Journal of Applied Logics - IfCoLog Journal of Logics and their Applications

Volume 6, Number 2

March 2019

Disclaimer

Statements of fact and opinion in the articles in Journal of Applied Logics - IfCoLog Journal of Logics and their Applications (JALs-FLAP) are those of the respective authors and contributors and not of the JALs-FLAP. Neither College Publications nor the JALs-FLAP make any representation, express or implied, in respect of the accuracy of the material in this journal and cannot accept any legal responsibility or liability for any errors or omissions that may be made. The reader should make his/her own evaluation as to the appropriateness or otherwise of any experimental technique described.

ISBN 978-1-84890-301-2
ISSN (E) 2631-9829
ISSN (P) 2631-9810

College Publications
Scientific Director: Dov Gabbay
Managing Director: Jane Spurr

`http://www.collegepublications.co.uk`

Editorial Board

Scope and Submissions

This journal considers submission in all areas of pure and applied logic, including:

- pure logical systems
- proof theory
- constructive logic
- categorical logic
- modal and temporal logic
- model theory
- recursion theory
- type theory
- nominal theory
- nonclassical logics
- nonmonotonic logic
- numerical and uncertainty reasoning
- logic and AI
- foundations of logic programming
- belief revision
- systems of knowledge and belief
- logics and semantics of programming
- specification and verification
- agent theory
- databases
- dynamic logic
- quantum logic
- algebraic logic
- logic and cognition
- probabilistic logic
- logic and networks
- neuro-logical systems
- complexity
- argumentation theory
- logic and computation
- logic and language
- logic engineering
- knowledge-based systems
- automated reasoning
- knowledge representation
- logic in hardware and VLSI
- natural language
- concurrent computation
- planning

This journal will also consider papers on the application of logic in other subject areas: philosophy, cognitive science, physics etc. provided they have some formal content.

Submissions should be sent to Jane Spurr (jane.spurr@kcl.ac.uk) as a pdf file, preferably compiled in LaTeX using the IFCoLog class file.

CONTENTS

ARTICLES

Preface

Ofer Arieli
School of Computer Science, The Academic College of Tel-Aviv, Israel
oarieli@mta.ac.il

Anna Zamansky
Department of Information Systems, University of Haifa, Israel
annazam@is.haifa.ac.il

This special issue of the *Journal of Applied Logics* contains revised and extended versions of selected papers presented at *the Israeli Workshop on Non-Classical Logics and Their Applications* (IsraLog'2017). The workshop took place on October 15–17, 2017 at the University of Haifa, Israel. It is the third edition of the IsraLog workshops, following the first meeting, held in Tel-Aviv University in November 2012, and the second meeting, which also took place in Haifa University, in October 2015. Post proceedings of these workshops have appeared as special issues of the *Journal of Logic and Computation* (Volume 26, Issue 1, February 2016) and the *Logic Journal of the IGPL* (Volume 24, Number 3, June 2016), respectively.

The main theme of the IsraLog workshops is the investigation of non-classical logics in general, and their application to computer science in particular. Triggered by a publication of a handbook on paraconsistent logics[1], the last edition of the meetings focused on reasoning with incomplete and inconsistent information. It featured four keynote talks by Didier Dubois (Paul Sabatier University), Michael Dunn (Indiana University), Edwin Mares (University of Wellington) and Daniele Mundici (University of Florence), as well as 33 selected presentations, given by world-wide experts in the areas from 15 countries and 4 continents.

Following the workshop, the papers in this volume were thoroughly reviewed and revised through a peer-refereeing process. A short summary of the accepted papers is provided below.

[1] A. Avron, O. Arieli, and A. Zamansky. *Theory of Effective Propositional Paraconsistent Logics*, Studies in Logic, volume 75 (sub-series: Mathematical Logic and Foundations), College Publications, 2018.

- Diana Costa, Manuel Martins and João Marcos show how quantified hybrid logic (an extension of modal logic which is expressive enough to allow referring to specific states of the associated possible worlds semantics) may be formalised by introducing Skolem functions. Then they introduce two variants of Herbrand's theorem for the "clausal-like" resulting formulas.

- Leonid Devyatkin considers a well-studied class of paraconsistent logics, obtained by extending the positive fragment of classical logic (consisting of conjunction, disjunction and implication), with a paraconsistent negation and possibly other operators. In particular, he investigates (countable) three- and (continual) four-valued matrices that are obtained in this way.

- Nissim Francez and Michael Kaminski present a systematic method of constructing natural deduction calculi for n-valued propositional logics from the truth tables for the connectives. The natural deduction systems are built from polysequents, which allow for explicitly referring to the truth-value of a formula. A general soundness and strong completeness theorem for the full consequence relation over polysequents is shown, and special cases for two 3-valued logics are considered.

- Oleg Grigoriev studies logical systems with two-dimensional truth values, which have ontological and epistemic parts. This induces a lattice with four elements, resembling the well-known Dunn-Belnap's structure. This structure is then extended to a 9-element distributive lattice, and a corresponding sound and complete axiomatization is provided, capturing the bipartite nature of truth values through two weak negations. A variation in which distribution scheme is omitted from the set of axiomatic schemata is also considered.

- Jesse Heyninck analyzes the relations between two paradigmatic formalisms to modelling defeasible reasoning: assumption-based argumentation and adaptive logic. It is shown how every (finite) assumption-based argumentation framework can be modelled as a sequential combination of adaptive logics, using a specific translation and a modal semantics to construct the appropriate adaptive logics. As a consequence, since it is already known that adaptive logics can be translated into assumption-based argumentation, the two frameworks are shown to be tightly related.

- Beata Konikowska and Arnon Avron study the logical aspects of covering-based rough sets, which are a generalization of the standard concept of Pawlak's rough sets, based on an arbitrary covering of the universe of objects instead of a partition. This is done by a three-valued logic with an analytic strongly sound and complete proof system based on a Gentzen-style sequent calculus.

- Daniele Mundici investigates some algebraic properties of de Finetti's notion of coherence, which is the basis of de Finetti's theory of probability. In particular, he studies preservation of coherence under quotients, definitions by fresh variables, and modifications of betting odds.

- Nenad Savić and Thomas Studer develop the logic RJ4 that combines the relevant logic R and the modal justification logic J4. It is shown that RJ4 overcomes some paradoxical situations that are obtained by justification logics that are based on classical logic. The logic is described in terms of a corresponding class of models and an axiomatic system, with respect to which it is sound and complete.

- Yaroslav Shramko provides another study in this volume of logics that emerge from Dunn-Belnap's four-valued logic. In particular, he examines logics in which three out of the four-values are designated (i.e., only pure falsity is left out), and their dual logics, in which only the value representing pure truth is designated. The dualization is carried out through corresponding proof systems, and soundness and completeness results are established with respect to a four-valued semantics.

We would like to thank the authors of the submitted papers for their valuable time and expertise devoted to writing contributions to this volume. It is also our pleasant duty to cordially thank all those who have acted as reviewers of the manuscripts submitted to this volume: Katalin Bimbó, Krysia Broda, Walter Carnielli, Davide Ciucci, Andrea Cohen, Marcelo Coniglio, Michael Dunn, Andreas Herzig, Jesse Heyninck, Hykel Hosni, Tomáš Kroupa, Roman Kuznets, A. Mani, Thiago Nascimento da Silva, Hitoshi Omori, Frederik Van De Putte, Revantha Ramanayake, Yaroslav Shramko, and Richard Zach.

The workshop, as well as the preparation of this volume, were supported by the Israel Science Foundation under grant number 817/15. We are also grateful to the Caesarea Rothschild Institute for Computer Science for their support with the organization of IsraLog'2017, and thank Dov Gabbay and Jane Spurr for their valuable help with the preparation of this volume.

 Received December 2018

On Herbrand's Theorem for Hybrid Logic

Diana Costa, Manuel A. Martins
CIDMA – Center for R&D in Mathematics and Applications, Department of Mathematics, University of Aveiro, Portugal
`{dianafcosta,martins}@ua.pt`

João Marcos
Department of Informatics and Applied Mathematics, Federal University of Rio Grande do Norte, Brazil
`jmarcos@dimap.ufrn.br`

Abstract

The original version of Herbrand's theorem [8] for first-order logic provided the theoretical underpinning for automated theorem proving, by allowing a constructive method for associating with each first-order formula χ a sequence of quantifier-free formulas $\chi_1, \chi_2, \chi_3, \cdots$ so that χ has a first-order proof if and only if some χ_i is a tautology. Some other versions of Herbrand's theorem have been developed for classical logic, such as the one in [6], which states that a set of quantifier-free sentences is satisfiable if and only if it is propositionally satisfiable. The literature concerning versions of Herbrand's theorem proved in the context of non-classical logics is meager. We aim to investigate in this paper two versions of Herbrand's theorem for hybrid logic, which is an extension of modal logic that is expressive enough so as to allow identifying specific sates of the corresponding models, as well as describing the accessibility relation that connects these states, thus being completely suitable to deal with relational structures [3]. Our main results state that a set of satisfaction statements is satisfiable in a hybrid interpretation if and only if it is propositionally satisfiable.

1 Introduction

Hybrid logics [3] are a breed of modal logics that provide appropriate syntax for referring to the associated possible-worlds semantics through the use of nominals. The latter, in particular, add to the modal description of relational structures the ability to refer to specific states. If modal logics have been successfully employed in specifying reactive systems, the hybrid component adds to them enough expressivity so as to refer to individual states and to reason about the system's local behavior at each of these states. Hybrid logics turn out thus to be strictly more expressive than their modal fragments. For example, irreflexivity ($i \to \neg\Diamond i$), asymmetry ($i \to \neg\Diamond\Diamond i$) or antisymmetry ($i \to \Box(\Diamond i \to i)$) are properties of the underlying transition structure which fail to be definable in standard modal logic (see [4]). Nonetheless, for the propositional case the satisfiability problem for hybrid logics is still decidable.

An important feature of hybrid logics that will play a central role in our approach is the fact that they allow for the specification of Robinson Diagrams [2]. Indeed, in these logics one may: (1) express equality between states named by i and j (note that $@_i j$ intends to affirm that the states named by i and j are identical, while $@_i \neg j$, being logically equivalent to $\neg @_i j$, intends to affirm that states i and j are distinct); (2) talk about accessibility between states through a modality (note that $@_i \Diamond j$ intends to affirm that the state named by j is a successor of the state named by i); (3) formulate satisfiability statements about a specific state (note that $@_i p$ intends to affirm that the proposition p is true at the state named by i, while $@_i \neg p$, being logically equivalent to $\neg @_i p$, intends to deny this). Consequently, within a hybrid logic one is able to completely describe the corresponding models using the rich underlying syntax.

Herbrand's theorem is a fundamental result of mathematical logic. It essentially allows a certain kind of reduction of first-order logic to propositional logic. While not aimed at providing an efficient procedure for (semi)decidability, Herbrand-like theorems are ordinarily used as useful intermediate steps in proving that some theorem-proving resolution-based method works as intended. Several versions of Herbrand's theorem are now available for classical logic; here we present two versions for hybrid logics, using the concepts of satisfiability and propositional satisfiability, following the approach described in [6].

Outline of the paper. In Section 2 we start by recalling the basic hybrid logic. Theorem 2.13, our first Herbrand-like theorem, states that hybrid satisfiability is equivalent to propositional satisfiability for sets of satisfaction statements containing the equality axioms. In Section 3 we discuss the quantified hybrid logic — a logic less known than the basic hybrid logic. The strategy to establish a Herbrand-like theorem in this case follows the one for the classical first-order version, by making use of Skolemization to eliminate the existential quantifiers on world variables. The main result here is stated on Theorem 3.26. Section 4 wraps up with some pointers for future investigation.

2 The Case of the Basic Hybrid Logic

The simplest form of hybrid logic is based on the *basic hybrid language*, which adds nominals and the satisfaction operator to the language of propositional modal logic. This simple upgrade of the usual modal language carries great power in terms of expressivity.

Definition 2.1. Let $\mathcal{L} = \langle \text{Prop}, \text{Nom} \rangle$ be a *hybrid signature*, where Prop is a countable set of *propositional symbols* and Nom is a countable set of symbols disjoint from Prop. We use p, q, r and so on to refer to the elements in Prop. The elements in Nom are called *nominals* and we typically write them as i, j, k, and so on. The hybrid formulas over $\mathcal{L}$, which we denote by $\text{Form}_{@}(\mathcal{L})$, are defined by the following grammar:

$$\varphi ::= i \mid p \mid \neg\varphi \mid \varphi_1 \wedge \varphi_2 \mid \Diamond\varphi \mid @_i\varphi$$
$$\text{where } i \in \text{Nom and } p \in \text{Prop}.$$

The formulas with prefix @ are called *satisfaction statements.* The connectives $\vee$, $\rightarrow$, and $\Box$ are defined as usual. ◀

Definition 2.2. Let $\mathcal{L} = \langle \text{Prop}, \text{Nom} \rangle$ be a hybrid signature. A *hybrid structure $\mathcal{M}$ over $\mathcal{L}$* is a tuple (W, R, N, V). Here, W is a non-empty set called *domain* whose elements are called *states* or *worlds*, $R \subseteq W \times W$ is called *accessibility relation*, $N : \text{Nom} \rightarrow W$ is a *hybrid nomination* and $V : \text{Prop} \rightarrow Pow(W)$ is a *hybrid valuation.* The pair $\langle W, R \rangle$ is called the *frame* underlying $\mathcal{M}$, and $\mathcal{M}$ is said to be a structure based on this frame. ◀

The satisfaction relation, which is defined next, is a generalization of Kripke-style satisfaction.

Definition 2.3. The *satisfaction relation* $\Vdash$ between a hybrid structure $\mathcal{M} = (W, R, N, V)$, a state $w \in W$, and a hybrid formula is recursively defined by:

- $\mathcal{M}, w \Vdash i$ iff $w = N(i)$;
- $\mathcal{M}, w \Vdash p$ iff $w \in V(p)$;
- $\mathcal{M}, w \Vdash \neg\varphi$ iff it is not the case that $\mathcal{M}, w \Vdash \varphi$;
- $\mathcal{M}, w \Vdash \varphi_1 \wedge \varphi_2$ iff $\mathcal{M}, w \Vdash \varphi_1$ and $\mathcal{M}, w \Vdash \varphi_2$;
- $\mathcal{M}, w \Vdash \Diamond\varphi$ iff $\exists w' \in W(wRw'$ and $\mathcal{M}, w' \Vdash \varphi)$;
- $\mathcal{M}, w \Vdash @_i\varphi$ iff $\mathcal{M}, w' \Vdash \varphi$, where $w' = N(i)$.

If $\mathcal{M}, w \Vdash \varphi$ we say that φ *is satisfied in* $\mathcal{M}$ *at* w. If φ is satisfied at all states in a structure $\mathcal{M}$, we write $\mathcal{M} \Vdash \varphi$. If φ is satisfied at all states in all structures based on a frame $\mathcal{F}$, then we say that φ *is valid on* $\mathcal{F}$ and we write $\mathcal{F} \Vdash \varphi$. If φ is valid on all frames, then we simply say that φ *is valid* and we write $\Vdash \varphi$. We say that a set Φ of hybrid formulas is *satisfiable* if there exists a model $\mathcal{M}$ and a world $w \in W$ such that $\mathcal{M}, w \Vdash \Phi$, i.e., $\mathcal{M}, w \Vdash \varphi$ for all $\varphi \in \Phi$. For $\Delta \subseteq \text{Form}_{@}(\mathcal{L})$, we say that $\mathcal{M}$ *is a model of* Δ if $\mathcal{M} \Vdash \delta$ for all $\delta \in \Delta$. ◀

Definition 2.4. Let $\mathcal{L}$ be a hybrid signature. The set $\text{At}(\mathcal{L})$ of *atomic satisfaction statements* (*atoms*, for short) *over* $\mathcal{L}$ is the set of $\mathcal{L}$-formulas of the forms $@_i p$, $@_i \Diamond j$, and $@_i j$ for $i, j \in \text{Nom}$ and $p \in \text{Prop}$. We use $\text{BCAt}(\mathcal{L})$ to denote the set of all (finite) Boolean combinations of atomic satisfaction statements over $\mathcal{L}$, i.e., $\text{BCAt}(\mathcal{L})$ is the smallest set containing $\text{At}(\mathcal{L})$ and closed under $\wedge$ and $\neg$. ◀

Definition 2.5. An $\mathcal{L}$-*truth assignment* is a mapping $v : \text{At}(\mathcal{L}) \to \{T, F\}$. Given an $\mathcal{L}$-truth assignment v, one may extend it to $\overline{v} : \text{BCAt}(\mathcal{L}) \to \{T, F\}$ through the truth-functional interpretation of the propositional connectives. In order to simplify notation, given that this extension is unique, we will use v in order to refer both to an $\mathcal{L}$-truth assignment and to its extension $\overline{v}$. Let $\Phi \subseteq \text{BCAt}(\mathcal{L})$. We say that Φ is *propositionally satisfiable* if there is an $\mathcal{L}$-truth assignment that simultaneously satisfies every member of Φ. We say that Φ is *propositionally unsatisfiable* if there is no such $\mathcal{L}$-truth assignment. ◀

We have now the basis to start investigating a first Herbrand-like theorem for hybrid logic:

Theorem 2.6. *Let* $\Phi \subseteq \text{BCAt}(\mathcal{L})$. *If* Φ *is propositionally unsatisfiable then* Φ *is unsatisfiable.*

Proof. Suppose that Φ is satisfiable: then there is a model $\mathcal{M}$ and a world $w \in W$ such that $\mathcal{M}, w \Vdash \Phi$, i.e., $\mathcal{M}, w \Vdash \varphi$ for all $\varphi \in \Phi$.

Define $v^{\mathcal{M}} : \text{At}(\mathcal{L}) \to \{T, F\}$ by setting $v^{\mathcal{M}}(\psi) = T$ iff $\mathcal{M}, w \Vdash \psi$.

Let us prove by induction on the structure of $\varphi \in \text{BCAt}(\mathcal{L})$ that $v^{\mathcal{M}}(\varphi) = T$ iff $\mathcal{M}, w \Vdash \varphi$.

- If $\varphi \in \text{At}(\mathcal{L})$, the result follows from the definition of $v^{\mathcal{M}}$.
- Suppose now, by Induction Hypothesis, (IH), that $\mathcal{M}, w \Vdash \varphi_i$ iff $v^{\mathcal{M}}(\varphi_i) = T$, for $i = 1, 2$.

– If $\varphi = \varphi_1 \wedge \varphi_2$, then

$$\begin{array}{lll} \mathcal{M}, w \Vdash \varphi & \text{iff} & \mathcal{M}, w \Vdash \varphi_1 \wedge \varphi_2 \\ & \text{iff} & \mathcal{M}, w \Vdash \varphi_1 \text{ and } \mathcal{M}, w \Vdash \varphi_2 \\ & \underset{\text{(IH)}}{\text{iff}} & v^{\mathcal{M}}(\varphi_1) = T \text{ and } v^{\mathcal{M}}(\varphi_2) = T \\ & \text{iff} & v^{\mathcal{M}}(\varphi_1 \wedge \varphi_2) = T \\ & \text{iff} & v^{\mathcal{M}}(\varphi) = T \end{array}$$

– If $\varphi = \neg\psi$, then

$$\begin{array}{lll} \mathcal{M}, w \Vdash \varphi & \text{iff} & \mathcal{M}, w \Vdash \neg\psi \\ & \text{iff} & \mathcal{M}, w \nVdash \psi \\ & \underset{\text{(IH)}}{\text{iff}} & v^{\mathcal{M}}(\psi) = F \\ & \text{iff} & v^{\mathcal{M}}(\neg\psi) = T \\ & \text{iff} & v^{\mathcal{M}}(\varphi) = T \end{array}$$

Since $\mathcal{M}, w \Vdash \Phi$, by assumption, we have that $v^{\mathcal{M}}(\varphi) = T$ for any $\varphi \in \Phi$. Therefore, Φ is propositionally satisfiable. ■

Example 2.7. Let $\mathcal{L} = \langle \{p, q\}, \{i, j\} \rangle$, and $\Phi = \{@_i p \vee @_i q, @_j \neg q, @_i j, @_i \Diamond j\}$.

The set Φ is satisfiable, as there is a model $\mathcal{M} = (W, R, N, V)$ such that $W = \{w\}$, $R = \{(i, i)\}$, $N(i) = N(j) = w$, $V(p) = \{w\}$ and $V(q) = \varnothing$, where $\mathcal{M}, w \Vdash \Phi$.

Define $v^{\mathcal{M}} : \mathrm{At}(\mathcal{L}) \to \{T, F\}$ by setting $v^{\mathcal{M}}(\psi) = T$ iff $\mathcal{M}, w \Vdash \psi$. This implies that $v^{\mathcal{M}}(@_i p) = T$, $v^{\mathcal{M}}(@_i \Diamond j) = T$, $v^{\mathcal{M}}(@_i j) = T$ and for all other atomic satisfaction statements in $\mathcal{L}$, $v^{\mathcal{M}}$ assigns F. The extension of $v^{\mathcal{M}}$ to $\overline{v^{\mathcal{M}}}$ is straightforward. Thus Φ is propositionally satisfiable. ♦

The converse of the previous theorem is not true in general. Here is a counterexample:

Example 2.8. Let $\mathcal{L} = \langle \{p\}, \{i, j\} \rangle$, and $\Phi = \{@_i j, @_i p, @_j \neg p\}$.

Note that Φ is propositionally satisfiable: take $v^{\mathcal{M}} : \mathrm{At}(\mathcal{L}) \to \{T, F\}$ to be such that $v^{\mathcal{M}}(@_i p) = T$, $v^{\mathcal{M}}(@_i j) = T$, and $v^{\mathcal{M}}$ assigns the value F to all other atomic satisfaction statements.

However, Φ is not satisfiable, as there is no model $\mathcal{M}$ such that $\mathcal{M}, w \Vdash \Phi$. Any model that satisfies the first formula in Φ has that $N(i) = N(j) = w$. From the second and the third formulas, one must have that $w \in V(p)$ and $w \notin V(p)$, respectively, which is a contradiction. ♦

As in the case of first-order logic with equality, the characteristic equality axioms need to be taken into consideration. In hybrid logic we do not have an explicit symbol of equality in the language; however, there are hybrid formulas that express the equality axioms over nominals in $\mathcal{L}$ (see [3]):

- *Reflexivity*: $@_i i$, for $i \in \mathrm{Nom}$;
- *Symmetry*: $@_i j \to @_j i$, for $i, j \in \mathrm{Nom}$;
- *Nom*: $(@_i \varphi \wedge @_i j) \to @_j \varphi$, for $i, j \in \mathrm{Nom}$ and $@_i \varphi$ an atomic satisfaction statement;
- *Bridge*: $(@_i \Diamond j \wedge @_j k) \to @_i \Diamond k$, for $i, j, k \in \mathrm{Nom}$.

The set of all equality axioms over the hybrid signature $\mathcal{L}$ is denoted by $\mathrm{Eq}(\mathcal{L})$. It is easy to check that these formulas are all valid in hybrid logic. Note that *Bridge* does not follow from the other axioms, as it is the only axiom where nominals are replaced in formula position.

Lemma 2.9. *Let $\mathcal{M}$ be a model and φ be a formula in* $\mathrm{BCAt}(\mathcal{L})$. *Then,*

$$\exists w \in W : \mathcal{M}, w \Vdash \varphi \text{ iff } \mathcal{M} \Vdash \varphi$$

Proof. We will check this result by induction on the structure of $\varphi \in \mathrm{BCAt}(\mathcal{L})$:
- For $\varphi = @_i \psi$ an atomic satisfaction statement:

$$\begin{array}{rll} \exists w \in W : \mathcal{M}, w \Vdash \varphi & \text{iff} & \exists w \in W : \mathcal{M}, w \Vdash @_i \psi \\ & \text{iff} & \mathcal{M}, w' \Vdash \psi, \text{ where } w' = N(i) \\ & \text{iff} & \mathcal{M} \Vdash @_i \psi \\ & \text{iff} & \mathcal{M} \Vdash \varphi \end{array}$$

- Suppose by (IH) that ψ and θ are such that the result holds. Then,
– For $\varphi = \neg\psi$:

$$\begin{array}{rll} \exists w \in W : \mathcal{M}, w \Vdash \varphi & \text{iff} & \exists w \in W : \mathcal{M}, w \Vdash \neg\psi \\ & \text{iff} & \exists w \in W : \mathcal{M}, w \nVdash \psi \\ & \underset{\text{(IH)}}{\text{iff}} & \mathcal{M} \nVdash \psi \\ & \text{iff} & \forall w \in W : \mathcal{M}, w \nVdash \psi \\ & \text{iff} & \forall w \in W : \mathcal{M}, w \Vdash \neg\psi \\ & \text{iff} & \mathcal{M} \Vdash \neg\psi \\ & \text{iff} & \mathcal{M} \Vdash \varphi \end{array}$$

– For $\varphi = \varphi_1 \wedge \varphi_2$:

For one implication:

$$\begin{array}{ll} \exists w \in W : \mathcal{M}, w \Vdash \varphi & \\ \quad \text{iff} & \exists w \in W : \mathcal{M}, w \Vdash \varphi_1 \wedge \varphi_2 \\ \quad \text{iff} & \exists w \in W : (\mathcal{M}, w \Vdash \varphi_1 \text{ and } \mathcal{M}, w \Vdash \varphi_2) \\ \quad \text{implies} & \exists w \in W : \mathcal{M}, w \Vdash \varphi_1 \text{ and } \exists w \in W : \mathcal{M}, w \Vdash \varphi_2 \\ \quad \underset{\text{(IH)}}{\text{iff}} & \mathcal{M} \Vdash \varphi_1 \text{ and } \mathcal{M} \Vdash \varphi_2 \\ \quad \text{iff} & \forall w \in W : \mathcal{M}, w \Vdash \varphi_1 \text{ and } \forall w \in W : \mathcal{M}, w \Vdash \varphi_2 \\ \quad \text{iff} & \forall w \in W : \mathcal{M}, w \Vdash \varphi_1 \text{ and } \mathcal{M}, w \Vdash \varphi_2 \\ \quad \text{iff} & \forall w \in W : \mathcal{M}, w \Vdash \varphi_1 \wedge \varphi_2 \\ \quad \text{iff} & \mathcal{M} \Vdash \varphi_1 \wedge \varphi_2 \\ \quad \text{iff} & \mathcal{M} \Vdash \varphi \end{array}$$

For the converse implication:

$$\begin{array}{lll} \mathcal{M} \Vdash \varphi & \text{iff} & \mathcal{M} \Vdash \varphi_1 \wedge \varphi_2 \\ & \text{iff} & \forall w \in W : \mathcal{M}, w \Vdash \varphi_1 \wedge \varphi_2 \\ & \underset{\text{(given that } W \neq \varnothing)}{\text{implies}} & \exists w \in W : \mathcal{M}, w \Vdash \varphi_1 \wedge \varphi_2 \\ & \text{iff} & \exists w \in W : \mathcal{M}, w \Vdash \varphi \end{array}$$

■

Let us consider next the binary relation $\sim$ defined on Nom by setting $i \sim j$ iff $v(@_i j) = T$.

Lemma 2.10. *The binary relation $\sim$ is an equivalence relation.*

Proof. [Reflexivity] is guaranteed by the homonymous axiom stated above, namely $@_i i$, for $i \in \text{Nom}$. Once $\text{Eq}(\mathcal{L}) \subseteq \Phi$, then $v(@_i i) = T$ implies $i \sim i$.

[Symmetry] holds due to the fact that if $i \sim j$, then $v(@_i j) = T$, and given that $\text{Eq}(\mathcal{L}) \subseteq \Phi$, we have $v(@_i j \to @_j i) = T$, which implies that $v(@_j i) = T$. So, $j \sim i$.

[Transitivity] follows from *Symmetry* and the axiom *Nom*. Suppose $i \sim j$ and $j \sim k$. By [Symmetry] it follows that $j \sim i$ and $j \sim k$, thus $v(@_j i) = T$ and $v(@_j k) = T$. Once more, since $\text{Eq}(\mathcal{L}) \subseteq \Phi$, we have in particular that $v((@_j i \wedge @_j k) \to @_i k) = T$. We conclude that $v(@_i k) = T$, thus $i \sim k$. ■

The above result is crucial in proving Herbrand's Theorem for languages containing equality. Next we show that if for a set Φ of Boolean combinations of atomic satisfaction statements with equality there is a valuation v that assigns the value true to all atomic satisfaction statements in Φ, then there is a hybrid structure that satisfies the equality axioms and where Φ is satisfiable.

Theorem 2.11. *Assume* $\text{Eq}(\mathcal{L}) \subseteq \Phi \subseteq \text{BCAt}(\mathcal{L})$. *If Φ is unsatisfiable then Φ is propositionally unsatisfiable.*

Proof. Suppose that Φ is propositionally satisfiable and let $v : \mathrm{At}(\mathcal{L}) \to \{T, F\}$ be such that $v(\varphi) = T$ for any $\varphi \in \Phi$.

Let $W = \mathrm{Nom}$. We define the hybrid structure $\mathcal{M} = (W_v, R_v, N_v, V_v)$ such that:

- $W_v = W/\sim$;
- $[i]R_v[j]$ iff $v\,(@_i\Diamond j) = T$, for $i, j \in \mathrm{Nom}$;
- $N_v(j) = [i]$ iff $v\,(@_i j) = T$, for $i, j \in \mathrm{Nom}$; and
- $[i] \in V_v(p)$ iff $v\,(@_i p) = T$, for $i \in \mathrm{Nom}$, $p \in \mathrm{Prop}$.

Claim I. R_v is well-defined.

We want to prove that if $i \sim j$ and $k \sim l$, then $[i]R_v[k]$ implies $[j]R_v[l]$.

– Suppose that $i \sim j, k \sim l$ and $[i]R_v[k]$. By definition, we know that $[i]R_v[k]$ means that $v\,(@_i\Diamond k) = T$, and $i \sim j$ means that $v\,(@_i j) = T$. It follows that $v\,(@_i\Diamond k \wedge @_i j) = T$. The axiom *Nom* let us conclude then that $v\,(@_j\Diamond k) = T$. We also know that $k \sim l$ means that $v\,(@_k l) = T$. From the axiom *Bridge*, since $v\,(@_j\Diamond k \wedge @_k l) = T$, it follows that $v\,(@_j\Diamond l) = T$. Therefore, by definition, $[j]R_v[l]$.

Claim II. V_v is well-defined.

We want to prove that if $i \sim j$ then $([i] \in V_v(p)$ iff $[j] \in V_v(p))$.

– Suppose that $i \sim j$ and $[i] \in V_v(p)$. By the definition of the equivalence relation $\sim$, $v(@_i j) = T$; and by the definition of V_v, $v(@_i p) = T$. Then $v(@_i p \wedge @_i j) = T$ and from *Nom* it follows that $v(@_j p) = T$. So, $[j] \in V_v(p)$. The converse direction is checked analogously in view of the symmetry of $\sim$.

All that is left to prove now is the satisfiability of Φ.

Claim III. For all $\varphi \in \mathrm{BCAt}(\mathcal{L})$, $(\mathcal{M} \Vdash \varphi$ iff $v(\varphi) = T)$.

Below you should recall that for Boolean combinations of atomic satisfaction statements, satisfiability at one state is equivalent to satisfiability at all states, by Lemma 2.9.

- $\varphi = @_i p$

$$\begin{aligned} \mathcal{M} \Vdash @_i p \ &\text{iff} \ \mathcal{M}, [i] \Vdash p \\ &\text{iff} \ [i] \in V_v(p) \\ &\text{iff} \ v(@_i p) = T \\ &\text{iff} \ v(\varphi) = T \end{aligned}$$

- $\varphi = @_i\Diamond j$

$$\begin{aligned}
\mathcal{M} \Vdash @_i\Diamond j \quad &\text{iff} \quad \mathcal{M}, [i] \Vdash \Diamond j \\
&\text{iff} \quad \exists k : [i]R_v[k] \text{ and } \mathcal{M}, [k] \Vdash j \\
&\text{iff} \quad \exists k : [i]R_v[k] \text{ and } [k] = [j] \\
&\text{iff} \quad [i]R_v[j] \\
&\text{iff} \quad v(@_i\Diamond j) = T \\
&\text{iff} \quad v(\varphi) = T
\end{aligned}$$

- $\varphi = @_i j$

$$\begin{aligned}
\mathcal{M} \Vdash @_i j \quad &\text{iff} \quad \mathcal{M}, [i] \Vdash j \\
&\text{iff} \quad [i] = [j] \\
&\text{iff} \quad v(@_i j) = T \\
&\text{iff} \quad v(\varphi) = T
\end{aligned}$$

- By (IH), let φ_1, φ_2 be such that $\mathcal{M} \Vdash \varphi_i$ iff $v(\varphi_i) = T$, for $i = 1, 2$.

  This part is similar to Theorem 2.6, so we omit the details.

  - Given $\varphi = \varphi_1 \wedge \varphi_2$, note that

$$\mathcal{M} \Vdash \varphi_1 \wedge \varphi_2 \quad \text{iff} \quad v(\varphi_1 \wedge \varphi_2) = T$$

  - Given $\varphi = \neg\varphi_1$, note that

$$\mathcal{M} \Vdash \neg\varphi_1 \quad \text{iff} \quad v(\neg\varphi_1) = T$$

Thus, in particular, $\mathcal{M} \Vdash \Phi$, and this means that Φ is satisfiable. ■

We finish this section by generalizing the above results to compound satisfaction statements. Let φ be any satisfaction statement. The following rules allow us to rewrite φ by recursively applying the following rules in order to obtain a semantically equivalent formula $\varphi^\circ \in \mathrm{BCAt}(\mathcal{L}^*)$, where $\mathcal{L}^*$ is an expansion of $\mathcal{L}$ obtained by the addition of new nominals to the initial hybrid signature. Observe that such extension is possible since we considered Nom to be a countable set.

Rewrite Rules:

1. $@_i@_j\varphi \twoheadrightarrow @_j\varphi$

2. $@_i\neg\varphi \twoheadrightarrow \neg @_i\varphi$

3. $@_i(\varphi \wedge \psi) \twoheadrightarrow @_i\varphi \wedge @_i\psi$

4. $@_i\Diamond\varphi \twoheadrightarrow @_i\Diamond k \wedge @_k\varphi$, for k a fresh nominal

As the above rules successively decrease the complexity of satisfaction statements, it is clear that the associated rewrite system is terminating. In fact, by using the Knuth-Bendix completion algorithm it is easy to see that the rewrite system is also confluent. In this respect, it is worth noting that the formula $@_i@_j\Diamond\varphi$ may rewrite in two ways, namely as $@_j\Diamond k_1 \wedge @_{k_1}\varphi$ and as $@_j\Diamond k_2 \wedge @_{k_2}\varphi$. These are the same, however, modulo the introduced fresh nominals. Moreover, we should point out that Areces and Gorín, in [1], have investigated labeled resolution calculi for hybrid logics with inference rules similar to the above rewrite rules; namely our rules 1., 3. and 4. correspond to their @, $\wedge$ and $\langle r\rangle$ rules, respectively.

Example 2.12. Consider the formula $\varphi = @_i@_j\Diamond(p \wedge \neg q)$ in $\mathcal{L}$. It is clear that φ is not a Boolean combination of atomic satisfaction statements of $\mathcal{L}$.

Applying the rewrite rules yields that:

$$\begin{array}{rcl} @_i@_j\Diamond(p \wedge \neg q) & \twoheadrightarrow & @_j\Diamond(p \wedge \neg q) \\ & \twoheadrightarrow & @_j\Diamond k \wedge @_k(p \wedge \neg q),\ k \text{ fresh} \\ & \twoheadrightarrow & @_j\Diamond k \wedge (@_k p \wedge @_k\neg q) \\ & \twoheadrightarrow & @_j\Diamond k \wedge (@_k p \wedge \neg @_k q) \end{array}$$

Thus $\varphi^\circ = @_j\Diamond k \wedge (@_k p \wedge \neg @_k q)$. Note that the new formula is in the hybrid signature $\mathcal{L}^*$ that expands $\mathcal{L}$ by the addition of the new nominal k. ♦

Theorem 2.13 (Herbrand-like). *Let Φ be a set of satisfaction statements such that* $\mathrm{Eq}(\mathcal{L}) \subseteq \Phi$. *Then Φ is propositionally unsatisfiable iff Φ is unsatisfiable.*

Proof. We exhaustively apply the previously introduced rules to the formulas of Φ and transform Φ into $\Phi^\circ := \{\varphi^\circ : \varphi \in \Phi\} \cup \mathrm{Eq}(\mathcal{L}^*)$. Note that Φ° is a subset of $\mathrm{BCAt}(\mathcal{L}^*)$, which contains the equality axioms in the expanded language, thus we may apply Theorems 2.6 and 2.11. ■

3 The Case of Quantified Hybrid Logic

In this section we introduce a hybrid logic enriched with operators over world variables, typically written as s, t, u and so on, distinct from both nominals and propositional variables. We will also resort to an algebraic similarity type in order to allow function symbols. This logic, which we will call Algebraic Strong Priorean Logic, shares some similarities with the logic $\mathcal{HLCOV}(@,\forall,\exists)$ found in [9], namely in the use of quantifiers and functions, but it differs in the definition of terms; in particular, while $\mathcal{HLCOV}(@,\forall,\exists)$ allows for quantification over both state variables and functional terms, the Algebraic Strong Priorean Logic restricts quantifications to state variables.

Definition 3.1. An *algebraic similarity type* Σ is a tuple (F, σ) such that F is a non-empty set of function symbols, and σ assigns to each function symbol its arity. An algebraic similarity type together with a countable set of world variables, WVar, and a countable set of nominals, Nom, induces the set $\mathrm{Term}(\Sigma, \mathrm{WVar}, \mathrm{Nom})$ of Σ-terms, whose elements are the algebraic terms given by the grammar:

$$t ::= i \mid s \mid f\big(t_1, \cdots, t_{\sigma(f)}\big)$$
$$\text{where } i \in \mathrm{Nom}, s \in \mathrm{WVar} \text{ and } f \in \mathrm{F}.$$
◀

We may now introduce a powerful hybrid language, $\mathcal{H}(\Sigma, @, \forall)$, whose grammar is defined below:

Definition 3.2. A *hybrid similarity type* L is a tuple (Prop, Nom, WVar), where Prop and Nom are as usual the set of propositional variables and the set of nominals of a hybrid signature, and WVar is a countable set of world variables. Let $\Sigma = \langle \mathrm{F}, \sigma \rangle$ be an algebraic similarity type. The well-formed formulas $\mathrm{Form}_{@,\forall}(L, \mathrm{Term}(\Sigma, \mathrm{WVar}, \mathrm{Nom}))$ over the hybrid similarity type L and the Σ-terms $\mathrm{Term}(\Sigma, \mathrm{WVar}, \mathrm{Nom})$ are defined by the following grammar:

$$\varphi ::= p \mid t \mid \neg\varphi \mid \varphi_1 \wedge \varphi_2 \mid \Diamond\varphi \mid @_t\varphi \mid \forall s\, \varphi \mid \exists s\, \varphi$$
$$\text{where } p \in \mathrm{Prop},\ t \in \mathrm{Term}(\Sigma, \mathrm{WVar}, \mathrm{Nom}) \text{ and } s \in \mathrm{WVar}.$$

Note that @ can make use of Σ-terms, i.e., world variables and functional terms. The connectives $\vee$, $\rightarrow$, and $\Box$ are defined as usual. ◀

The earlier definition of a 'hybrid structure' is now upgraded as follows:

Definition 3.3. Let $L = \langle \mathrm{Prop}, \mathrm{Nom}, \mathrm{WVar} \rangle$ and $\Sigma = \langle \mathrm{F}, \sigma \rangle$ be, respectively, a hybrid and an algebraic similarity types. A *hybrid structure* $\mathcal{H}$ *over* $\langle L, \Sigma \rangle$ is a tuple $(W, R, \left(f^W\right)_{f \in \mathrm{F}}, N, V)$, where W, R, N and V are the domain, accessibility relation, hybrid nomination and valuation as introduced in Definition 2.2, and $\left(f^W\right)_{f \in \mathrm{F}}$ is a family containing for each $f \in \mathrm{F}$ an interpretation $f^W : W^{\sigma(f)} \rightarrow W$. ◀

As we need a mechanism for coping with the terms introduced in the above grammars, we consider now a *world assignment* $g : WVar \rightarrow W$. Two world assignments

g and g' are called *s-variant* iff $g(u) = g'(u)$, for all $u \in$ WVar such that $u \neq s$; in such case we write $g \overset{s}{\sim} g'$. We extend g to Term$(\Sigma, WVar)$ in the following way:

$$\bar{g}(t) = \begin{cases} g(t), & \text{if } t \in \text{WVar} \\ N(t), & \text{if } t \in \text{Nom} \\ f^W\Big(\bar{g}(t_1), \ldots, \bar{g}(t_{\sigma(f)})\Big), & \text{if } t = f(t_1, \ldots, t_{\sigma(f)}), \text{ for some } f \in F \end{cases}$$

In order to simplify notation, we will use g to denote both a world assignment and its extension.

The notion of satisfaction is now defined in the following way:

Definition 3.4. The *satisfaction relation* $\Vdash$ between a hybrid structure $\mathcal{H} = \left(W, R, \left(f^W\right)_{f \in \mathrm{F}}, N, V\right)$, a state $w \in W$, a world assignment g and a hybrid formula is recursively defined by:

- $\mathcal{H}, g, w \Vdash p$ iff $w \in V(p)$, for $p \in$ Prop;
- $\mathcal{H}, g, w \Vdash t$ iff $w = g(t)$, for $t \in$ Term$(\Sigma, \text{WVar}, \text{Nom})$;
- $\mathcal{H}, g, w \Vdash \neg\varphi$ iff it is not the case that $\mathcal{H}, g, w \Vdash \varphi$;
- $\mathcal{H}, g, w \Vdash \varphi_1 \wedge \varphi_2$ iff $\mathcal{H}, g, w \Vdash \varphi_1$ and $\mathcal{H}, g, w \Vdash \varphi_2$;
- $\mathcal{H}, g, w \Vdash \Diamond\varphi$ iff $\exists w' \in W(wRw'$ and $\mathcal{H}, g, w' \Vdash \varphi)$;
- $\mathcal{H}, g, w \Vdash @_t\varphi$ iff $\mathcal{H}, g, w' \Vdash \varphi$, where $w' = g(t)$, for $t \in$ Term$(\Sigma, \text{WVar}, \text{Nom})$;
- $\mathcal{H}, g, w \Vdash \forall s\, \varphi$ iff $\mathcal{H}, g', w \Vdash \varphi$ for all g' such that $g' \overset{s}{\sim} g$;
- $\mathcal{H}, g, w \Vdash \exists s\, \varphi$ iff $\mathcal{H}, g', w \Vdash \varphi$ for some g' such that $g' \overset{s}{\sim} g$.

Here, $\mathcal{H}, g, w \Vdash \varphi$ is read as saying that φ *is satisfied at the state w in the hybrid structure $\mathcal{H}$ under the world assignment g.* ◀

We shall use the appellation *Algebraic Strong Priorean Logic* to refer to the logic induced by the above notion of satisfaction. It is worth pointing out that the Algebraic Strong Priorean Logic contains the logic of the hybrid language with a binder, as $\downarrow s.\, \varphi$ is expressible here by $\exists s\,(s \wedge \varphi)$. Such logic is very expressive. The algebraic structure over the set of worlds may be useful in several contexts. Here are some examples: on *trees*, one can consider a functional symbol for referring to the first common ancestor of two given nodes; on the *graph representations of maps*, one can consider a functional symbol for referring to an intermediate city that minimizes the distance between two other given cities; on *temporal frames*, one can consider functional symbols that allow pointing to a specific time after or before the current moment, or a function that allows one to say that something happens periodically.

Definition 3.5. A set Φ of formulas in $\mathrm{Form}_{@,\forall}(L, \mathrm{Term}(\Sigma, \mathrm{WVar}, \mathrm{Nom}))$ is said to be *satisfiable* if there exists a hybrid structure $\mathcal{H}$ over $\langle L, \Sigma \rangle$, a $w \in W$ and a world assignment g such that $\mathcal{H}, g, w \Vdash \varphi$ for all $\varphi \in \Phi$. We say that $\varphi \in \mathrm{Form}_{@,\forall}(L, \mathrm{Term}(\Sigma, \mathrm{WVar}, \mathrm{Nom}))$ is *satisfiable* if the singleton $\{\varphi\}$ is satisfiable. ◀

Definition 3.6. A *literal* in $\mathcal{H}(\Sigma, @, \forall)$ is a formula of the form: $@_a p$, $@_a \neg p$, $@_a b$, $@_a \neg b$ $@_a \Diamond b$, $@_a \neg \Diamond b$, where $p \in \mathrm{Prop}$, and $a, b \in \mathrm{Term}(\Sigma, \mathrm{WVar}, \mathrm{Nom})$. ◀

Lemma 3.7 (Labelling). *Let φ be a formula in* $\mathrm{Form}_{@,\forall}(L, \mathrm{Term}(\Sigma, \mathrm{WVar}, \mathrm{Nom}))$. *Then*

$$\varphi \textit{ is satisfiable iff } @_i\varphi \textit{ is satisfiable},$$

where i is a fresh nominal.

Proof.

$$\begin{array}{lll} \varphi \text{ is satisfiable} & \text{iff} & \exists \mathcal{H}, \exists g, \exists w : \mathcal{H}, g, w \Vdash \varphi \\ & \text{iff} & \exists \tilde{\mathcal{H}}, \exists g, \exists w : \tilde{\mathcal{H}}, g, w \Vdash \varphi, \; w = \tilde{N}(i) \\ & \text{iff} & \exists \tilde{\mathcal{H}}, \exists g, \exists \tilde{w} : \tilde{\mathcal{H}}, g, \tilde{w} \Vdash @_i\varphi \\ & \text{iff} & @_i\varphi \text{ is satisfiable} \end{array}$$

■

Our goal in what follows is to study the satisfiability of a formula in the Algebraic Strong Priorean Logic. Since the satisfiability problem of a formula φ is equivalent to the satisfiability problem of a formula $@_i\varphi$ — where i does not occur in φ — we will prove satisfiability of the latter. In order to do so, it will be convenient to rearrange formulas so that we end up with a formula in Prenex Conjunctive Normal Form, i.e., a formula in which quantifiers appear on the left, prefixing a quantifier-free part that is a conjunction of clauses, where clauses are disjunctions of literals.

Definition 3.8. A formula is said to be *rectified* if no world variable occurs both bound and free and if all quantifiers in the formula refer to different world variables. ◀

The renaming of bound world variables follows the same approach as in first-order logic, whose proof is standard:

Lemma 3.9. *It is always possible to perform a systematic renaming of bound (world) variables such that the result is a rectified formula, equivalent to the original one in the following way: if s occurs bounded in a formula φ and u does not occur at all,* *then φ is equivalent to the formula obtained by replacing all occurrences of s in the scope of a quantifier in φ with u.*

Given a formula φ as input, we will refer to the formula $\tilde{\varphi}$ produced by the above renaming procedure *the rectified version of φ*.

Definition 3.10. Let $s_1, \ldots, s_n$ be the world variables occurring free in φ. The *[rectified] existential closure of φ* is the formula which results from rectifying φ and then existentially bounding its free variables, i.e., it is the formula $\exists s_1 \ldots \exists s_n \tilde{\varphi}$, where $\tilde{\varphi}$ is the rectified version of φ. ◀

Lemma 3.11. *A formula φ and its existential closure ψ are equisatisfiable.*

Proof.

$$
\begin{array}{ll}
\multicolumn{2}{l}{\psi \text{ is satisfiable}} \\
\text{iff} & \exists \mathcal{H}, \exists g, \exists w : \mathcal{H}, g, w \Vdash \exists s_1 \ldots \exists s_n \varphi \\
\text{iff} & \exists \mathcal{H}, \exists g, \exists w : \mathcal{H}, g_1, w \Vdash \exists s_2 \ldots \exists s_n \varphi, \text{ for some } g_1 \overset{s_1}{\sim} g \\
\text{iff} & \exists \mathcal{H}, \exists g, \exists w : \mathcal{H}, g_2, w \Vdash \exists s_3 \ldots \exists s_n \varphi, \text{ for some } g_2 \overset{s_2}{\sim} g_1 \overset{s_1}{\sim} g \\
\text{iff} & \cdots \\
\text{iff} & \exists \mathcal{H}, \exists g, \exists w : \mathcal{H}, g_n, w \Vdash \varphi, \text{ for some } g_n \overset{s_n}{\sim} g_{n-1} \overset{s_{n-1}}{\sim} \cdots \overset{s_2}{\sim} g_1 \overset{s_1}{\sim} g \\
\text{iff} & \exists \mathcal{H}, \exists g_n, \exists w : \mathcal{H}, g_n, w \Vdash \varphi \\
\text{iff} & \varphi \text{ is satisfiable}
\end{array}
$$

■

Let us apply the latter two results in the following examples:

Example 3.12. Let $\varphi_1 = @_i(\Diamond p \wedge \neg @_s p)$.
– This formula is rectified.
– The existential closure of φ_1 is the formula $\psi_1 = \exists s\, @_i(\Diamond p \wedge \neg @_s p)$.
It is easy to check that φ_1 and ψ_1 are equisatisfiable. ♦

Example 3.13. Let $\varphi_2 = @_i\, (\neg\, (\forall s @_s \neg p \wedge \exists s\, @_s p) \wedge @_s \neg p)$.
– This formula is not rectified.
The renaming of variables leads to $@_i\, (\neg\, (\forall t @_t \neg p \wedge \exists u @_u p) \wedge @_s \neg p)$,
which is equivalent to φ_2.
– The (rectified) existential closure of φ_2 is the formula
$\psi_2 = \exists s\, @_i\, (\neg\, (\forall t @_t \neg p \wedge \exists u @_u p) \wedge @_s \neg p)$.
The formulas φ_2 and ψ_2 are equisatifiable. ♦

Example 3.14. Let $\varphi_3 = @_i(\forall s \exists t @_s \Diamond t)$.

– This formula is rectified.

– Since φ_3 does not have free world variables, it coincides with its existential closure, ψ_3. ♦

The following theorem allows us to convert a formula into an equivalent formula in Prenex Conjunctive Normal Form.

Theorem 3.15. *Let $L = \langle \text{Prop}, \text{Nom}, \text{WVar} \rangle$ be a hybrid similarity type, Σ be an algebraic similarity type, and $\mathcal{H}$ be a hybrid structure over $\langle L, \Sigma \rangle$. For each formula of the form $@_i\varphi$, where $\varphi \in \text{Form}_{@,\forall}(L, \text{Term}(\Sigma, \text{WVar}, \text{Nom}))$ and $i \in \text{Nom}$ does not occur in φ, its existential closure ψ is equivalent to a formula in Prenex Conjunctive Normal Form.*

Proof. Let ψ be a formula in the conditions of the theorem.

Step 1: Use the double negation law, the De Morgan's laws, the duality equivalences $\forall s\, \varphi \equiv \neg \exists s\, \neg \varphi$ and $\Diamond \varphi \equiv \neg \Box \neg \varphi$, and the following rewrite rules until no further transformations apply.

$$\begin{array}{ll}
@_a(\theta_1 \wedge \theta_2) \twoheadrightarrow @_a\theta_1 \wedge @_a\theta_2 & @_a(\theta_1 \vee \theta_2) \twoheadrightarrow @_a\theta_1 \vee @_a\theta_2 \\
\neg @_a\theta \twoheadrightarrow @_a\neg\theta & @_a @_b\theta \twoheadrightarrow @_b\theta \\
@_a\Diamond\theta \twoheadrightarrow \exists u(@_a\Diamond u \wedge @_u\theta) & @_a\exists s\, \theta \twoheadrightarrow \exists s\, @_a\theta \\
@_a\Box\theta \twoheadrightarrow \forall u(@_a\Box\neg u \vee @_u\theta) & @_a\forall s\theta \twoheadrightarrow \forall s @_a\theta
\end{array}$$

where $a, b \in \text{Term}(\Sigma, \text{WVar}, \text{Nom})$ and $u \in \text{WVar}$ does not occur in ψ.

Step 2: Flush all quantifiers to the prefix position, as usual, and the result is a formula in Prenex Normal Form (since the variables added in **Step 1** are new, the formula remains rectified). Apply the associative and distributive laws as necessary in order to reach a formula in Prenex Conjunctive Normal Form.

Due to the rectified nature of the formulas over which the transformations have been applied, the resulting formulas are equivalent to the original ones. ∎

We return to the previous examples and apply the latter result:

Example 3.16. Let $\psi_1 = \exists s\, @_i(\Diamond p \wedge \neg @_s p)$:

Step 1:

$$\begin{array}{rcl}
\exists s\, @_i\, (\Diamond p \wedge \neg @_s p) & \twoheadrightarrow & \exists s\, (@_i \Diamond p \wedge @_i \neg @_s p) \\
 & \twoheadrightarrow & \exists s\, (\exists u\, (@_i \Diamond u \wedge @_u p) \wedge @_i @_s \neg p) \\
 & \twoheadrightarrow & \exists s\, (\exists u\, (@_i \Diamond u \wedge @_u p) \wedge @_s \neg p)
\end{array}$$

Step 2: $\exists s \exists u\, (@_i \Diamond u \wedge @_u p \wedge @_s \neg p)$ ♦

Example 3.17. Let $\psi_2 = \exists s\, @_i\, (\neg\, (\forall t @_t \neg p \land \exists u @_u p) \land @_s \neg p)$.
Step 1:

$$\begin{aligned}
&\exists s\, @_i\, (\neg\, (\forall t @_t \neg p \land \exists u @_u p) \land @_s \neg p) \\
&\quad \twoheadrightarrow\ \exists s\, @_i\, ((\neg \forall t @_t \neg p \lor \neg \exists u @_u p) \land @_s \neg p) \\
&\quad \twoheadrightarrow\ \exists s\, @_i\, ((\exists t \neg @_t \neg p \lor \forall u \neg @_u p) \land @_s \neg p) \\
&\quad \twoheadrightarrow\ \exists s\, (@_i\, (\exists t @_t \neg\neg p \lor \forall u @_u \neg p) \land @_i @_s \neg p) \\
&\quad \twoheadrightarrow\ \exists s\, ((@_i \exists t @_t p \lor @_i \forall u @_u \neg p) \land @_s \neg p) \\
&\quad \twoheadrightarrow\ \exists s\, ((\exists t @_i @_t p \lor \forall u @_i @_u \neg p) \land @_s \neg p) \\
&\quad \twoheadrightarrow\ \exists s\, ((\exists t @_t p \lor \forall u @_u \neg p) \land @_s \neg p)
\end{aligned}$$

Step 2: $\exists s \exists t \forall u\, ((@_t p \lor @_u \neg p) \land @_s \neg p)$ ♦

Example 3.18. Let $\psi_3 = @_i(\forall s \exists t @_s \Diamond t)$.
Step 1:

$$\begin{aligned}
@_i(\forall s \exists t\, @_s \Diamond t) \ &\twoheadrightarrow\ \forall s \exists t\, (@_i @_s \Diamond t) \\
&\twoheadrightarrow\ \forall s \exists t\, (@_s \Diamond t)
\end{aligned}$$

Step 2: $\forall s \exists t\, (@_s \Diamond t)$ ♦

Analogously to the corresponding construction in first-order logic, we can also resort to Skolemization in the Algebraic Strong Priorean Logic.

Lemma 3.19 (Skolemization in $\mathcal{H}(\Sigma, @, \forall)$). *Let φ be a sentence of the form $\forall s_1 \ldots \forall s_n \exists s_{n+1}\, G(s_1, \ldots, s_n, s_{n+1})$ of $\mathcal{H}(\Sigma, @, \forall)$, where the existentially quantified variable s_{n+1} is preceded by n universally quantified variables. In case $n = 0$, augment the underlying hybrid similarity type with a new nominal c and form the sentence $G(c)$; otherwise, augment the underlying hybrid similarity type with a new n-ary function symbol f and form the sentence $\forall s_1, \ldots, s_n G(s_1, \ldots, s_n, f(s_1, \ldots, s_n))$. Let φ' denote this new sentence, formed after the appropriate augmentation of the language. Then, there is an extension $\mathcal{H}'$ of the model $\mathcal{H}$ such that:*

$$\mathcal{H}, g, w \Vdash \varphi \ \textit{ iff } \ \mathcal{H}', g, w \Vdash \varphi'.$$

The (standard) proof of the latter result shows how to build the mentioned extension of the original model.

We now apply Skolemization to the previous examples.

Example 3.20. $\overline{\psi_1} = @_i \Diamond c_1 \land @_{c_1} p \land @_{c_2} \neg p$ ♦

Example 3.21. $\overline{\psi_2} = \forall u\, ((@_{c_2} p \lor @_u \neg p) \land @_{c_1} \neg p)$ ♦

Example 3.22. $\overline{\psi_3} = \forall s\,(@_s \Diamond f(s))$ ♦

Definition 3.23. A formula of $\mathcal{H}(\Sigma, @, \forall)$ is in *conjunctive Skolem form* if it is in Prenex Conjunctive Normal Form and its prefix contains only universal quantifiers. ◀

For a given formula φ, its Skolem Form is the result of applying labelling (Lemma 3.7), followed by the rectification and existential closure of the new formula (Lemma 3.11), then putting it in Prenex Conjunctive Normal Form (Theorem 3.15) and finally performing Skolemization (Lemma 3.19).

With conjunctive Skolem forms defined, we can state the following result:

Theorem 3.24. *A set Φ of formulas in $\mathcal{H}(\Sigma, @, \forall)$ is satisfiable iff the set of conjunctive Skolem forms of formulas in Φ is satisfiable.*

Proof. In view of Lemma 3.7, we know that the satisfiability of Φ is preserved when one considers the set $\{@_i\varphi \mid \varphi \in \Phi\}$, with i not occurring in any formula φ. Recall that such nominal is always possible to find, as we assumed Nom to be a countable set.

From Lemma 3.11, the satisfiability problem for $\{@_i\varphi \mid \varphi \in \Phi\}$ is the same as for $\{\overline{@_i\varphi} \mid \varphi \in \Phi\}$ where $\overline{@_i\varphi}$ represents the existential closure of $@_i\varphi$. This step is possible to accomplish since we also assumed WVar to be a countable set.

Furthermore, we can use the procedure employed in the proof of Theorem 3.15 in order to put formulas in Prenex Conjuntive Normal Form, and this is a procedure that strictly preserves the satisfiability of formulas. Thus we can deal with the satisfiability problem of $\{\mathrm{PCNF}\left(\overline{@_i\varphi}\right) \mid \varphi \in \Phi\}$ where $\mathrm{PCNF}(\psi)$ is the result of applying the steps in the proof of Theorem 3.15 to the formula ψ. Next we apply Skolemization to all formulas. Beware of the fact that the Skolem symbols introduced in each formula are to be disjoint. Let us call the resulting set $\tilde{\Phi}$. Clearly, by Lemma 3.19, the satisfiability problem for $\tilde{\Phi}$ is the same as for Φ. ■

The above relatively straightforward proof contrasts with proofs of the analogous result in first-order logic (see, e.g., [5]), which are often involved.

Definition 3.25. A *ground instance* of a sentence $\forall s_1 \ldots \forall s_n\, G(s_1, \ldots, s_n)$, with $G(s_1, \ldots, s_n)$ a quantifier-free formula of $\mathcal{H}(\Sigma, @, \forall)$, is a formula of the form $G(i_1, \ldots, i_n)$ which results from substituting all occurrences of $s_1, \ldots, s_n$ in G with nominals $i_1, \ldots, i_n$. ◀

Before presenting our Herbrand-like result for hybrid logic with quantifiers, we find it worth pointing out that hybrid logic can be translated into first-order logic with equality, and (a fragment of) first-order logic with equality can be translated back into (a fragment of) hybrid logic (cf. [3]). Both translations are truth-preserving. First-order logic is compact, which means that a set of first-order sentences is satisfiable if and only if every finite subset of it is satifiable. Furthermore, from our earlier Herbrand-like result (Theorem 2.6), we know that for a set of Boolean combinations of atomic satisfaction statements, satisfiability implies propositional satisfiability.

Theorem 3.26 (Herbrand-like). *Let L and Σ be, respectively, a hybrid and an algebraic similarity type, and let $\Phi \subseteq \mathrm{Form}_{@,\forall}(L, \mathrm{Term}(\Sigma, \mathrm{WVar}, \mathrm{Nom}))$. Then Φ is unsatisfiable iff some finite set Φ^* of ground instances of Skolem forms of $\Phi \cup \mathrm{Eq}(L)$ is propositionally unsatisfiable.*

Proof. By Theorem 3.24 the set Φ is unsatisfiable iff the set Ψ of conjunctive Skolem forms of formulas in Φ is unsatisfiable. So, in the present proof we will deal with Ψ.

Let us now prove the right-to-left direction of the theorem. First observe that, from Theorem 2.6, if a set Φ^* of ground instances of $\Psi \cup \mathrm{Eq}(L)$ is propositionally unsatisfiable then it is unsatisfiable. Furthermore, notice that a ground instance of a universal sentence τ is a logical consequence of τ. Therefore, if a set Φ^* of ground instances of $\Psi \cup \mathrm{Eq}(L)$ is unsatisfiable, then $\Psi \cup \mathrm{Eq}(L)$ is unsatisfiable, which yields that Ψ is unsatisfiable. It follows from the previous paragraph that Φ is unsatisfiable.

For the left-to-right direction of the theorem we prove the contrapositive: if every finite set of ground instances of Skolem forms of $\Phi \cup \mathrm{Eq}(L)$, i.e., ground instances of $\Psi \cup \mathrm{Eq}(L)$, is propositionally satisfiable, then Φ is satisfiable. Let Φ_0 be the set of all ground instances of $\Psi \cup \mathrm{Eq}(L)$. From the assumption that every finite subset of Φ_0 is propositionally satisfiable, it follows from compactness that the entire set Φ_0 is propositionally satisfiable. From Theorem 2.11, we conclude that Φ_0 is satisfiable. Thus $\Psi \cup \mathrm{Eq}(L)$ is satisfiable, from which Ψ is satisfiable, which finally implies that Φ is satisfiable. ■

4 Conclusion

We have proposed two versions of Herbrand's theorem in the context of hybrid logic, with a restriction to satisfaction statements, by making use of rules that rewrite each satisfaction statement as a Boolean combination of atomic satisfaction statements, and making use also of the fact that each model can be described by its diagram. We proved that a set of satisfaction statements is propositionally unsatisfiable if and only if it is unsatisfiable.

Formulas with quantifiers over objects constitute a challenge. In fact, allowing non-rigidity introduces a new set of problems: when dealing with non-rigid terms, i.e. terms that can designate different things at different possible worlds, the act of designation and the act of passing to an alternative world need not commute. For an example of how this has been dealt with elsewhere, it is worth to point out Fitting's version (cf. [7]) of Herbrand's theorem for the modal logic K with varying domains. Following the standard steps for Herbrand-like theorems, after going through Skolemization one gets non-rigid designators for some formulas and the above mentioned difficulty concerning non-commutativity ensues. In order to overcome this issue, Fitting resorted to the concepts of predicate abstraction and validity functional form. In short, if φ is a formula, then $\langle \lambda x.\varphi \rangle$ is a predicate abstraction that is to be applied to terms; loosely speaking, for $\langle \lambda x.\varphi \rangle(t)$ to be true at a world w, φ should be true in that world provided we take the value of x to be whatever the term t designates at w. The predicate abstraction mechanism does not have an important role to play in classical logic because all the classical connectives and quantifiers are 'transparent' to it. On the other hand, $\langle \lambda x.\Box\varphi \rangle(t)$ and $\Box\langle \lambda x.\varphi \rangle(t)$ may have very different meanings, from a semantical viewpoint. Fitting defines as modal Herbrand transform of a formula X the formula X' such that $X \to X'$ can be derived from a certain calculus that he presents. He later proves equivalence between the validity problem for a closed formula φ and for one of its modal Herbrand expansions, a notion built over that of modal Herbrand transforms. We are confident that within the hybrid scenario something similar is to be done: by adding just nominals and the satisfaction operator, and assuming that nominals are rigid, it would seem that @ is to behave as classical connectives and quantifiers do when interacting with the predicate abstraction mechanism, namely, that $\langle \lambda x.@_i\varphi \rangle(t)$ and $@_i\langle \lambda x.\varphi \rangle(t)$ are to share the same meaning. If the addition of nominals proves not to be worrisome, then updating the concept of modal Herbrand transform into hybrid Herbrand transform, after proper adjustments to the calculus proposed by Fitting in order to incorporate the hybrid machinery, should be rather trouble-free. The details need to be checked, of course, and we propose that as future work.

As in [1], we have here investigated a direct path towards the proofs of our main (Herbrand-like) results, without taking an indirect approach through first-order translations of the hybrid formulas. However, for a more straightforward comparison with the standard formulation of the Herbrand Theorem and its numerous applications, it might be worth exploring the connection of our present results concerning Hybrid Logic to the more long winded route going through its translation into classical first-order logic. For space reasons, though, we have to leave details of this reconnaissance to a future opportunity.

Acknowledgments. This work was supported in part by the Portuguese Foundation for Science and Technology (FCT) through CIDMA within project UID/MAT/04106/2019 and Dalí project POCI-01-0145-FEDER-016692, and in part by the Marie Curie project PIRSES-GA-2012-318986 funded by EU-FP7. Diana Costa also thanks the support of FCT via the Ph.D. scholarship PD/BD/105730/2014. João Marcos acknowledges partial support by CNPq and by the Humboldt Foundation. Comments by Raquel Oliveira, Cláudia Nalon and two anonymous referees have helped us to improve the paper.

References

[1] Carlos Areces and Daniel Gorín. Resolution with order and selection for hybrid logics. *Journal of Automated Reasoning*, 46(1):1–42, Jan 2011.

[2] Patrick Blackburn. Internalizing labelled deduction. *Journal of Logic and Computation*, 10(1):137–168, 2000.

[3] Patrick Blackburn. Representation, reasoning, and relational structures: A hybrid logic manifesto. *Logic Journal of the IGPL*, 8(3):339–365, 2000.

[4] Patrick Blackburn, Maarten de Rijke, and Yde Venema. *Modal logic.* Cambridge University Press, 2001.

[5] Alan Bundy. *The Computer Modelling of Mathematical Reasoning.* Academic Press Professional, Inc., San Diego, CA, USA, 1985.

[6] Stephen Cook and Phuong Nguyen. *Logical Foundations of Proof Complexity.* Cambridge University Press, New York, NY, USA, 1st edition, 2010.

[7] Melvin Fitting. A modal Herbrand theorem. *Fundamenta Informaticae*, 28(1-2):101–122, 1996.

[8] Jacques Herbrand. *Logical Writings.* Dordrecht, Holland, D. Reidel Pub. Co., 1971.

[9] Evangelos Tzanis. Algebraizing hybrid logic. Master's thesis, University of Amsterdam, Institute of Logic, Language and Computation, 2005.

Received 8 February 2018

Many-Valued Paraconsistent Extensions of Classical Positive Propositional Calculus

Leonid Devyatkin
RAS Institute of Philosophy, 12/1 Goncharnaya Str., Moscow, 109240, Russian Federation
deviatkin@iph.ras.ru

Abstract

This paper is devoted to several infinite classes of paraconsistent matrices possessing a number of desirable logical properties. Many-valued matrices have been an invaluable tool in many fields of logic, including the study of paraconsistency. In the latter case, the widespread approach to construction of logical matrices is to supplement the classically behaving conjunction, disjunction and implication with a paraconsistent negation, possibly with addition of extra operators. We show how to obtain countable sets of three-valued matrices and continual sets of four-valued matrices of this kind.

1 Introduction

Paraconsistent logics are logics where contradictions do not necessarily lead to triviality. In terms of logical consequence, such logics are defined as ones lacking *the principle of explosion*. That is, for some formulae α and β it is not the case that $\alpha, \neg\alpha \vdash \beta$. Clearly, this way paraconsistency is determined by the properties of negation alone, without regard for other connectives.

At the same time, in construction of paraconsistent logics, researchers usually aim to retain as much of classical propositional calculus (**CPC**) as possible. This results in the $\{\wedge, \vee, \supset\}$-fragment of classical logic ($\mathbf{CPC}^+$) being left intact.

For instance, in da Costa's calculus $\mathbf{C}_1$ 'all schemata and rules of deduction of the classical positive propositional calculus are true' [19, Th. 3]. The same is true for A.I. Arruda's system **V1** aimed at formalization of N.A. Vasiliev's philosophical ideas [5], and for the infinite sequence of paraconsistent logics obtained by generalization of Arruda's approach [40]. J. Ciuciura axiomatized Jaśkowski's discursive

The author is indebted to the anonymous referees whose comments helped to substantially improve the paper.

logic $\mathbf{D_2}$ as an extension of $\mathbf{CPC}^+$[15]. Other examples include but are not limited to paraconsistent extensional propositional logics of D. Batens [7] and annotated logics $\mathbf{P}\mathcal{T}$ [20].

The situation is no different in the many-valued realm. Starting from Sette's logic $\mathbf{P}^1$, in matrices of many prominent paraconsistent logics, including $\mathbf{J}_3$ (**CLuNs**, **LFI**1, **MPT**, **SP3B**) and **PAC** ($\mathbf{RM_3}$), the binary operations have been designed to behave classically. These and other examples are explored in [2], [14] and [13, § 4.4]. R.A. Lewin and I.F. Mikenberg have defined an infinite class of literal-paraconsistent matrices with the same property [32]. More examples can be found in [10], [9], and [25], where three-valued paraconsistent logics without *the principle of non-contradiction* ($\neg(\alpha \wedge \neg\alpha)$) are investigated.

It is then hardly surprising that projects aimed at generalized construction of 'good' many-valued paraconsistent logics converge on preservation of classical conjunction, disjunction and implication. J. Marcos proposed a class of 8,192 three-valued matrices which induce paraconsistent logics with a number of desirable properties [35]. This class is the result of generalization of theoretical considerations laid out in [37]. It consists of matrices for the language in the signature $\{\wedge, \vee, \supset, \neg\}$ where operations can vary as long as they preserve the classical truth-values $\{0, 1\}$, $\neg$ behaves as appropriate for paraconsistent negation, and the binary operations conform with the *standard conditions* of J.B. Rosser and A.R. Turquette [41, p. 26], which makes them essentially classical. In [14] this set of operations is supplemented with the unary operators $\circ$ and $\bullet$ to obtain a class of **LFI**s (*logics of formal inconsistency*), and their various properties are investigated.

Another generalized approach to many-valued paraconsistent logics is presented by O. Arieli, A. Avron and A. Zamansky in [3] and [2]. In the first paper, the authors set out to determine what properties an 'ideal propositional paraconsistent logic' is supposed to have. The resulting definition of such an ideal logic requires it to include a classically behaving implication. Given the restrictions placed by the authors on negation, this implies the definability of classical conjunction and disjunction as well. It is made explicit in the second latter paper, where the authors see only the appropriate negation as the necessary basic operation, and refer to operations for conjunction, disjunction and implication as 'possibly definable'. In [3, Th. 3], the authors also provide an algorithm for construction of an infinite sequence of 'ideal paraconsistent logics' defined by matrices with steadily increasing numbers of truth-values.

The prominence of the approach laid out above merits the question, how many $\{0, 1\}$-preserving three-valued paraconsistent extensions of $\mathbf{CPC}^+$ are there in total? In the sequel, we will demonstrate that the answer is: infinitely many.

The remaining part of the paper is structured as follows.

In the next section, the central problem is addressed in a more formal manner. We introduce the necessary basic definitions, explore the classes of matrices from [14] and [3], and determine the further steps required to answer the principal question of the paper for the three-valued case.

In the third section, we use elements of clone theory to demonstrate that the set of three-valued paraconsistent extensions of $\mathbf{CPC}^{+}$ that are **LFI**s is uncountably infinite, and the set of such extensions that are not **LFI**s is at least countably infinite.

The final part is devoted to four-valued paraconsistent extensions of $\mathbf{CPC}^{+}$. In particular, we deal with such extensions that are both paraconsistent and paracomplete at the same time. We generalize the three-valued truth-table schemata from the second section to obtain a class of four-valued matrices and utilize it to prove that two kinds of four-valued extensions of $\mathbf{CPC}^{+}$ — those that are **LFI**s and those that are not — form continuum sets.

2 Preliminaries

Let $\mathcal{L} = \langle L, F\rangle$ be a *propositional language* treated as an absolutely free algebra. We assume that the free generators of $\mathcal{L}$ form a countable set $Var = \{p_1, p_2, \dots\}$ and for each $i \leq n$ the arity of $F_i \in F$ equals k_i.

A *logical matrix* is a structure $\mathcal{M} = \langle \mathcal{A}, D\rangle$ where $\mathcal{A} = \langle A, F\rangle$ is an algebra and $D \subseteq A$. In this case, a homomorphism h from $\mathcal{L}$ into $\mathcal{A}$ is called a *valuation* for $\mathcal{L}$ in $\mathcal{M}$. We denote as $Val(\mathcal{M})$ the set of all the valuations over $\mathcal{M}$.

If $\mathcal{L}$ is a propositional language and $\mathcal{M}$ is a matrix for $\mathcal{L}$,

$$Cn(\mathcal{M}) = \{\langle X, \alpha\rangle \in \wp(L) \times L | \forall h \in Val(\mathcal{M})(h(X) \subseteq D \Rightarrow h(\alpha) \in D)\}$$

is said to be a *matrix consequence* induced by $\mathcal{M}$. A pair $\mathbf{L} = \langle \mathcal{L}, Cn(\mathcal{M})\rangle$ is then called a *many-valued (propositional) logic*[1].

Let $\mathcal{M} = \langle A, \wedge, \vee, \rightarrow, \neg, D\rangle$ be a logical matrix. Suppose its basic operations satisfy the *standard conditions* of J.B. Rosser and A.R. Turquette [41, p. 26]:

- $x \wedge y \in D \Leftrightarrow x \in D$ and $y \in D$;
- $x \vee y \notin D \Leftrightarrow x \notin D$ and $y \notin D$;
- $x \rightarrow y \notin D \Leftrightarrow x \in D$ and $y \notin D$;
- $\neg x \in D \Leftrightarrow x \notin D$.

[1] We are following D.J. Shoesmith and T.J. Smiley in this definition [45], [46, § 13.1].

Then $\mathcal{M}$ induces the classical consequence in the corresponding language: **CPC** $= \langle \mathcal{L}, Cn(\mathcal{M}) \rangle$ [46, § 18.3].

We say that a matrix $\mathcal{M}_k = \langle \{0, 1, \ldots, k\}, F, D \rangle$ is $\{0,1\}$-*preserving* iff $\mathcal{M}_2 = \langle \{0, 1\}, F, \{1\} \rangle$ is its *submatrix*, i.e. $\langle \{0, 1\}, F \rangle$ is a subalgebra of $\langle \{0, 1, \ldots, k\}, F \rangle$ and $1 \in D$. If $\mathcal{M} = \langle A, F, D \rangle$ is $\{0,1\}$-preserving, then $Cn(\mathcal{M}_k) \subseteq Cn(\mathcal{M}_2)$.

A matrix $\mathcal{M} = \langle A, f_1, \ldots, f_n, \neg, D \rangle$ for $\mathcal{L}$ is said to be *paraconsistent* (w.r.t. $\neg$) iff $\langle \{\alpha, \neg\alpha\}, \beta \rangle \notin Cn(\mathcal{M})$ for some $\alpha, \beta \in L$. Obviously, the necessary and sufficient condition for this is the existence of $a \in D$, such that $\neg a \in D$ [2, § 4].

Given the above, if one intends to define a three-valued paraconsistent matrix and at the same time diverge from the classical logic as little as possible, it seems rather natural to take a $\{0,1\}$-preserving three-valued matrix with 'standard' basic operations $\{\wedge, \vee, \rightarrow, \neg\}$ and modify the negation to obtain paraconsistency: $2 \in D$ and $\neg 2 \in D$. This approach results in matrices of the form

$$\mathcal{M} = \langle \{0, 1, 2\}, \wedge, \vee, \rightarrow, \neg, \{1, 2\} \rangle,$$

where basic operations correspond to the truth-table schemata below [2].

$\wedge$	0	1	2
0	0	0	0
1	0	1	1 ≀ 2
2	0	1 ≀ 2	1 ≀ 2

$\vee$	0	1	2
0	0	1	1 ≀ 2
1	1	1	1 ≀ 2
2	1 ≀ 2	1 ≀ 2	1 ≀ 2

$\rightarrow$	0	1	2
0	1	1	1 ≀ 2
1	0	1	1 ≀ 2
2	0	1 ≀ 2	1 ≀ 2

	$\neg x$
0	1
1	0
2	1 ≀ 2

Here and elsewhere in the paper the notation '1 ≀ 2' signifies that either 1 or 2 should be picked for the resulting truth-table. This way we obtain 2^3 conjunctions, 2^5 disjunctions, 2^4 implications, 2 negations, and, as a result, $2^{13} = 8,192$ matrices. We will label the class of all such matrices as NAT. As pointed out in [2], each logic defined by a matrix from NAT is different from the others.

In [14, p. 77–79], the schemata are supplemented with two more operations:

	$\circ x$
0	1
1	1
2	0

	$\bullet x$
0	0
1	0
2	1

We denote as LFI the class of all matrices of the form

$$\mathcal{M} = \langle \{0, 1, 2\}, \wedge, \vee, \rightarrow, \neg, \circ, \bullet, \{1, 2\} \rangle.$$

By construction, LFI contains 8,192 matrices as well, and they also define pairwise distinct logics [14, p. 78, Th. 130].

Although, there are two observations to be made. First, as noted in [2], while all matrices in NAT define distinct logics, some of them have equivalent expressive power. Obviously, the same goes for LFI. At the same time, as shown by P. Wojtylak, whenever two matrices which differ only in the sets of basic operations have equivalent expressive power, the logics they define can be conservatively translated into each other [52], [53]. In other words, they can be considered linguistic variants of the same logic (see [51, § 1.8]).

To identify logics that are genuinely different, we should partition the matrices into equivalence classes with respect to mutual definability of basic operation sets. An example of such approach can be found in [49], where this work is done for implicative extensions of Weak Kleene Logic and Paraconsistent Weak Kleene Logic (see [16], [11], [48] regarding the latter). The author has shown that 24 matrices obtained from the matrix of **PWK** can be partitioned into just 7 classes of equivalence. If we adopt this approach, we will find that NAT generates much fewer negative extensions of $\mathbf{CPC}^+$ as well.

Second, the definition yields matrices for the fixed language in the signature $\{\wedge, \vee, \supset, \neg\}$. However, most results regarding such matrices hold for their $\{0,1\}$-preserving extensions as well. At the same time, some of those extensions can not be defined within NAT. Consider the matrix $\langle\{0,1,2\}, \wedge, \vee, \rightarrow, \neg, *, \{1,2\}\rangle$.

$\wedge$	0	1	2
0	0	0	0
1	0	1	1
2	0	1	1

$\vee$	0	1	2
0	0	1	1
1	1	1	1
2	1	1	1

$\rightarrow$	0	1	2
0	1	1	1
1	0	1	1
2	0	1	1

	$\neg x$
0	1
1	0
2	1

	$*x$
0	0
1	0
2	2

Evidently, there is no matrix $\langle\{0,1,2\}, \wedge, \vee, \rightarrow, \neg, \{1,2\}\rangle$ from NAT of equivalent expressive power. So there are more paraconsistent extensions of $\mathbf{CPC}^+$ than NAT generates.

In light of the observations presented above, we can conclude that the question regarding the number of three-valued paraconsistent extensions of $\mathbf{CPC}^+$ has not yet been answered decisively. To frame this question in more precise terms, let us first focus on some facts concerning NAT and LFI.

Notice that $CPC = \langle\{0,1\}, \wedge, \vee, \rightarrow, \neg, \{1\}\rangle$, the usual matrix of **CPC**, is a submatrix of every $M \in$ NAT. As it follows from A.V. Makarov's results [34], this entails that in every M one can define at least one of the following operations:

$\downarrow_1$	0	1	2
0	1	0	1
1	0	0	0
2	1	0	1

$\downarrow_2$	0	1	2
0	1	0	0
1	0	0	0
2	0	0	0

$\downarrow_3$	0	1	2
0	1	0	0
1	0	0	0
2	0	0	2

If $\downarrow_1$ or $\downarrow_2$ are definable in $\mathcal{M}$, then $\mathcal{M}$ is an extension of Sette's matrix $\mathcal{P}^1$ [43]. It follows from the results in [14] (see Theorem 136 in particular) and the fact that $x \downarrow_1 x$ defines the Sette negation, and $x \downarrow_2 x$ defines the classical negation[2].

Both $\circ$ and $\bullet$ are definable in $\mathcal{P}^1$, so every extension of $\mathcal{P}^1$ defines an **LFI**, a logic that is paraconsistent, but *gently explosive* [14, pp. 19–21].

If $\downarrow_3$ is definable in $\mathcal{M}$, then $\mathcal{M}$ is an extension of Sobociński's matrix $\mathcal{A}_1$ (as described in [2, § 5.2]). In fact, $\downarrow_3$ is a Pierce arrow for $\mathcal{A}_1$, as demonstrated by the following identities: $\sim x =: x \downarrow_3 x$; $x \cap y =: \sim x \downarrow_3 \sim y$; $x \rightarrow y =: \sim (\sim x \downarrow_3 y)$; $x \Rightarrow y =: x \rightarrow\sim (y \rightarrow\sim x)$; $x \twoheadrightarrow y =: (x \Rightarrow y) \cap (\sim y \Rightarrow\sim x)$; $x \otimes y =\sim (x \twoheadrightarrow\sim y)$, where $\otimes$ is Sobociński's conjunction.

A matrix is called $\{2\}$-preserving iff for each its operation $f(x_1, \ldots, x_n)$ it holds that $f(2, \ldots, 2) = 2$. If $\mathcal{M}$ is $\{2\}$-preserving, then for every set of formulas $\bigcirc(p)$ we have $h(\{\bigcirc(p), p, \neg(p)\}) \subseteq D$ whenever $h(p) = 2$. Therefore, $\langle\{\bigcirc(p), p, \neg(p)\}, q\rangle \notin Cn(\mathcal{M})$, and the logic induced by $\mathcal{M}$ is not an **LFI** (cf. Example 17 and Theorem 25 in [14, pp. 17–25]). In particular, this is the case for $\mathcal{A}_1$ or any of its $\{2\}$-preserving extensions. Notice that a more general claim also holds: for every $X \subseteq \mathcal{L}$, if $Var(X) \neq Var(\mathcal{L})$, then there is $\alpha \in \mathcal{L}$, such that $X \nvdash \alpha$. The logics with such a property are called *non-exploding* [1, Def. 26].

Now consider the following truth-tables:

$\wedge$	0	1	2
0	0	0	0
1	0	1	2
2	0	2	2

$\vee$	0	1	2
0	0	1	2
1	1	1	1
2	2	1	2

$\rightarrow_1$	0	1	2
0	1	1	1
1	0	1	1
2	0	1	1

$\rightarrow_2$	0	1	2
0	1	1	1
1	0	1	2
2	0	1	2

	$\neg x$
0	1
1	0
2	2

The matrix $\langle\{0, 1, 2\}, \wedge, \vee, \rightarrow_1, \neg, \{1, 2\}\rangle$ defines the logic **MPT** [17] (also studied as **TLP** in [47, § 2.1]). Numerous linguistic variants of this logic are known in the literature: $\mathbf{J}_3$, **CLuNs**, **LFI1** (see [14, p. 18–19], [13, § 4.4.3] and references therein). Recent new version of this logic has appeared in [9] under the name **SP3B**.

[2]Interestingly enough, whenever $\downarrow_1$ is definable in $\mathcal{M} \in \mathsf{NAT}$, so is $\downarrow_2$, and vice versa, as $\{\downarrow_1, \downarrow_2\}$ constitutes a base for $\mathcal{P}^1$. See [28] and [29] for detailed elaborations on this theme.

The matrix $\langle\{0,1,2\},\wedge,\vee,\rightarrow_2,\neg,\{1,2\}\rangle$ defines the logic **PAC**. This logic is also known in the literature under other names, such as **PCont** [42], **RM**$_3$, **PI**^{s} (see [2, § 5.4], [14, p. 17–18] and references therein).

We will label the two matrices as $\mathcal{MPT}$ and $\mathcal{PAC}$ respectively. Observe that both $\mathcal{MPT}$ and $\mathcal{PAC}$ belong to NAT. As it follows from [6, Th. 2.21], every $\{0,1\}$-preserving operation is definable in $\mathcal{MPT}$, and in $\mathcal{PAC}$ one can define any operation that is both $\{0,1\}$-preserving and $\{2\}$-preserving at the same time.

Hence, all $\{0,1\}$-preserving three-valued paraconsistent extensions of **CPC**$^+$ that are **LFI**s are defined by matrices that are extensions of $\mathcal{P}^1$ and are definable in $\mathcal{MPT}$. All $\{0,1\}$-preserving three-valued paraconsistent extensions of **CPC**$^+$ that are non-exploding are defined by matrices that are extensions of $\mathcal{A}_1$ and are definable in $\mathcal{PAC}$.

Now we can refine the central question of this paper for the three-valued case in the following way:

- How many matrices with expressive power between $\mathcal{P}^1$ and $\mathcal{MPT}$ are not mutually definable?
- How many matrices with expressive power between $\mathcal{A}_1$ and $\mathcal{PAC}$ are not mutually definable?

The number of matrices can be used to determine the number of many-valued logics they induce. Still, there are points related to the connection between matrices and calculi that require clarification.

Consider the set $E_k = \{0,1,\ldots,k-1\}$. Let us denote the set of all n-ary functions on E_k as P_k^n. For any fixed $n \in \mathbb{N}$ the total number of n-ary functions definable on E_k equals k^{k^n}, so P_k^n is always finite. Now define the set $P_k = \bigcup_{n\geq 1} P_k^n$. Since $\mathbb{N}$ is countably infinite, so is P_k. Therefore, P_k contains the countably infinite number of finite subsets and the uncountably infinite (cardinality of $\aleph_1$) number of infinite subsets.

When we define a k-valued logic, we establish a one-to-one correspondence between basic connectives of a given propositional language and a set of functions of a k-valued algebra. In other words, a set of basic connectives is put into correspondence with a subset of P_k.

If the arity of the basic connectives of $\mathcal{L}$ is not allowed to exceed some fixed m (for example, all basic connectives have to be either unary or binary), only finitely many k-valued logics in the language $\mathcal{L}$ can be defined for a given k. On the other hand, if there is no restriction regarding the arity of basic connectives, the amount of possible k-valued logics becomes at least countably infinite.

Some authors (e.g. R. Wòjcicki [51, p. 12]) define propositional languages as having strictly finitely many basic connectives. In this case only the countably infinite number of k-valued logics can be defined for a given k. However, it is also rather common in the newer literature to not make the finiteness of the set of basic connectives a hard requirement (e.g. [23, pp. 2–4], [24, p. 15], [27, pp. 51–52]). Once we allow the set of basic connectives to be infinite, we are enabled to define continuum-many k-valued logics.

A three-valued version of $\mathbf{CPC}^+$ is obtained when we pick three binary functions ($f_\wedge$, $f_\vee$, $f_\supset$) on $\{0, 1, 2\}$, which satisfy the 'standard conditions' of Rosser and Turquette for conjunction, disjunction and implication respectively. If we supplement the syntax of $\mathbf{CPC}^+$ with new connectives $\S_1, \ldots, \S_r$ and assign to them the functions $f_{\S_1}, \ldots, f_{\S_r}$ of corresponding arities, we get a three-valued linguistic extension of $\mathbf{CPC}^+$. The number of three-valued linguistic extensions of $\mathbf{CPC}^+$ then coincides with the number of supersets of $\{f_\wedge, f_\vee, f_\supset\}$ in P_3.

We have already established that P_k contains countably many finite subsets and uncountably many infinite subsets for every k. Trivially, P_3 contains countably many finite subsets and uncountably many infinite subsets, which contain $\{f_\wedge, f_\vee, f_\supset\}$ — we obtain them by simply supplementing $\{f_\wedge, f_\vee, f_\supset\}$ with one subset of $P_3 \setminus \{f_\wedge, f_\vee, f_\supset\}$ at a time. Although the matter becomes not as trivial when we focus on not just any subsets of P_3, but only the ones closed w.r.t. superposition.

In the literature, a number of *sufficient* conditions is known under which a closed subset of P_3 has finitely, countably or uncountably many closed supersets (see e.g. [57]). However, no universal procedure to determine the number of closed supersets of a given closed subset of P_3 is known to the author of the present paper. In what follows, we demonstrate that some closed subsets of P_3 generated by sets of the form $\{f_\wedge, f_\vee, f_\supset\}$ are in fact contained in infinitely many pairwise distinct subsets of P_3 — and that is how the infinite number of three-valued pairwise distinct linguistic extensions of $\mathbf{CPC}^+$ is shown to be possible.

3 Three-Valued Extensions of CPC$^+$

In this section, we will show that there is a uncountably infinite set of closed sets of functions, each of which can be used to define a functional extension of $\mathcal{P}^1$, and a countably infinite chain of closed sets of functions, each of which can be used to define a functional extension of $\mathcal{A}_1$. First, a number of definitions and notational remarks is in order.

Let A be a finite set with at least two elements. We will denote the set of all functions on A as P_A. The so-called *Mal'tsev operations* on P_A are defined as follows[3]:

- $(\zeta f)(x_1, \ldots, x_n) = f(x_2, x_3, \ldots, x_n, x_1)$,
- $(\tau f)(x_1, \ldots, x_n) = f(x_2, x_1, x_3, \ldots, x_n)$,
- $(\Delta f)(x_1, \ldots, x_{n-1}) = f(x_1, x_1, x_2, \ldots, x_{n-1})$ in $n \geq 2$,
- $\zeta f = \tau f = \Delta f = f$ if $n = 1$,
- $(\nabla f)(x_1, \ldots, x_{n+1}) = f(x_2, x_3, \ldots, x_{n+1})$,
- $(f \star g)(x_1, \ldots, x_{m+n-1}) = f(g(x_1, \ldots, x_m), x_{m+1}, \ldots, x_{m+n-1})$.

The algebra $\mathcal{P}_A = \langle P_A, \zeta, \tau, \nabla, \Delta, \star\rangle$ is called *iterative full function algebra* on A.

A function $f \in P_A$ is called a *superposition* over $F \subseteq P_A$ iff f can be obtained from the functions of F by a finite number of applications of Mal'tsev operations. The set of all superpositions over $F \subseteq P_A$ is called the *closure* of F and is denoted by $[F]$. A set $F \subseteq P_A$ satisfying $[F] = F$ is called a *closed set* or a *subclass* of P_A.

Denote as P_3 the iterative full function algebra on $\{0, 1, 2\}$. In the previous section we have defined three-valued matrices on the same set. This way, the set of all operations definable in a three-valued matrix $\mathcal{M}$ coincides with some subclass M of P_3. If $\mathcal{M}_1$ and $\mathcal{M}_2$ are three-valued matrices with the same designated values, $\mathcal{M}_1$ is definable in M_2 iff M_1 is a subclass of M_2. Such matrices are equivalent in expressive power iff their basic operations generate the same subclass of P_3.

The basic operations of $\mathcal{MPT}$ generate the maximal subclass T_{01} of P_3 which consists of all $\{0, 1\}$-preserving functions. The basic operations of $\mathcal{P}^1$ generate the subclass $P_{3,2}$ of P_3 which consists of all functions with the domain in $\{0, 1, 2\}$ and the image in $\{0, 1\}$ [44].

The amount of matrices with expressive power between $\mathcal{P}^1$ and $\mathcal{MPT}$ that are not mutually definable then equals the amount of subclasses of P_3 that contain $P_{2,3}$ and are contained in T_{01}.

Theorem 1. *There are continuum-many subclasses of P_3 that contain $P_{3,2}$ and are contained in T_{01}.*

Proof. Consider the $\{0, 1\}$-preserving function $r_n(x_1, \ldots, x_n)$, where $n \in \mathbb{N} \setminus \{1\}$:

$$r_n(x_1, \ldots, x_n) = \begin{cases} 1, & \text{if for some } i \in \{1, \ldots, n\}\ x_i = 2 \text{ and} \\ & x_j = 1 \text{ for all } j \neq i, \\ 2, & \text{if } x_1 = \cdots = x_n = 2, \\ 0 & \text{otherwise.} \end{cases}$$

[3]For more detailed exposition of this material see [30, P. II, Ch. 1]

It is known that the system $\{r_i | i \geq 2\}$ constitutes a countable *basis* for $[\{r_i | i \geq 2\}]$ [30, p. 426, Lemma 14.10.4]. This entails $r_k \notin [\{r_i | i \geq 2\} \setminus \{r_k\}]$ for every $r_k \in \{r_i | i \geq 2\}$. Let us denote $[\{r_i | i \geq 2\} \setminus \{r_k\}]$ as R_k.

Now suppose $r_k(x_1, \ldots, x_k) = \Phi(\Phi_1, \ldots, \Phi_m)$, where $\Phi \in R_k \cup P_{3,2}$, $\Phi_1, \ldots, \Phi_m$ are either variables or functions from $R_k \cup P_{3,2}$.

Notice that functions from R_k produce the value 2 iff all their arguments take this value. At the same time, functions from $P_{3,2}$ never produce the value 2. Consequently, if there was even one instance of a function from $P_{3,2}$ in $\Phi, \Phi_1, \ldots, \Phi_m$, then $\Phi(\Phi_1, \ldots, \Phi_m) \neq 2$ would hold. By definition, $r_k(x_1, \ldots, x_k) = 2$ iff $x_1 = \cdots = x_k = 2$, so $\Phi, \Phi_1, \ldots, \Phi_m$ can contain no functions from $P_{3,2}$.

As it follows from the above, $\Phi, \Phi_1, \ldots, \Phi_m$ contain only variables and functions from R_k. But then $\Phi(\Phi_1, \ldots, \Phi_m) \in [R_k]$. Again, there is a contradiction, and we conclude that $r_k(x_1, \ldots, x_k) \notin [R_k \cup P_{3,2}]$.

Denote two arbitrary subsets of $\{r_i | i \geq 2\}$ as R' and R''. Since $r_k(x_1, \ldots, x_k) \notin [R_k \cup P_{3,2}]$, it is also the case that $[R' \cup P_{3,2}] \neq [R'' \cup P_{3,2}]$ whenever $R' \neq R''$. Now, $\{r_i | i \geq 2\}$ is countable, so it contains continuum-many (possibly infinite) subsets. This means that there are also continuum-many different subclasses of P_3 generated by systems of the form $R \cup P_{3,2}$, where $R \subseteq \{r_i | i \geq 2\}$. □

This theorem provides a recipe to obtain continuum-many paraconsistent linguistic extensions of **CPC**$^+$ that are **LFI**s. Consider Sette's matrix

$$\mathcal{P}^1 = \langle \{0, 1, 2\}, \wedge, \vee, \supset, \neg, \{1, 2\} \rangle$$

for the language $\mathcal{L} = \langle L, \dot{\wedge}, \dot{\vee}, \dot{\supset}, \dot{\neg} \rangle$. It defines the logic $\mathbf{P}^1$ which is already known to be a paraconsistent extension of **CPC**$^+$ and an **LFI**. Consequently, so are all linguistic extensions of $\mathbf{P}^1$.

Let us supplement the basic operations of $\mathcal{P}^1$ with some (possibly infinite) $R \subseteq \{r_i | i \geq 2\}$. We obtain the matrix $\mathcal{P}^1_R = \langle \{0, 1, 2\}, \wedge, \vee, \supset, \neg, r_{j_1}, r_{j_2}, \ldots \{1, 2\} \rangle$ for the language $\mathcal{L} = \langle L, \dot{\wedge}, \dot{\vee}, \dot{\supset}, \dot{\neg}, \dot{r}_{j_1}, \dot{r}_{j_2}, \ldots \rangle$. Clearly, $\mathcal{P}^1_R$ defines a linguistic extension of $\mathbf{P}^1$.

By virtue of Theorem 1, such matrices $\mathcal{P}^1_{R'}$ and $\mathcal{P}^1_{R''}$ constructed using subsets R' and R'' of $\{r_i | i \geq 2\}$ are equivalent in expressive power iff $R' = R''$. Therefore, the set of all logics of the type $\mathbf{P}^1_R$ that are not linguistic variants of each other has the power of continuum. However, if we limit ourselves to languages with finite amount of basic connectives, this set becomes just countable, since only finite subsets of $\{r_i | i \geq 2\}$ would be available in this case.

Now we will use the similar strategy to estimate the amount of matrices with expressive power between $\mathcal{A}_1$ and $\mathcal{PAC}$. We were able to prove that this amount is at least countably infinite. Whether there are exactly that many is an open question. Let us introduce some necessary concepts and auxillary lemmata.

Notice that F is a subclass of P_A iff $\langle F, \zeta, \tau, \nabla, \Delta, \star\rangle$ is a subalgebra of $\mathcal{P}_A$. Let $\mathcal{P}_A$ and $\mathcal{P}_{A'}$ be iterative full function algebras on A and A' respectively, $\mathcal{F} = \langle F, \zeta, \tau, \nabla, \Delta, \star\rangle$ a subalgebra of $\mathcal{P}_A$, and $\mathcal{G} = \langle G, \zeta', \tau', \nabla', \Delta', \star'\rangle$ a subalgebra of $\mathcal{P}_{A'}$. We say that $F \subseteq P_A$ is a *homomorphic inverse image* of $G \subseteq P_{A'}$ iff there exists a homomorphism from $\mathcal{F}$ onto $\mathcal{G}$.

The following lemmata belong to S.V. Yablonskii [54, §8]. In English, similar topics are treated in [30, P. II, §§5.2.3, 12.1].

Lemma 2. *Let $A' \subseteq A$. We say that a subclass F of P_A* preserves a set A' *iff for every $f \in F$ it is true that $f(x_1, \ldots, x_n) \in A'$ whenever $x_i \in A'$ for each i $(1 \leq i \leq n)$. If F preserves A', it is a homomorphic inverse image of a subclass G of $P_{A'}$ with respect to the following homomorphism φ: for $f^n \in F$ and $g^m \in G$ let $\varphi(f^n) = g^m$ iff $m = n$ and $f(x_1, \ldots, x_n) = g(x_1, \ldots, x_n)$ whenever $x_i \in A'$ for each $1 \leq i \leq n$.*

Lemma 3. *A subclass F of P_A is said to* preserve a partition $\pi : A = A_0 + A_1 + \cdots + A_n$ *iff for every $f \in F$ the following holds:*

if $f(x_1, \ldots, x_{i-1}, x_i, x_{i+1}, \ldots, x_m) \in A_j$ and $x_i \in A_k$ $(j, k \in \{0, 1, \ldots, n\}$, $1 \leq i \leq m)$, then $f(x_1, \ldots, x_{i-1}, x'_i, x_{i+1}, \ldots, x_m) \in A_j$ whenever $x'_i \in A_k$.

Suppose F preserves π, and $A' = \{a_0, a_1, \ldots, a_n\}$. Then F is a homomorphic inverse image of a subclass G of $P_{A'}$ with respect to the following homomorphism ψ: for $f^n \in F$ and $g^m \in G$ let $\psi(f^n) = g^m$ iff $m = n$ and $q(f(x_1, \ldots, x_n)) = g(q(x_1), \ldots, q(x_n))$, where $q(A_j) = a_j$.

Using the lemmata above, A.V. Makarov has proposed a method which allows one to show that certain subclasses of P_3 form countably infinite chains [33].

Let $\{f_\mu(x_1, \ldots, x_{\mu+1}) | \mu \in \mathbb{N} \setminus \{1, 2\}\}$ be a system of functions from P_3, where

$$f_\mu(x_1, \ldots, x_{\mu+1}) = \begin{cases} 2, & \text{if at least } \mu \text{ variables} \\ & \text{take the value 2,} \\ 0 & \text{otherwise.} \end{cases}$$

Lemma 4. *Closed sets of the form $F_\mu = [\{f_\mu(x_1, \ldots, x_{\mu+1})]$ constitute a countably infinite chain.*

Proof. Notice that $F_3 = [\{f_\mu(x_1, \ldots, x_{\mu+1}) | \mu \in \mathbb{N} \setminus \{1, 2\}\}]$ preserves the partition with blocks $A_0 = \{0, 1\}$, $A_1 = \{2\}$. By Lemma 3, F_3 is an inverse homomorphic image of a subclass of P_2 with respect to ψ, where $q(\{0,1\}) = 0$, $q(\{2\}) = 1$. For each f_μ, $\psi(f_\mu) = h_\mu$, where

$$h_\mu(x_1,\dots,x_{\mu+1}) = \begin{cases} 1, & \text{if at least } \mu \text{ variables} \\ & \text{take the value 1,} \\ 0 & \text{otherwise.} \end{cases}$$

It is known that $[h_\mu(x_1,\dots,x_{\mu+1})] = F_6^\mu$ for $\mu \geq 3$. Moreover, it is known that $h_q(x_1,\dots,x_{q+1}) \notin F_6^p$, where $q \geq 3$, $p > q$. Therefore, there is an infinite chain of subclasses of P_2 [55, Ch. IV, § 9, Lemma 35][4]:

$$F_6^3 \supset F_6^4 \supset \dots \supset F_6^p \supset \dots$$

Since F_3 is an inverse homomorphic image of F_6^3, it is also the case that $[f_\mu(x_1,\dots,x_{\mu+1})] = F_\mu$, $f_q(x_1,\dots,x_{q+1}) \notin F_p$, if $q \geq 3$, $p > q$, and there is an infinite chain of subclasses of P_3:

$$F_3 \supset F_4 \supset \dots \supset F_p \supset \dots$$

□

Denote as T_2 the subclass of P_3 which consists of $\{2\}$-preserving functions. The closed set of all functions definable in $\mathcal{PAC}$ then coincides with $T_{01} \cap T_2$. Moreover, the class of all functions definable in $\mathcal{A}_1$ (i.e. $[\downarrow_3]$) is contained in T_2, and so is each F_m $(3 \leq m \leq p)$ from Lemma 4. This leads to the following theorem.

Theorem 5. *The set of subclasses of P_3 which contain $[\downarrow_3]$ and are contained in $T_{01} \cap T_2$ is at least countably infinite.*

Proof. Consider the system of functions $\{\downarrow_3, \{f_\mu(x_1,\dots,x_{\mu+1}) | \mu \in \mathbb{N} \setminus \{1,2\}\}\}$.

Notice that $\downarrow_3$ preserves the partition with blocks $A_0 = \{0,1\}$, $A_1 = \{2\}$. By Lemmata 3 and 4, F_3^* is an inverse homomorphic image of a subclass of P_2 with respect to ψ, where $q(\{0,1\}) = 0$, $q(\{2\}) = 1$.

As shown in Lemma 4, $\psi(f_\mu) = h_\mu$. By the definition of ψ, $\psi(x_1 \downarrow_3 x_2) = x_1 \wedge x_2$, where '$\wedge$' denotes Boolean conjunction.

It is known[5] that $h_\mu(x_1,\dots,x_{\mu+1}) = \bigvee_{i=1}^{\mu+1} x_1 \wedge \dots \wedge x_{i-1} \wedge x_{i+1} \wedge \dots \wedge x_{\mu+1}$. Consequently, $x_1 \wedge x_2$ can be obtained from $h_\mu(x_1,\dots,x_{\mu+1})$ by identification of variables, so $x_1 \wedge x_2 \in [h_\mu(x_1,\dots,x_{\mu+1})]$ for every μ.

As a result, F_3^* is an inverse homomorphic image of F_6^3,

[4] We are following [55] and [33] in use of notation. In [30] this matter is treated in P. II, Ch. 3, and $T_{0,\mu} \cap M \cap T_1$ is used instead of F_6^μ.

[5] See [55, Ch. IV, § 6, Lemma 32] or [30, p. 146].

$$[\{\downarrow_3, f_\mu(x_1, \ldots, x_{\mu+1})\}] = F^*_\mu,\ f_q(x_1, \ldots, x_{q+1}) \notin F^*_p,$$

if $q \geq 3$, $p > q$, and there is an infinite chain of subclasses of P_3:

$$F^*_3 \supset F^*_4 \supset \cdots \supset F^*_p \supset \ldots$$

□

Theorem 5 provides a recipe to obtain countably many paraconsistent linguistic extensions of **CPC**$^+$ that are non-exploding. Together, Theorems 1 and 5 then answer the two questions raised at the end of the previous section. Although the answer regarding the amount of functional extensions of $\mathcal{A}_1$ is only definitive regarding the matrices for languages with finitely many basic connectives. Nevertheless, in a more general case, it can be shown that the set of many-valued paraconsistent linguistic extensions of **CPC**$^+$ has the power of continuum. To do that, we will take the investigation a step further and explore the four-valued case.

4 Four-Valued Extensions of CPC$^+$

The approach used to obtain three-valued paraconsistent logics from the previous sections can be easily generalized to an arbitrary number of values. This section deals with one such generalization, a class of four-valued matrices of logics which are simultaneously paraconsistent and paracomplete, which we will label as NAT_4.

According to A. Loparić and N.C.A. da Costa [31], 'a logical system is paracomplete if it can function as the underlying logic of theories in which there are (closed) formulas such that these formulas and their negations are simultaneously false'.

In terms of logical consequence, a logic is said to be *paracomplete* iff for some theory T and formulas α, β it holds that $\{T, \alpha\} \vdash \beta$, $\{T, \neg\alpha\} \vdash \beta$, $T \nvdash \beta$ [1][6].

A logic is called *paranormal* iff it is both paraconsistent and paracomplete [8]. As noted in [1], the necessary condition for paranormality in a four-valued matrix is the existence of $a \in D$, such that $\neg a \in D$, and $b \notin D$, such that $\neg b \notin D$. Together with Rosser and Turquette's standard conditions for binary operations (the designated values are $\{1, 3\}$ in this case) and the requirement of $\{0, 1\}$-preservation for all connectives, this leads to the following schemata.

$\wedge$	0	1	2	3
0	0	0	$0 \wr 2$	$0 \wr 2$
1	0	1	$0 \wr 2$	$1 \wr 3$
2	$0 \wr 2$	$0 \wr 2$	$0 \wr 2$	$0 \wr 2$
3	$0 \wr 2$	$1 \wr 3$	$0 \wr 2$	$1 \wr 3$

$\vee$	0	1	2	3
0	0	1	$0 \wr 2$	$1 \wr 3$
1	1	1	$1 \wr 3$	$1 \wr 3$
2	$0 \wr 2$	$1 \wr 3$	$0 \wr 2$	$1 \wr 3$
3	$1 \wr 3$	$1 \wr 3$	$1 \wr 3$	$1 \wr 3$

[6]Another approach to paracompleteness is to regard it as dual to paraconsistency and handle it within the multiple-conclusion framework (see [38], [36], [18]).

$\to$	0	1	2	3
0	1	1	$1 \wr 3$	$1 \wr 3$
1	0	1	$0 \wr 2$	$1 \wr 3$
2	$1 \wr 3$	$1 \wr 3$	$1 \wr 3$	$1 \wr 3$
3	$0 \wr 2$	$1 \wr 3$	$0 \wr 2$	$1 \wr 3$

	$\neg x$
0	1
1	0
2	$0 \wr 2$
3	$1 \wr 3$

As in the three-valued case, the most 'expressive' matrix we can obtain is the one where every $\{0, 1\}$-preserving operation can be defined. A. Avron and O. Arieli have shown that such a property is possessed by the matrix

$$\mathcal{M}_{CC} = \langle \{0, 1, 2, 3\}, \wedge, \vee, \to, \neg, -, \{1, 3\} \rangle,$$

where operations are defined the following way [1, Th. 11]:

$\wedge$	0	1	2	3
0	0	0	0	0
1	0	1	2	3
2	0	2	2	0
3	0	3	0	3

$\vee$	0	1	2	3
0	0	1	2	3
1	1	1	1	1
2	2	1	2	1
3	3	1	1	3

$\to$	0	1	2	3
0	1	1	1	1
1	0	1	2	3
2	1	1	1	1
3	0	1	2	3

	$\neg x$
0	1
1	0
2	2
3	3

	$-x$
0	0
1	1
2	3
3	2

Obviously, $\{\wedge, \vee, \to, \neg\}$ satisfy the schemata above. However, an additional operation '$-$' is added to obtain the necessary expressive power. Let's show that we can replace the implication in such a way that no extra operations will be necessary. Consider the following schema:

$\Rightarrow$	0	1	2	3
0	1	1	$1 \wr 3$	$1 \wr 3$
1	0	1	0	1
2	3	$1 \wr 3$	1	$1 \wr 3$
3	2	$1 \wr 3$	$0 \wr 2$	1

As before, the use of $\wr$ signifies that one of two values must be chosen for the actual truth-table. Since the schema for $\Rightarrow$ requires it to be $\{0, 1\}$-preserving, no matter what choices we make, the resulting version of $\Rightarrow$ will be representable in $\mathcal{M}_{CC}$ by a formula containing only connectives from $\{\wedge, \vee, \to, \neg, -\}$ [1, Th. 11]. Therefore, the following identities are sufficient to demonstrate that the matrix $\mathcal{M}'_{CC} = \langle \{0, 1, 2, 3\}, \wedge, \vee, \Rightarrow, \neg, \{1, 3\} \rangle$ is equivalent to $\mathcal{M}_{CC}$ in its expressive power: $\sim x =: x \Rightarrow \neg(x \Rightarrow x)$; $-x = \neg \sim x$; $\diamond x =: (x \Rightarrow x) \Rightarrow x$; $x \supset y =: \sim (x \wedge \sim y)$; $x \to y = \diamond x \supset y$.

We will now address the number of four-valued matrices which induce paranormal logics extending $\mathbf{CPC}^+$ that can be defined in $\mathcal{M}_{CC}$. As in the third-valued case, we will be dealing with closed sets of functions. The first example, which we will label as $\mathcal{I}^1\mathcal{P}^1$, is similar to $\mathcal{P}^1$ in a sense that all operations produce classical values exclusively:

$\wedge$	0	1	2	3
0	0	0	0	0
1	0	1	0	1
2	0	0	0	0
3	0	1	0	1

$\vee$	0	1	2	3
0	0	1	0	1
1	1	1	1	1
2	0	1	0	1
3	1	1	1	1

$\rightarrow$	0	1	2	3
0	1	1	1	1
1	0	1	0	1
2	1	1	1	1
3	0	1	0	1

	$\neg x$
0	1
1	0
2	0
3	1

The logic $\mathbf{I}^1\mathbf{P}^1$ defined by $\mathcal{I}^1\mathcal{P}^1$ has been explored in [21, § 4.3] and [22][7].

$\mathbf{I}^1\mathbf{P}^1$ is an **LFI**. The following operation is definable in $\mathcal{I}^1\mathcal{P}^1$: $\circ x = \neg(x \wedge y) \vee \neg\neg x$. Add the appropriate connective to the language of $\mathbf{I}^1\mathbf{P}^1$: $h(\dot{\circ}p) = \circ(h(p))$ for every $h \in Val(\mathcal{I}^1\mathcal{P}^1)$. Verify that $p, \dot{\circ}p \nvdash q$, $\neg p, \dot{\circ}p \nvdash q$, $p, \neg p, \dot{\circ}p \vdash q$ in $\mathbf{I}^1\mathbf{P}^1$. as a corollary, not only is $\mathbf{I}^1\mathbf{P}^1$ an **LFI**, but so are all of its linguistic extensions (cf. [14, pp. 20–21]). Let us now demonstrate that there are continuum-many such extensions.

By abuse of notation, denote the class of all functions definable in $\mathcal{M}_{CC}$ as T_{01} as in the three-valued case. The class of all functions definable in $\mathcal{I}^1\mathcal{P}^1$ coincides with $P_{4,2}$ (this follows from the results presented in [50]).

Theorem 6. *There are continuum-many subclasses of P_4 that contain $P_{4,2}$ and are contained in T_{01}.*

Proof. Consider the $\{0,1\}$-preserving function $r'_n(x_1,\dots,x_n) \in P_4$, where $n \in \mathbb{N} \setminus \{1\}$:

$$r'_n(x_1,\dots,x_n) = \begin{cases} 1, & \text{if for some } i \in \{1,\dots,n\}\ x_i = 2 \text{ and} \\ & x_j = 1 \text{ for all } j \neq i, \\ 2, & \text{if } x_1 = \dots = x_n = 2, \\ 3, & \text{if } x_1 = \dots = x_n = 3, \\ 0 & \text{otherwise.} \end{cases}$$

[7] As a side note, a similar matrix can be found in [32] and [26]. It differs in the definition of negation: $\neg 2 = 2$ and $\neg 3 = 3$.

Since r'_n preserves $\{0,1,2\}$, it is an inverse homomorphic image of of r_n from Theorem 1. Therefore, the system $\{r'_i | i \geq 2\}$ constitutes a countable basis for $[\{r'_i | i \geq 2\}]$. This entails $r'_k \notin [\{r'_i | i \geq 2\} \setminus \{r'_k\}]$ for every $r'_k \in \{r'_i | i \geq 2\}$, so $[\{r'_i | i \geq 2\}]$ contains continuum-many closed subclasses. Notice that functions from $\{r'_i | i \geq 2\}$ produce the value 2 iff all their arguments take this value. At the same time, functions from $\mathcal{I}^1\mathcal{P}^1$ never produce the value 2. The remainder of the proof is as in Theorem 1. □

We have just shown that the set of all matrices with expressive power between $\mathcal{I}^1\mathcal{P}^1$ and $\mathcal{M}_{CC}$ has the cardinality of continuum. Consequently, there are as many four-valued paraconsistent extensions of **CPC**$^+$ that are **LFI**s. Now we turn to matrices from NAT_4 that induce non-exploding logics.

First, let us find a condition that is sufficient to estimate a lower boundary of the number of non-exploding matrices definable in NAT_4.

If $\mathcal{M} \in \mathsf{NAT}_4$ is $\{3\}$-preserving, then $\mathcal{M}$ does not induce an **LFI**. Let $\bigcirc(p)$ be a set of formulas depending on exactly one variable p. Suppose $\mathcal{M}$ is $\{3\}$-preserving. Then $h(\neg p) = h(\bigcirc(p)) = 3$ for every $h \in Val(\mathcal{M})$, such that $h(p) = 3$. Consequently, $p, \neg p, \bigcirc(p) \nvdash q$ in $\mathcal{M}$ and it does not induce an **LFI**. In more general terms, $X \nvdash q$ whenever $q \notin Var(X)$, so $\mathcal{M}$ is non-exploding.

Let T_{01} and T_3 stand for the subclasses of P_4 which consist of all $\{0,1\}$-preserving and $\{3\}$-preserving functions respectively. Consider a matrix

$$\mathcal{M}_{013} = \langle \{0,1,2,3\}, F, \{1,3\} \rangle,$$

such that $[F] = T_{01} \cap T_3$. Since $[F] \subset T_{01}$, $\mathcal{M}_{013}$ is definable in $\mathcal{M}_{CC}$. Since $[F] \subset T_3$, each matrix definable in $\mathcal{M}_{013}$ is non-exploding.

One of the matrices definable in $\mathcal{M}_{013}$ is quite similar to Sobociński's matrix $\mathcal{A}_1$. Consider the matrix $\mathcal{A}'_1 = \langle \{0,1,2,3\}, \wedge, \vee, \rightarrow, \neg, \{1,3\} \rangle$.

$\wedge$	0	1	2	3
0	0	0	0	0
1	0	1	0	1
2	0	0	0	0
3	0	1	0	3

$\vee$	0	1	2	3
0	0	1	0	1
1	1	1	1	1
2	0	1	0	1
3	1	1	1	3

$\rightarrow$	0	1	2	3
0	1	1	1	1
1	0	1	0	1
2	1	1	1	1
3	0	1	0	3

	$\neg x$
0	1
1	0
2	0
3	3

Denote as A'_1 the closed set of all functions definable in $\mathcal{A}'_1$. The number of non-exploding matrices definable in NAT_4 can not be lower than the number of closed subclasses of P_4 between $T_{01} \cap T_3$ and $\mathcal{A}'_1$.

Theorem 7. *There are continuum-many subclasses of P_4 that contain A_1' and are contained in $T_{01} \cap T_3$.*

Proof. As in Theorem 6. □

Theorems 7 and 6 allow us to produce uncountably many four-valued **LFI**s and non-exploding logics, or countably many of them, if we only consider languages with finite amount of basic connectives. However, there is a caveat: none of such logics are maximally paraconsistent.

A logic is said to be *maximally paraconsistent* iff it is paraconsistent, but none of its deductive extensions[8] are [1, Def. 16]. It has been shown that all logics induced by three-valued matrices from NAT, as well as their linguistic extensions, possess this property [4, Ex. 3.8]. In the four-valued case, the matrix $\mathcal{M}_{CC}$ and several others have been demonstrated to induce maximally paraconsistent logics [1]. The following lemma, which is similar to Theorem 17 in the paper just cited, provides a necessary condition for maximal paraconsistency in logics induced by matrices from NAT_4.

Lemma 8. *If $\mathcal{M}_4 \in \mathsf{NAT}_4$ is $\{0,1,3\}$-preserving, it does not induce a maximally paraconsistent logic.*

Proof. If $\mathcal{M}_4 \in \mathsf{NAT}_4$ is $\{0,1,3\}$-preserving, then it has a three-valued submatrix, which is isomorphic to some matrix $\mathcal{M}_3$ from NAT. Let $\mathbf{L}_4$ and $\mathbf{L}_3$ be logics induced by $\mathcal{M}_4$ and $\mathcal{M}_3$. Then $\mathbf{L}_4 \subsetneq \mathbf{L}_3$. Verify that $p \vee \neg p$ is necessary valid in $\mathbf{L}_3$, but can not be valid in $\mathbf{L}_4$. Therefore, $\mathbf{L}_3$ is a paraconsistent proper extension of $\mathbf{L}_4$, and the latter is not maximally paraconsistent. □

Since every matrix obtained via Theorems 6 and 7 is $\{0,1,3\}$-preserving, they can not induce maximally paraconsistent logics. Although, some four-valued linguistic extensions of those logics can be shown to be maximally paraconsistent. The following lemmata, which are adapted from [4, Th. 3.2], provide the sufficient conditions for maximality in such extensions.

Lemma 9. *If $\mathcal{M}$ is a functional extension of $\mathcal{I}^1\mathcal{P}^1$ and is not $\{0,1,3\}$-preserving, the logic $\mathbf{L} = \langle \mathcal{L}, \vdash \rangle$ induced by $\mathcal{M}$ is maximally paraconsistent w.r.t. $\neg$ of $\mathcal{I}^1\mathcal{P}^1$.*

Proof. Suppose $\mathbf{L}' = \langle \mathcal{L}, \Vdash \rangle$ is a logic in the language of $\mathbf{L}$ that is strictly stronger than $\mathbf{L}$. Then there are X and α, such that $h^*(X) \subseteq \{1,3\}$ and $h^*(\alpha) \in \{0,2\}$

[8]That is, extensions that are obtained by addition of axioms or inference rules without changing the language.

for some $h^* \in Val(\mathcal{M})$, but $X \Vdash \alpha$. Since $\mathcal{M}$ is not $\{0,1,3\}$-preserving, there is a formula β, such that $h'(q) \in \{0,1,3\}$ for each $q \in Var(\beta)$ and $h'(\beta) = 2$. For every $q \in Var(\beta)$ and $p \in Var(X \cup \{\alpha\})$ define the substitutions:

$$e_0(q) = \begin{cases} \neg\neg p_0, \text{ if } h'(q) = 0, \\ \neg p_0, \text{ if } h'(q) = 1, \\ p_0, \text{ if } h'(q) = 3. \end{cases} \quad e_1(p) = \begin{cases} \neg\neg p_0, \text{ if } h^*(p) = 0, \\ \neg p_0, \text{ if } h^*(p) = 1, \\ e_0(\beta), \text{ if } h^*(p) = 2, \\ p_0, \text{ if } h^*(p) = 3. \end{cases}$$

For every $h \in Val(\mathcal{M})$, if $h(p_0) = 3$, then $h(e_1(X)) \subseteq \{1,3\}$ and $h(e_1(\alpha)) \in \{0,2\}$. Therefore, the following holds: (1) $p_0, \neg p_0 \vdash e_1(\gamma)$ for each $\gamma \in X$; (2) $p_0, \neg p_0, e_1(\beta) \vdash q_0$. Recall that $\mathbf{L}'$ is strictly stronger than $\mathbf{L}$. So $p_0, \neg p_0 \Vdash e_1(\gamma)$ for each $\gamma \in X$ and $p_0, \neg p_0, e_1(\beta) \Vdash q_0$. Together with $e_1(X) \Vdash e_1(\alpha)$, this entails $p_0, \neg p_0 \Vdash q_0$, and $\mathbf{L}'$ is not paraconsistent. □

Lemma 10. *If $\mathcal{M}$ is a $\{3\}$-preserving functional extension of $\mathcal{A}'_1$ and is not $\{0,1,3\}$-preserving, the logic $\mathbf{L} = \langle \mathcal{L}, \vdash \rangle$ induced by $\mathcal{M}$ is maximally paraconsistent w.r.t. $\neg$ of $\mathcal{A}'_1$.*

Proof. Start as in Lemma 9, then define the substitutions:

$$e_0(q) = \begin{cases} (q_0 \wedge \neg q_0), \text{ if } h'(q) = 0, \\ \neg(q_0 \wedge \neg q_0), \text{ if } h'(q) = 1, \\ p_0, \text{ if } h'(q) = 3. \end{cases} \quad e_1(p) = \begin{cases} (q_0 \wedge \neg q_0), \text{ if } h^*(p) = 0, \\ \neg(q_0 \wedge \neg q_0), \text{ if } h^*(p) = 1, \\ e_0(\beta), \text{ if } h^*(p) = 2, \\ p_0, \text{ if } h^*(p) = 3. \end{cases}$$

For every $h \in Val(\mathcal{M})$, if $h(p_0) = 3$ and $h(q_0) \in \{0,1,2\}$, then $h(e_1(X)) \subseteq \{1,3\}$ and $h(e_1(\alpha)) \in \{0,2\}$. Moreover, since $\mathcal{M}$ is $\{3\}$-preserving, if $h(p_0) = h(q_0) = 3$, then $h(e_1(\alpha)) = 3$ and $h(e_1(\gamma)) = 3$ for every $\gamma \in X$. Therefore, the following holds: (1) $p_0, \neg p_0 \vdash e_1(\gamma)$ for each $\gamma \in X$; (2) $p_0, \neg p_0, e_1(\beta) \vdash q_0$. The remainder of the proof is as in Lemma 9. □

Now we will use Lemmata 9 and 10 to 'recover' maximal paraconsistency in classes of matrices provided by Theorems 6 and 7. Supplement each of them with the following operator:

⊛	0	1	2	3
0	0	0	0	2
1	0	0	0	0
2	0	0	0	0
3	2	0	0	3

Lemma 11. *For every $r'_k \in \{r'_i | i \geq 2\}$ the following holds:*

$$r'_k \notin [(\{r'_i | i \geq 2\} \setminus \{r'_k\}) \cup P_{4,2} \cup \{\circledast\}].$$

Proof. Let $\Phi(\Phi_1, \ldots, \Phi_m)$ be a formula, where $\Phi \in \{r'_i | i \geq 2\} \cup P_{4,2} \cup \{\circledast\}$ and $\Phi_1, \ldots, \Phi_m$ are either variables or functions from $\{r'_i | i \geq 2\} \cup P_{4,2} \cup \{\circledast\}$. By definitions of r'_n, $\mathcal{I}^1\mathcal{P}^1$, $\circledast$:

1. If $x_1 = \cdots = x_n = 2$, then $r'_n(x_1, \ldots, x_n) = 2$;
2. If $x_i \neq 2$ for some $x_i \in \{x_1, \ldots, x_n\}$, then $r'_n(x_1, \ldots, x_n) \neq 2$;
3. If $x_i \neq 3$ for each $x_i \in \{x_1, \ldots, x_n\}$, then $f(x_1, \ldots, x_n) \neq 2$ for every $f \in [P_{4,2} \cup \{\circledast\}]$.

Suppose $\Phi(\Phi_1, \ldots, \Phi_m)$ contains at least one instance of a function from $P_{4,2} \cup \{\circledast\}$. By virtue of (1)–(3), $x_1 = \cdots = x_n = 2$ entails $\Phi(\Phi_1, \ldots, \Phi_m) \neq 2$. Therefore, $r'_k \notin [(\{r'_i | i \geq 2\} \setminus \{r'_k\}) \cup P_{4,2} \cup \{\circledast\}]$ for every $r'_k \in \{r'_i | i \geq 2\}$. □

Lemma 12. *For every $r'_k \in \{r'_i | i \geq 2\}$ the following holds:*

$$r'_k \notin [(\{r'_i | i \geq 2\} \setminus \{r'_k\}) \cup A'_1 \cup \{\circledast\}].$$

Proof. Same as in Lemma 11. □

By virtue of Lemmata 9 and 11, the set of four-valued matrices that define maximally paraconsistent **LFI**s and are pairwise distinct in expressive power has cardinality of continuum. By virtue of Lemmata 10 and 12, the set of four-valued matrices that define maximally paraconsistent non-exploding logics and are pairwise distinct in expressive power has cardinality of continuum as well. That concludes this contribution to the study of many-valued paraconsistent extensions of $\mathbf{CPC}^+$. However, the study itself is far from over. In the final section we lay out some open problems and directions for further investigations.

5 Conclusion

We have given several individual examples of infinite subclasses of P_3 and P_4 that could be used to produce infinite sets of many-valued paraconsistent extensions of $\mathbf{CPC}^+$. Although our results raise more questions than they answer.

First of all, it is yet not known whether it is possible to define continuum-many functional extensions $\mathcal{A}_1$ which are definable in $\mathcal{PAC}$.

Second, while we have demonstrated the existence of uncountably infinitely many subclasses of P_3 between $P_{3,2}$ and T_{01}, the example we have given is clearly not the only one possible. At least one different way to expand $P_{3,2}$ is readily available if we modify the definition of r_n by replacing all 1s with 0s and vice versa.

Quite a lot of other similar subsets of P_k that have continuum cardinality is known in the literature (cf. [30, §§ 12.3 14.10], [57]). This makes case-by-case analysis a futile task.

Therefore, to provide the definitive answers regarding functional extensions of $\mathcal{A}_1$ and $\mathcal{P}_1$, we would need to produce complete description of respective sublattices of the lattice of all subclasses of P_3 ($\mathbb{L}_3$) — the one between A_1 and $T_{01} \cap T_2$ as well as the one $P_{3,2}$ and T_{01} — which would be not unlike the well known description of subclasses of P_2 (cf. [30, p. 149]).

At the same time, the existence of subclasses with countable bases, which is the source of the results presented in our paper, is the very reason for the lack of such a general description for $\mathbb{L}_3$. The known descriptions of sublattices of $\mathbb{L}_3$ usually deal with 'manageable' sets of subclasses that are at most countable [30, § 8.3].

The important exception can be found in D. N. Zhuk's paper [56]. The author isolated the 'unmanageable' segments of the lattice of all clones of self-dual functions of P_3 and was therefore able to present a comprehensible description of this lattice. This result hints that a similar description could be possible for the lattices of clones related to the topic of our paper.

The four-valued case raises several additional questions. The first group of problems is related to our generalization of NAT for the four values. We have identified some sufficient conditions for formal inconsistency, non-explosiveness and maximal paraconsistency in NAT_4. However, the general description of necessary and sufficient conditions for such properties in NAT_4 is yet to be found.

Next, further research of the expressive power of matrices in NAT_4 is in order. In the three-valued case, the upper and lower boundaries of subclasses of NAT is known for both the subclass of matrices that define **LFI**s and the subclass of matrices that define non-explosive logics. In NAT_4, on the other hand, the situation is more complicated. The 'strongest' matrix has been identified as $\mathcal{M}_{CC}$. We have also established that the class of operations definable in the strongest $\{3\}$-preserving matrix coincides with $T_{01} \cap T_3$. While such a matrix is definable in $\mathcal{M}_{CC}$, it is still not known whether $T_{01} \cap T_3$ has a basis of the form $\{\wedge, \vee, \supset, \neg\}$. The same is also the case for the strongest matrix where every function from $T_{01} \cap T_{23}$ is definable.

The matrix $\mathcal{I}^1\mathcal{P}^1$ is one of the 'minimal' ones (i.e. no other matrix from NAT_4 is definable in it), since the set of all operations definable in it coincides with $P_{4,2}$. The matrix $\mathcal{A}_1'$ should also be minimal, but this has to be proved. In any case, the two matrices we discussed do not exhaust the set of minimal matrices in NAT_4. Neither

$\mathcal{I}^1\mathcal{P}^1$ nor $\mathcal{A}'_1$ preserve $\{2,3\}$, so there has to be a minimal matrix with this property. The problem is directly related to the question of necessary and sufficient conditions for maximality raised above. Notice that the relevant lemmata in our paper deal with extensions of particular minimal matrices, and this is what facilitates their proofs.

The second group of problems stems from the fact that there is more than one way to generalize NAT. In NAT_4, we deal with the class of designated valued $D = \{1,3\}$. But we could also take $D = \{1,2,3\}$, and this would lead to a different class of matrices and a different class of logics, ones that are only paraconsistent, but not paracomplete. All questions raised in this paper would be relevant for such a class as well.

While the three-valued and four valued cases already provide plenty of material, it is also of interest to consider the many-valued extensions of $\mathbf{CPC}^+$ from a generalized finite-valued viewpoint. In [3, § 5] the way to obtain an infinite sequence of logics with gradually increasing number of truth-values is described. It seems worth investigating whether it is possible to design a similar sequence, such that every k-valued logic in the sequence would have continuum-many pairwise distinct k-valued linguistic extensions. Moreover, the questions regarding the conditions of formal inconsistency, maximality and other issues discussed above would best answered for the generalized k-valued case.

Last but not least, all of the above is applicable to the paracompleteness property. As pointed out by J. Marcos, 'Any definition involving paraconsistency can immediately be converted into a definition involving its dual, paracompleteness' [39] (see also [12]). Such a conversion essentially doubles the field of proposed research. In addition to the problems of formal inconsistency, non-explosiveness and maximal paraconsistency, the problems of formal uncertainty, non-implosiveness and maximal paracompleteness arise [36]. Such problems are already relevant to the matrices discussed in Section 4, since all the logics they define are paracomplete. But what is perhaps more interesting, paraconsistency also motivates the study of linguistic extensions of a different fragment of **CPC**, the dual-positive one.

We have established in Section 2 that the schemata of NAT yield 8,192 matrices with $D = \{1,2\}$ for the language $\langle L, \wedge, \vee, \rightarrow, \neg\rangle$. At the same time, as noted in [37], a similar construction for $D = \{1\}$ would yield only 1,024 such matrices. This seems to contradict the symmetry between paraconsistency and paracomleteness pointed out above. The source of this discrepancy is in the standard condition for the implication operator:

$D = \{1\}$

$\to$	0	1	2
0	1	1	1
1	0	1	$0 \wr 2$
2	1	1	1

$D = \{1, 2\}$

$\to$	0	1	2
0	1	1	$1 \wr 2$
1	0	1	$1 \wr 2$
2	0	$1 \wr 2$	$1 \wr 2$

$\leftarrow$	0	1	2
0	0	1	$0 \wr 2$
1	0	0	$0 \wr 2$
2	$0 \wr 2$	1	$0 \wr 2$

$\leftarrow$	0	1	2
0	0	1	$1 \wr 2$
1	0	0	0
2	0	0	0

For $D = \{1, 2\}$ there are 2^4 options, but for $D = \{1\}$ there are only two. The symmetry is restored once we take into consideration the operation $\leftarrow$, where the 'standard condition' is obtained from those for $\neg$ and $\to$ via the following identity: $x \leftarrow y =: \neg(\neg x \to \neg y)$. Since $\wedge$ is dual to $\vee$, $\{\wedge, \vee, \leftarrow\}$ is dual to $\{\wedge, \vee, \to\}$. Notice that in the classical two-valued case, the set of operations $\{\wedge, \vee, \to\}$ preserves the value 1 and the set of operations $\{\wedge, \vee, \leftarrow\}$ preserves the value 0. Let us call the fragment of **CPC** in the language $\{L, \wedge, \vee, \leftarrow\}$ *dual-positive* and denote it as $\mathbf{CPC}^-$. If we set out to explore the many-valued negative extensions of non-negative fragments of **CPC** and preserve the symmetry between paraconsistency and paracompleteness while doing that, in addition to the titular many-valued paracconsistent extensions of classical positive propositional calculus, we need to at least study the many-valued paracomplete extensions of classical dual-positive propositional calculus as well.

In light of the contents of this section, we conclude that the results presented in the paper should be seen not as final answers, but rather as motivating examples for research devoted to a rather sizeable subfield of non-classical logics.

References

[1] Arieli, O., Avron, A.: Four-valued paradefinite logics. Stud. Log. 105(6), 1087–1122 (2017) https://doi.org/10.1007/s11225-017-9721-4

[2] Arieli, O., Avron, A.: Three-valued paraconsistent propositional logics. In: Beziau, J.-Y., Chakraborty, M., Dutta, S. (eds.) New Directions in Paraconsistent Logic, 91–129. Springer (2015) https://doi.org/10.1007/978-81-322-2719-9_4

[3] Arieli, O., Avron, A., Zamansky, A.: Ideal paraconsistent logics. Stud. Log. 99(1–3), 31–60 (2011) https://doi.org/10.1007/s11225-010-9296-9

[4] Arieli, O., Avron, A., Zamansky, A.: Maximal and premaximal paraconsistency in the framework of three-valued semantics. Stud. Log. 97(1), 31–60 (2011)

[5] Arruda, A. I.: On the imaginary logic of N. A. Vasil'év. In: Arruda, A. I., da Costa, N. C. A., Chuaqui, R. (eds.) Non-classical Logics, Model Theory, and Computability, 3–24. Elsevier (1977) https://doi.org/10.1016/S0049-237X(08)70642-6

[6] Avron, A.: On the expressive power of three-valued and four-valued languages. J. Log. Comput. 9(6), 977–994 (1999) https://doi.org/10.1093/logcom/9.6.977

[7] Batens D. Paraconsistent extensional propositional logics. Log. Anal., Nouv. Ser. 23(90–91), 195–234 (1980)

[8] Beziau, J.-Y.: The future of paraconsistent logic. Log. Issled. 2, 1–23 (1999)

[9] Beziau, J.-Y.: Two genuine 3-valued paraconsistent logics. In: Akama, S. (ed.), Towards Paraconsistent Engineering, Intelligent Systems Reference Library 110. Springer (2016) https://doi.org/10.1007/978-3-319-40418-9_4

[10] Beziau, J.-Y., Franceschetto, A.: Strong three-valued paraconsistent logics. In: Beziau, J.-Y., Chakraborty, M., Dutta, S. (eds.) New Directions in Paraconsistent Logic, 131–145. Springer (2015) https://doi.org/10.1007/978-81-322-2719-9_5

[11] Bonzio, S., Gil-Férez, J., Paoli, F., Peruzzi, L.: On paraconsistent weak Kleene logic: axiomatisation and algebraic analysis. Stud. Log., 105(2), 253–297 (2017) https://doi.org/10.1007/s11225-016-9689-5

[12] Brunner, A. B., Carnielli, W. A. . Anti-intuitionism and paraconsistency. J. Appl. Log., 3(1), 161–184 (2005) https://doi.org/10.1016/j.jal.2004.07.016

[13] Carnielli, W., Coniglio, M. E.: Paraconsistent Logic: Consistency, Contradiction and Negation. Springer (2016) https://doi.org/10.1007/978-3-319-33205-5

[14] Carnielli, W., Coniglio, M. E., Marcos, J.: Logics of formal inconsistency. In Gabbay, D., Guenthner, F. (eds.), Handbook of Philosophical Logic, 2nd edn., vol. 14, 1–93. Springer (2007) https://doi.org/10.1007/978-1-4020-6324-4_1

[15] Ciuciura, J.: Frontiers of the discursive logic. Bull. Sect. Log., 37(2), 81–92 (2008).

[16] Ciuni, R., Carrara, M.: Characterizing logical consequence in Paraconsistent Weak Kleene. In: Felline, L., Ledda, A., Paoli, F., Rossanese, E. (eds.) New Developments in Logic and the Philosophy of Science, 165–176. College, London (2016)

[17] Coniglio, M. E., Silvestrini, L. H.: An alternative approach for quasi-truth. Log. J. IGPL, 22(2), 387–410 (2014). doi:10.1093/jigpal/jzt026

[18] Cobreros, P.: Vagueness: Subvaluationism. Philos. Compass. 8(5), 472–485 (2013) doi:10.1111/phc3.12030

[19] da Costa, N. C. A.: On the theory of inconsistent formal systems. Notre Dame J. Formal Logic. 15(4), 497–510 (1974) doi:10.1305/ndjfl/1093891487

[20] da Costa, N. C. A., Subrahmanian, V. S., Vago, C.: The paraconsistent logics $\mathcal{PT}$. Z. Math. Logik Grundlagen Math. 37(2), 139–148 (1991) doi:10.1002/malq.19910370903

[21] Fernández V. L.: Semântica de Sociedades para Lógicas n-valentes. Campinas: IFCH-UNICAMP (2001)

[22] Fernández V. L., Coniglio, M. E.: Combining valuations with society semantics. J. Appl. Non-Class. Log. 13(1), 21–46 (2003) https://doi.org/10.3166/jancl.13.21-46

[23] Font, J. M.: Abstract algebraic logic: An introductory textbook. London: College Publications (2016)

[24] Gottwald, S.: A treatise on many-valued logics. Baldock: Research Studies Press (2001)

[25] Hernández-Tello, A., Ramírez, J. A., Galindo, M. O. The pursuit of an implication for the logics L3A and L3B. Log. Univers. 11, 507–524 (2017) https://doi.org/10.1007/s11787-017-0182-3

[26] Hirsh, E., Lewin, R. A.: Algebraization of logics defined by literal-paraconsistent or literal-paracomplete matrices. Math. Log. Q. 54(2), 153–166 (2008) doi:10.1002/malq.200710021

[27] Humberstone, L. The connectives. Cambrige: MIT Press (2011)

[28] Karpenko, A. S.: A maximal paraconsistent logic: the combination of two three-valued isomorphs of classical propositional logic. In: Batens, D., Mortensen, C., Priest, G., van Bendegem, J.-P. (eds.) Frontiers of Paraconsistent Logic, 181–187. Baldock Research Studies Press (2000)

[29] Karpenko, A. S., Tomova, N. E.: Bochvar's three-valued logic and literal paralogics: their lattice and functional equivalence. Log. and Log. Philos. 26(2), 207–235 (2017) http://dx.doi.org/10.12775/LLP.2016.029

[30] Lau, D.: Function Algebras on Finite Sets: Basic Course on Many-Valued Logic and Clone Theory. Springer (2006) doi:10.1007/3-540-36023-9

[31] Loparic, A., da Costa, N. C. A.: Paraconsistency, paracompleteness, and valuations. Logique Anal., Nouv. Sér. 27(106), 119–131 (1984)

[32] Lewin, R. A., Mikenberg, I. F.: Literal-paraconsistent and literal-paracomplete matrices. Math. Log. Quart. 52(5), 478–493 (2006) doi:10.1002/malq.200510044

[33] Makarov, A. V.: On homomorphisms of functional systems of many-valued logics. Mat. Vopr. Kibern. 4, 5–29 (1992); in Russian.

[34] Makarov, A. V.: Description of all minimal classes in the partially ordered set $\mathcal{L}_2^3$ of closed classes of the three-valued logic that can be homomorphically mapped onto the two-valued logic. Mosc. Univ. Math. Bull. 70(1), 48 (2015); translation from Vestn. Mosk. Univ., Ser. I 70(1), 65–66 (2015) https://doi.org/10.3103/S0027132215010106

[35] Marcos, J.: 8K solutions and semi-solutions to a problem of da Costa. Unpublished draft.

[36] Marcos, J.: Nearly every normal modal logic is paranormal. Log. Anal., Nouv. Sér. 48(189/192), 279–300 (2005)

[37] Marcos, J.: On a problem of da Costa. In: G. Sica (ed.) Essays on the Foundations of Mathematics and Logic 2, 53–69. Polimetrica (2005)

[38] Marcos, J.: On negation: pure local rules. J. Appl. Log. 3(1), 185–219 (2005) https://doi.org/10.1016/j.jal.2004.07.017

[39] Marcos, J.: Ineffable inconsistencies. In: Beziau, J.-Y., Carnielli, W., Gabbay, D. (eds.) Studies in Logic and Cognitive Systems 9. Handbook of Paraconsistency, 341–354. College: London (2007)

[40] Popov, V. M., Shangin V. O.: On sublogics in Vasiliev fragment of the logic definable with A. Arruda's calculus V1. In: Markin, V. I., Zaitsev, D. V. (eds.) The Logical Legacy of Nikolai Vasiliev and Modern Logic, 181–188. Springer (2017) https://doi.org/10.1007/978-3-319-66162-9_13

[41] Rosser, J. B., Turquette, A. R.: Many-valued logics. Amsterdam: North-Holland. (1952)

[42] Rozonoer, L. I.: On interpretation of inconsistent theories. Inf. Sci. 47(3), 243–266 (1989) https://doi.org/10.1016/0020-0255(89)90003-0

[43] Sette, A. M.: On propositional calculus P^1. Math. Japon. 18, 173–180 (1973)

[44] Shestakov, V. I.: On one fragment of D. A. Bochvar's calculus. Information issues of semiotics, linguistics and automatic translation 1, 102–115 (1971); In Russian.

[45] Shoesmith, D. J., Smiley, T. J.: Deducibility and many-valuedness. J. Symb. Log. 36(4), 610–622 (1971) https://doi.org/10.2307/2272465

[46] Shoesmith, D. J., Smiley, T. J.: Multiple-Conclusion logic. CUP Archive (1978)

[47] Silvestrini, L. H.: Uma Nova Abordagem Para a Noção de Quase-Verdade (A new approach to the Notion of Quasi-Truth, in Portuguese). PhD thesis, IFCH, State University of Campinas, Brazil (2011).

[48] Szmuc, D. E.: Defining LFIs and LFUs in extensions of infectious logics. J. Appl. Non-Class. Log. 26(4), 286–314 (2016) https://doi.org/10.1080/11663081.2017.1290488

[49] Tomova, N. E.: A lattice of implicative extensions of regular Kleene's Logics. Rep. on Math. Logic 47, 173–182 (2012) doi:10.4467/20842589RM.12.008.0689

[50] Tomova, N. E., Nepeivoda, A. N.: Functional properties of four-valued paralogics. Log.-Philos. Stud. 16(1–2), 130–132 (2018)

[51] Wójcicki, R.: Theory of Logical Calculi: Basic Theory of Consequence Operations. Dordrecht: Kluwer (1988) doi:10.1007/978-94-015-6942-2

[52] Wojtylak, P.: Mutual interpretability of sentential logic I. Rep. Math. Logic 11, 69–89 (1981)

[53] Wojtylak, P.: Mutual interpretability of sentential logic II. Rep. Math. Logic 12, 51–66 (1981)

[54] Yablonskii, S. V.: Functional constructions in a k-valued logic. Trudy Mat. Inst. Steklov., 51, 5–142 (1958); in Russian.

[55] Yablonskii, S. V., Gavrilov, G. P., Kudryavtsev, V. B.: Functions of the algebra of logic and Post classes. Moscow: Nauka (1964); in Russian.

[56] Zhuk, D. N.: The lattice of all clones of self-dual functions in three-valued logic. J. Mult.-Val. Log. Soft Comput. 24(1–4), 251–316 (2015)

[57] Zhuk, D. N.: The cardinality of the set of all clones containing a given minimal clone on three elements. Algebra Univers., 68(3–4), 295–320 (2012) https://doi.org/10.1007/s00012-012-0207-y

 Received 28 February 2018

On Poly-logistic Natural-deduction for Finitely-valued Propositional Logics

Nissim Francez and Michael Kaminski
Computer Science dept., the Technion-IIT, Haifa, Israel.
{francez,kaminski}@cs.technion.il

Abstract

The paper presents a systematic construction of natural-deduction proof-systems for multi-valued logics from the truth-tables for the connectives. The construction is based on poly-sequents of the form $\Gamma_1|\cdots|\Gamma_n : \Delta_1|\cdots|\Delta_n$, $n \geq 2$, improving on a previous approach by Baaz et. al. [2] Poly-sequents allow to speak explicitly about the truth-value of a formula, and have in I/E-rules both assumptions and conclusion that have any truth-value. Soundness and strong completeness are proved. The generality of the construction is exemplified by *retrieving* within the constructed ND-system a host of well-known ND-systems for multi-valued logics.

1 Introduction

Our point of departure is the construction of a natural-deduction (ND) proof-system for many-valued logics out of the truth-tables for the connectives, as presented in Baaz et. al. In contrast to [2], our construction relies directly on the truth-tables *only*, while Baaz et. al. rely on *consequences of the truth-table* which are not easy to establish in general (see [2]) as it appeals to *all interpretations.* This reliance on consequences of the truth-table is then shown to be eliminable.

A talk based on this paper was presented at ISRALOG17, Haifa, October 2017.

Our central purpose is to present simpler and more intuitive construction, based on a more advantageous extension of Gentzen's logistic-ND to poly-logistic *ND, an extension referred to as* poly-sequents, *where both contexts of a sequent are* poly-contexts, *a structure originating from sequent calculi. The central idea of using poly-contexts is to allow for different kinds of assumptions and conclusions in an ND-system, having arbitrary truth-values.*

In more detail, our main aim is to devise a *uniform* construction of natural-deduction proof-systems, denoted by $\mathcal{N}^n$, for any n-valued logic, $n \geq 2$, constructed in a systematic way (only!) from the truth-tables for the connectives in the object language. In particular, unlike [2], the construction is *direct*, without an intermediate passage through a mediating sequent-calculus. This directness of construction is facilitated by the following two characteristics of poly-sequents:

1. While only disjunction is present in the meta-language defining satisfaction in Baaz et. al. [2] we have in the meta-language all the classical connectives: disjunction, conjunction, implication and (in a certain form) negation.
2. Our poly-sequents provide a clear separation of assumptions and conclusions, thereby being more faithful to the general concept of natural deduction and to Gentzen's original sequents.

We denote the truth-values generically as $\mathcal{V} = \{v_1, \cdots v_i, \cdots v_n\}$. Mnemonic names for truth-values are used where appropriate.

Poly-sequents should be carefully distinguished from some notationally-related generalizations of sequents, found in the literature. In the sequel, Γ_i (assumptions) and Δ_i (conclusions), $1 \leq i \leq n$, are (finite, possibly empty) sets of formulas in some object language.

poly-sequents: These are structures of the form

$$\Pi = \Gamma_1 | \cdots | \Gamma_n : \Delta_1 | \cdots | \Delta_n, \; n \geq 2 \tag{1.1}$$

Each Γ_i and Δ_i, $i \in \hat{n}$, are called (corresponding) compartments (of Π).

In a multi-valued logic, poly-sequents are interpreted, relative to some truth-value assignment[1] σ, *conjunctively* on the *assumptions* (Γ_is) and *disjunctively*

[1]Assignments are also called valuations in the literature.

on the *conclusions* (Δ_js): *if, for every* $1 \leq i \leq n$, *every* $\varphi \in \Gamma_i$ *has truth-value* v_i *under* σ, *then, for some* $1 \leq j \leq n$, *some* $\psi \in \Delta_j$ *has truth-value* v_j *under* σ.

A remark[2] about notation: Typically, the Γ_is, as well as the Δ_is are presented in ascending order of i. However, it is often convenient not to adhere to this strict order of display. In such cases, however, if *the name* Γ_i itself is not present, there might be ambiguity in determining the intended compartment index. For example, for $n = 2$, consider $\alpha|\beta : \gamma|\delta$. This could mean either

$$\Gamma_1 = \{\alpha\}, \Gamma_2 = \{\beta\}, \Delta_1 = \{\gamma\} \text{ and } \Delta_2 = \{\delta\}$$

but also

$$\Gamma_2 = \{\alpha\}, \Gamma_1 = \{\beta\}, \Delta_2 = \{\gamma\} \text{ and } \Delta_1 = \{\delta\}$$

To disambiguate, we use remnants of the name of the compartment where needed in the form of Λ_i, an empty compartment (of v_i-valued formulas).

Thus, for the first reading,

$$\Lambda_1, \alpha|\Lambda_2, \beta : \Lambda_1, \gamma|\Lambda_2, \delta$$

while for the second reading we have

$$\Lambda_2, \alpha|\Lambda_1, \beta : \Lambda_2, \gamma|\Lambda_1, \delta$$

many-sided sequents: These are structures of the form

$$\Delta_1|\cdots|\Delta_n, \quad n \geq 2 \tag{1.2}$$

In a multi-valued logic, n-sided sequents are interpreted, relative to some truth-value assignment σ, disjunctively: for *some* $1 \leq i \leq n$, and *some* $\varphi \in \Delta_i$, φ has truth-value v_i under σ (see [2]).

hyper-sequents: These were introduced by Avron [1] (and, independently, also presented by Pottinger [21]) and have the following form:

$$\Gamma_1 : \Delta_1 \mid \cdots \mid \Gamma_n : \Delta_n \tag{1.3}$$

Hyper-sequents are also interpreted differently than poly-sequents, as follows: If, for *some* $1 \leq i \leq n$, every $\psi \in \Gamma_i$ is true, then some $\psi \in \Delta_i$ is true. This interpretation appeals to two truth-values only.

[2]We thank an anonymous referee for drawing our attention to the need for this remark.

The following advantages of using our poly-sequents as the building blocks of ND-systems for multi-valued logics, in comparison to the use of n-sided sequents in [2], are:

- As stated above, poly-sequents keep the formal separation of assumptions and conclusions. In particular, poly-sequents allow for assumptions and conclusions of the form[3]
$$\overline{\Gamma_{\overline{\{i,j\}}}}|\Gamma_i, \varphi_1|\Gamma_j, \varphi_2 : \overline{\Delta_{\overline{\{k,l\}}}}|\Delta_k, \psi_1|\Delta_l, \psi_2 \tag{1.4}$$
For example, anticipating what follows, we can state the (single) premise for an I-rule for the truth of the (bivalent) classical implication, using mnemonics $\{t, f\}$ instead $\{1, 2\}$, as
$$\overline{\Gamma} : \Lambda_f, \varphi|\Lambda_t, \psi$$
namely, either φ is false or ψ is true.

- The following property, essential in object languages lacking negation, can be proved (cf. Theorem 3.1)
$$\overline{\Lambda} : \Lambda_1, \varphi|\cdots|\Lambda_n, \varphi$$
expressing[4] the claim that every φ has one of the n truth-values (extending the bivalent claim that every formula is either true or false). This property is taken as an axiom in [2].

Our contributions go beyond those in [2] in:

- Providing an extended theory of poly-sequents.
- Providing a strong completeness theorem for the full consequence relation over poly-sequents.
- Establish the harmony and stability of the constructed I/E-rules, a major ingredient for Proof-Theoretic semantics (see Section 3.6).
- Studying the derivability of specific ND-systems for multi-valued logics from the literature within our general systems. See the remark on p. 21 for an important point regarding this issue.

[3]As described below, the notation $\Gamma_{\overline{\{i,j\}}}$ abbreviates (assuming $i < j$) $\Gamma_1|\cdots|\Gamma_{i-1}|\Gamma_{i+1}|\cdots|\Gamma_{j-1}|\Gamma_{j+1}|\cdots|\Gamma_n$ and similarly for $\Delta_{\overline{\{k,l\}}}$.

[4]Here $\overline{\Lambda}$ are empty Γs, defined below.

We mention on passing that poly-sequents were used for other purposes too, besides formulating ND-systems for multi-valued logics.

- In [9], poly-sequents were used for expressing *multilateralism*, a natural generalization of *bilateralism*. The lateral is an approach to logic based on having *assertion* and *denial* as two independent speech acts, allowing a "nice" proof-theory for classical logic (see [29] for more details). In the case of multilateralism, one can formulate *abstract positions* in which an agent can have any of $n \geq 2$ independent *stances* towards a proposition.
- In [12], a theory of *transparent truth-value assignment* predicates was formulated. It is based on predicates $T_i(\hat{\varphi})$ (where $\hat{\varphi}$ is a unique name for φ), asserting, when true, that φ has the truth-value v_i. It generalizes the known disqoutational use of the transparent truth-predicate $T(\hat{\varphi})$ expressed by $T(\hat{\varphi}) \leftrightarrow \varphi$.

Traditionally, a *specific* multi-valued logic is defined not only by its truth-tables for its connectives, but also by a collection $\mathcal{V}_d \subseteq \mathcal{V}$ of *designated* truth-values, the preservation of which defines the consequence relation of the logic. We would like to draw the reader's attention, and emphasize, that our construction of an ND-system is *not* over formulas, but over poly-sequents, endowed with their own consequence relation (not depending on designated truth-values). For another approach to construct ND-systems from multi-valued truth-tables that *is* based on designation of some truth-values the reader is referred to [8]. There a reductive approach is employed, reducing the construction to a bivalent one.

We defer the treatment of the consequence relation over formulas to Section 3.7.

2 Poly-sequents

Consider an arbitrary object language for a multi-valued logic. A truth-value assignment σ is a mapping of the formulas in the object language to $\mathcal{V} = \{v_1, \cdots, v_n\}$ assigning to each formula φ a truth-value $\sigma[\![\varphi]\!] \in \mathcal{V}$. The value of σ for atomic sentences is arbitrary, and the extension to arbitrary formulas respects the truth-tables of the connectives.

Definition 2.1 (poly-sequents). *A* poly-sequent Π *is a structure of the form*

$$\Pi = \Gamma_1 | \cdots | \Gamma_n : \Delta_1 | \cdots | \Delta_n, \; n \geq 2 \tag{2.5}$$

where each Γ_i *and* Δ_i *is a (finite, possibly empty) set of formulas in the object language. We denote by* $\mathbf{\Pi}$ *a (possibly empty) set of poly-sequents (over the same* $\mathcal{V}$*).*

Definition 2.2 (support, satisfaction, validity, consequence).

support:

a-support: *A truth-value assignment* σ a-supports *a poly-context* $\overline{\Gamma}$ *iff for every* $1 \leq i \leq n$ *and every* $\varphi \in \Gamma_i$ $\sigma[\![\varphi]\!] = v_i$.

c-support *A truth-value assignment* σ c-supports *a poly-context* $\overline{\Delta}$ *iff for some* $1 \leq i \leq n$ *and some* $\varphi \in \Delta_i$ $\sigma[\![\varphi]\!] = v_i$.

satisfaction: *A truth-value assignment* σ satisfies *a poly-sequent* $\Pi = \overline{\Gamma} : \overline{\Delta}$*, denoted* $\models_\sigma \Pi$*, iff the following holds: If* σ *a-supports* $\overline{\Gamma}$ *then* σ *c-supports* $\overline{\Delta}$.

Spelled out, the satisfaction condition is as follows.
If for every $1 \leq i \leq n$ *and every* $\varphi \in \Gamma_i$, $\sigma[\![\varphi]\!] = v_i$*, then for some* j, $1 \leq j \leq n$*, and some* $\psi \in \Delta_j$, $\sigma[\![\psi]\!] = v_j$.
If σ does not *a-support* $\overline{\Gamma}$ *(i.e.,* $\sigma[\![\psi]\!] \neq v_j$ *for some* $\psi \in \Gamma_j$, $1 \leq j \leq n$ *) we say that* σ *satisfies* Π vacuously.

σ *satisfies* $\mathbf{\Pi}$*, denoted* $\models_\sigma \mathbf{\Pi}$*, iff* $\models_\sigma \Pi$ *for every* $\Pi \in \mathbf{\Pi}$.

validity: Π *is* valid*, denoted* $\models \Pi$*, iff* $\models_\sigma \Pi$ *for every truth-value assignment* σ.

consequence: Π *is a* consequence *of a set of poly-sequents* $\mathbf{\Pi}$*, denoted* $\mathbf{\Pi} \models \Pi$*, iff for every assignment* σ*, if* $\models_\sigma \mathbf{\Pi}$ *then* $\models_\sigma \Pi$.

Note again that unlike consequence relations in multi-valued logics defined over formulas, the consequence relation over poly-sequents does not depend on any designation of some of the truth-values. *All* the truth-values take part in the definition of consequence. Designation will be considered in Section 3.7, and also later, when we discuss the retrievability of specific formula-logics within our poly-sequent logics (see Section 4).

Observe that under these definitions of the semantic notions, a poly-sequent implicitly embodies, via its description in the meta-language, the classical connectives of conjunction, disjunction and implication. In a forthcoming paper [13], this property is exploited for the presentation of a family of deductively equivalent (to the currently constructed system) proof systems exhibiting certain *dualities and symmetries.*

A precursor to multi-context sequents, without this kind of interpretation, can be found in [26, 27].

Notational and terminological conventions

- Let, for $n \geq 2$, $\hat{n} =^{\text{df.}} \{1, \cdots, n\}$.

- We abbreviate $\Gamma_1|\cdots|\Gamma_n$ to $\overline{\Gamma}$, and similarly for $\overline{\Delta}$. As usual, $\overline{\Gamma}, \overline{\Gamma}'$ is defined by $\Gamma_1 \cup \Gamma'_1|\cdots|\Gamma_n \cup \Gamma'_n$. If (in 2.5) $\varphi \in \Gamma_i$ (resp. Δ_i), we say that φ is a *resident formula* of Γ_i (resp. Δ_i).

- For $J = \{j_1, \cdots, j_m\} \subseteq \hat{n}$, $\overline{\Gamma}_J$ abbreviates $\Gamma_{j_1}|\cdots|\Gamma_{j_m}$, and $\Gamma_{j_1}, \varphi \mid \cdots \mid \Gamma_{j_m}, \varphi$ abbreviates to $\overline{\Gamma}_J, \varphi$. Similarly for $\overline{\Delta}_J, \varphi$. The notation is naturally extended to mnemonic indices. For example, for a four-valued logic with $\mathcal{V} = \{f, n, b, t\}$, standing, respectively, for false, neither false nor true, both false and true and true, if $\overline{\Gamma} = \Gamma_f \mid \Gamma_n \mid \Gamma_b \mid \Gamma_t$, then $\overline{\Gamma}_{\{b,t\}}$ denotes $\Gamma_b \mid \Gamma_t$.

  One special case, of the form $\Gamma_1 \mid \cdots \mid \Gamma_{i-1} \mid \Gamma_{i+1} \ldots |\Gamma_n$, where Γ_i is excluded, occurs often (for various values of i). We abbreviate it to $\overline{\Gamma}_{\overline{i}}$. Similarly for $\overline{\Gamma}_{\overline{J}}$, for $J \subseteq \hat{n}$. Analogous abbreviations apply to the Δs.

- An empty compartment is denoted by Λ. We abbreviate $\Lambda|\cdots|\Lambda$ to $\overline{\Lambda}$. Occasionally, for readability, we use the hybrid notation Λ_i to indicate that the empty compartment is the one corresponding to Γ_i or Δ_i (cf. the remark about notation above). Similarly, Λ_N when N is a mnemonic index of a compartment. For example, Λ_f in the bivalent case.

- Two poly-sequents Π, Π' of the form

  $$\Pi = \overline{\Gamma} : \Delta_1|\cdots|\Delta_i, \varphi|\cdots| \Delta_n \text{ and } \Pi' = \overline{\Gamma}' : \Delta'_1|\cdots|\Delta'_j, \varphi|\cdots|\Delta'_n, \quad i \neq j \tag{2.6}$$

  are called *incompatible*, expressing a conclusion of two different truth-values to the same formula φ.

As mentioned above, for multi-valued logics the ith compartment of a poly-sequent Π is intended to hold resident formulas evaluating, under a given truth-value assignment satisfying Π, to v_i, the ith truth-value. Thus, truth-values, that often raise philosophical problems regarding their nature, are reduced merely to *positions in a context of a poly-sequent.*

3 Saturated poly-logistic ND-systems

3.1 The general primitive rules

The idea is to construct systematically *directly* from any n-valued truth-tables a poly-sequent ND-system, say $\mathcal{N}^n$, over some signature[5], that *fully* reflects the whole truth-tables of the connectives in the signature, thereby being sound and strongly complete by construction.

initial poly-sequents: For every $1 \leq i \leq n$:

$$\overline{\Gamma}_{\bar{i}}|\Gamma_i, \varphi : \overline{\Delta}_{\bar{i}}|\Delta_i, \varphi \tag{3.7}$$

This makes the structural rules of (WL_i, WR_i) (*Weakening*)

$$\frac{\overline{\Gamma} : \overline{\Delta}}{\overline{\Gamma}_{\bar{i}}|\Gamma_i, \varphi : \overline{\Delta}} (WL_i) \qquad \frac{\overline{\Gamma} : \overline{\Delta}}{\overline{\Gamma} : \overline{\Delta}_{\bar{i}}|\Delta_i, \varphi|} (WR_i) \tag{3.8}$$

admissible. Alternatively, one can restrict the initial sequents to

$$\overline{\Lambda}_{\bar{i}}|\Lambda_i, \varphi : \overline{\Lambda}_{\bar{i}}|\Lambda_i, \varphi$$

and admit (WL_i, WR_i) as primitive.

shifting rules: For every $1 \leq i, j \leq n$:

$$\frac{\overline{\Gamma}_{\bar{i}}|\Gamma_i, \varphi : \overline{\Delta}}{\overline{\Gamma} : \Delta_i|\overline{\Delta}_{\bar{i}}, \varphi} (\overrightarrow{s}_i) \qquad \frac{\overline{\Gamma} : \overline{\Delta}_{\bar{i}}|\Delta_i, \varphi}{\overline{\Gamma}_{\bar{j}}|\Gamma_j, \varphi : \overline{\Delta}} (\overleftarrow{s}_{i,j}), \ j \neq i \tag{3.9}$$

coordination

$$\frac{\overline{\Gamma} : \overline{\Delta}_{\bar{i}}|\Delta_i, \varphi \quad \overline{\Gamma}' : \overline{\Delta}'_{\bar{j}}|\Delta'_j, \varphi}{\overline{\Gamma}, \overline{\Gamma}' : \overline{\Delta}, \overline{\Delta}'} (c_{i,j}), \ i \neq j \tag{3.10}$$

The rule $(c_{i,j})$ expresses a *coherence* requirement, namely a formula φ cannot be proved to have two different truth-values, v_i (from the one premise) and v_j (from the other). Those two premises are incompatible. This is a generalization of the traditional *resolution* rule.

[5]In the sequel, we will mainly be concerned with signatures which are a subset of $\{\neg, \wedge, \vee, \rightarrow\}$, where '$\rightarrow$' represents some generic implication.

operational rules: The guiding lines for the construction are the following, expressed in terms of a generic p-ary operator, say '$*$'.

$(*I)$: Such rules introduce a conclusion $\overline{\Gamma} : \overline{\Delta}_{\overline{k}}|\Delta_k, *(\varphi_1, \cdots, \varphi_p)$.

- In general, if in the truth-table for '$*$' the values v_{i_j} for φ_j, $1 \leq j \leq p$, yield the value v_k for $*(\varphi_1, \cdots, \varphi_p)$, then there is a rule

$$\frac{\{\overline{\Gamma} : \overline{\Delta}_{\overline{i_j}}|\Delta_{i_j}, \varphi_j,\ 1 \leq j \leq p\}}{\overline{\Gamma} : \overline{\Delta}_{\overline{k}}|\Delta_k, *(\varphi_1, \cdots, \varphi_p)}\ (*I_{i_1,\cdots,i_p,k}) \tag{3.11}$$

The rule $(*I_{i_1,\cdots,i_p,k})$ has, thus, p premises.

$(*E)$: Such rules have a major premise $\overline{\Gamma} : \overline{\Delta}_{\overline{k}}|\Delta_k, *(\varphi_1, \cdots, \varphi_p)$.

- Let I_k be the collection of all $\{i_1, \cdots, i_p\} \subseteq \hat{n}^p$ such that in the truth-table for '$*$' the values v_{i_j} for φ_j yield the value v_k for $*(\varphi_1, \cdots, \varphi_p)$. Then, there is a rule

$$\frac{\overline{\Gamma} : \overline{\Delta}_{\overline{k}}|\Delta_k, *(\varphi_1, \cdots, \varphi_p) \qquad \{\overline{\Gamma}_{\overline{\{i_1,\cdots,i_p\}}}|\Gamma_{i_1}, \varphi_1| \cdots |\Gamma_{i_p}, \varphi_p : \overline{\Delta}\ \mathbf{s.t.}\ \langle i_1, \cdots, i_p\rangle \in I_k\}}{\overline{\Gamma} : \overline{\Delta}}\ (*E_{I_k}) \tag{3.12}$$

Thus, the rule $(*E_{I_k})$ has $|I_k| + 1$ premises.

See an important Corollary 3.2 regarding the E-rules.

Digression

The pattern (3.12) of an E-rule are a generalization of the pattern known as *general elimination rules (GE)*.[6] Originally, GE-rules emerged from a concern regarding the relationship between Cut-free derivations in sequent calculi and normal derivations in ND-systems; see, for example, [28] and [20]. Later, they emerged (see [11], [24]) as inducing *harmony* between I-rules and E-rules, a criterion for an ND-system to qualify as *meaning-conferring* according to the *proof-theoretic semantics* theory of meaning (see [10]).
(end of digression)

[6] The similarity to the GE-form would be easier recognized had we formulated the rule as

$$\frac{\overline{\Gamma} : \overline{\Delta}_{\overline{k}} \mid \Delta_k, *(\varphi_1, \cdots, \varphi_p) \quad \overline{\Gamma}_{\overline{i_1,\cdots,i_p}}|\{\Gamma_{i_1}, \varphi_1| \cdots |\Gamma_{i_p}, \varphi_p : \overline{\Delta}_{\overline{l}} \mid \Delta_l, \chi\ \mathbf{s.t.}\ \langle i_1, \cdots, i_p\rangle \in I_k\}}{\overline{\Gamma} : \overline{\Delta}_{\overline{l}} \mid \Delta_l, \chi}\ (*E_{I_k,l})$$

However, in a multi-conclusion right context an empty succedent has to be accounted for too, hence the formulation is as given.

Two remarkable properties of the operational rules for $\mathcal{N}^n$ when constructed by the above schemes are the following.

- *Purity* ([6]): no rule features a connective different from the connective it introduces/eliminates. This is only possible due to the ability to have assumptions of any truth value! For example, in bivalent logic, one obtains pure rules for implication (not referring to $\perp$), because one can have a premise with φ being false.
- The premises of the I-rules manipulate only the Δs, while the minor premises of the E-rules manipulate only the Γs. In particular, discharged assumption never emerge!

Tree-shaped $\mathcal{N}^n$*-derivations* over poly-sequents, ranged over by $\mathcal{D}$, are defined recursively as usual, iterating applications of rules, starting from assumption poly-sequents or initial poly-sequents. *Derivability* (i.e., existence of a $\mathcal{N}^n$-derivation) of a poly-sequent Π from assumption poly-sequents $\mathbf{\Pi}$ is denoted by $\mathbf{\Pi} \vdash_{\mathcal{N}^n} \Pi$.

3.2 Some properties of $\mathcal{N}^n$

Definition 3.3 (composition of derivations). *Let* $\begin{array}{c}\mathbf{\Pi}, \Pi \\ \mathcal{D} \\ \Pi^*\end{array}$ *and* $\begin{array}{c}\mathbf{\Pi}' \\ \mathcal{D}' \\ \Pi\end{array}$ *be two* $\mathcal{N}^n$*-derivations. Their* composition $\begin{array}{c}\mathbf{\Pi}, \mathbf{\Pi}' \\ \mathcal{D}'' \\ \Pi^*\end{array}$ *is obtained as usual by replacing the leaf* Π *with the sub-tree* $\mathcal{D}'$.

Refer to Π *as the* anchor *of the composition.*

It is convenient to denote the above composition by $\mathcal{D}[\Pi := \begin{array}{c}\mathcal{D}' \\ \Pi\end{array}]$. This notation is conveniently extended to parallel multiple compositions. For example, for two anchors, Π_1 and Π_2, we get

$$\mathcal{D}[\Pi_1, \Pi_2 := \begin{array}{c}\mathcal{D}'_1 \\ \Pi_1\end{array}, \begin{array}{c}\mathcal{D}'_2 \\ \Pi_2\end{array}] \tag{3.13}$$

Remark: It is convenient to extend the definition to the case where the anchor Π does not occur in $\mathcal{D}$ as a *vacuous composition*, resulting in $\mathcal{D}$ itself, left unchanged. This is like a substitution for a variable not occurring in the term into which a substitution takes place.

Proposition 3.1 (closure of $\mathcal{N}^n$ under composition of derivations). *If* $\begin{array}{c}\boldsymbol{\Pi}, \Pi\\ \mathcal{D}\\ \Pi^*\end{array}$ *and* $\begin{array}{c}\boldsymbol{\Pi}'\\ \mathcal{D}'\\ \Pi\end{array}$ *are two $\mathcal{N}^n$-derivations, so is* $\mathcal{D}[\Pi := \begin{array}{c}\mathcal{D}'\\ \Pi\end{array}]$.

The proof is standard, like for other ND-systems, and omitted. See [17].

Definition 3.4 (simple poly-sequents). *Those are limit cases of poly-sequents.*

- *A poly-sequent* $\Pi = \overline{\Lambda} : \overline{\Delta}$ *is called* a-simple.
- *A poly-sequent* $\Pi = \overline{\Gamma} : \overline{\Lambda}$ *is called* c-simple.

Thus, a-simple poly-sequents have empty assumption compartments, while c-simple poly-sequents have empty conclusion compartments.

The following notation is handy for the next definition. Let $\Delta \cup \overline{\Gamma}$ abbreviate $\Delta \cup \Gamma_1 \cdots \cup \Gamma_n$.

Definition 3.5 (simplification). *The simplification* $\Pi_{\overline{\Gamma} \to \overline{\Delta}}$ *of a poly-sequent* $\Pi = \overline{\Gamma} : \overline{\Delta}$ *is defined as*

$$\Pi_{\overline{\Gamma} \to \overline{\Delta}} =^{\text{df.}} \overline{\Lambda} : \Delta_1 \cup \overline{\Gamma}_{\overline{1}} | \cdots | \Delta_n \cup \overline{\Gamma}_{\overline{n}} \tag{3.14}$$

Abbreviate $\Delta_1 \cup \overline{\Gamma}_{\overline{1}} | \cdots | \Delta_n \cup \overline{\Gamma}_{\overline{n}}$ *to* $conc(\Pi_{\overline{\Gamma} \to \overline{\Delta}})$.

Proposition 3.2 (simplification equivalence).

$$\Pi \dashv\vdash_{\mathcal{N}^n} conc(\Pi_{\overline{\Gamma} \to \overline{\Delta}}) \tag{3.15}$$

Proof: Each direction is proved by iterating the appropriate shifting rule.

Thus, when convenient, it is always possible to assume w.l.o.g that poly-sequents are a-simple.

The following proposition establishes the relationship of our poly-sequents to the many-sided sequents of Baaz et. al. [2].

Proposition 3.3. *Let* $\Delta = \Delta_1|\cdots|\Delta_n,\ n \geq 2$.

- Δ *is valid under the semantics of [2] iff* $\overline{\Lambda} : \overline{\Delta}$ *is valid under our semantics.*
- $\Pi = \Gamma : \Delta$ *is valid under our semantics iff* $conc(\Pi_{\overline{\Gamma}\to\overline{\Delta}})$ *is valid under the semantics of [2].*

3.3 Some useful derived and admissible rules

The structural rules below are useful for conveniently establishing some properties of $\mathcal{N}^n$.

Weakening: Those follow from the conjunctive reading of antecedents and disjunctive reading of succedents and are easily shown admissible. Still, their use in derivations is often convenient.

$$\frac{\overline{\Gamma} : \overline{\Delta}}{\overline{\Gamma}_{\overline{i}}|\Gamma_i, \varphi : \overline{\Delta}}\ (WL_i) \qquad \frac{\overline{\Gamma} : \overline{\Delta}}{\overline{\Gamma} : \overline{\Delta}_{\overline{i}}|\Delta_i, \varphi}\ (WR_i) \tag{3.16}$$

(Cut): For every $1 \leq i \leq n$:

$$\frac{\overline{\Gamma} : \overline{\Delta}_{\overline{i}}|\Delta_i, \varphi \quad \overline{\Gamma}'_{\overline{i}}|\Gamma'_i, \varphi : \overline{\Delta}'}{\overline{\Gamma}, \overline{\Gamma}' : \overline{\Delta}, \overline{\Delta}'}\ (cut_i) \tag{3.17}$$

Proof of (cut)**-derivability**: We show by induction on m the derivability of each of

$$\frac{\overline{\Gamma} : \overline{\Delta}_{\overline{i}}|\Delta_i, \varphi \quad \overline{\Gamma}'_{\overline{i}}|\Gamma'_i, \varphi : \overline{\Delta}'}{\overline{\Gamma}, \overline{\Gamma}' : \overline{\Delta}, \overline{\Delta}', \overline{\Lambda}_{m+1,\cdots,n}, \varphi} \tag{3.18}$$

The result follows for $m = n$. W.l.o.g., assume $i = 1$.

Basis: $m = 1$.

$$\cfrac{\cfrac{\overline{\Gamma}'_{\overline{1}}|\Gamma'_1, \varphi : \overline{\Delta}'}{\overline{\Gamma}' : \overline{\Delta}', \overline{\Lambda}_{2,\cdots,n}, \varphi}\ (\overrightarrow{s}_1)}{\overline{\Gamma}, \overline{\Gamma}' : \overline{\Delta}, \overline{\Delta}', \overline{\Lambda}_{2,\cdots,n}, \varphi}\ (WLs, WRs) \tag{3.19}$$

Induction step: Assume (3.18) for m.

$$\frac{\overline{\Gamma}_{\overline{1}}|\Gamma_1, \varphi : \overline{\Delta} \quad \overline{\Gamma}, \overline{\Gamma}' : \overline{\Delta}, \overline{\Delta}', \overline{\Lambda}_{m+1,\cdots,n}, \varphi}{\overline{\Gamma}, \overline{\Gamma}' : \overline{\Delta}, \overline{\Delta}', \overline{\Lambda}_{m+2,\cdots,n}, \varphi}\ (c_{1,m+1}) \tag{3.20}$$

Elsewhere [13], we present a stronger result, that *both* (*cut*) and the $(c_{i,j})$s are jointly eliminable.

Redundancy rules

$$\frac{\{\overline{\Gamma}_{\bar{i}}|\Gamma_i, \varphi : \overline{\Delta}, \quad i \in \hat{n}\}}{\overline{\Gamma} : \overline{\Delta}} (\overleftarrow{st}) \qquad \frac{\{\overline{\Gamma} : \overline{\Delta}|\Delta_i, \quad \varphi, i \in \hat{n}\}}{\overline{\Gamma} : \overline{\Delta}} (\overrightarrow{st}) \tag{3.21}$$

The derivation of $(\overleftarrow{st})$ is by $(\overrightarrow{s}_1)$ followed by (*cut*)s. $(\overrightarrow{st})$ is just a special case of $(c_{i,j})$.

3.4 Is $\mathcal{N}^n$ a natural-deduction proof system?

Traditionally, there is a basic distinguishing feature between natural-deduction and sequent-calculi, even when the former are formulated in Gentzen's logistic style (using sequents, not formulas): Natural-deduction rules manipulate only the succedent of a sequent, while sequent-calculi rules manipulate both the succedent and the antecedent of a sequent.

In view of the $(\overleftarrow{s}_{i,j})$ shifting rules, that manipulates also the antecedent of a sequent, one might wonder whether $\mathcal{N}^n$ still qualifies as an ND-system.

We claim it does.

The reason, expressed in terms of Schroeder-Heister's [30] terminology, is the following. Schroeder-Heister distinguishes between two ways of introducing an assumption into an antecedent of a sequent: *specific*, i.e., according to the assumptions meaning, and *non-specific*. The sequent-calculi left-rules all introduce assumption into an antecedent specifically: each logical operator has its own L-rule, reflecting its meaning. On the other hand, the shifting rules introduce an assumption into an antecedent non-specifically, independently of the form (and, hence, of the meaning) of that assumption.

Thus, the shifting rules are more like a structural rule, not disrupting the general characteristic of ND-system listed above.

3.5 Further properties of $\mathcal{N}^n$

The following theorem is the natural generalization of bivalency in bivalent logics. It expresses the fact that every φ must have *some* truth-value. It is the natural generalization of the bivalent excluded-middle law, but expressed without appealing to negation.

Theorem 3.1.

$$\vdash_{\mathcal{N}^n} \overline{\Lambda} : \Lambda_1, \varphi | \cdots | \Lambda_n, \varphi \tag{3.22}$$

Proof: The derivation is

$$\frac{\overline{\Lambda}_{\bar{i}} | \Lambda_i, \varphi \ : \Lambda_i, \varphi | \overline{\Lambda}_{\bar{i}}}{\overline{\Lambda} : \Lambda_1, \varphi | \cdots | \Lambda_n, \varphi} \ (\overrightarrow{s}_i) \tag{3.23}$$

In the sequel, we refer to the poly-sequent $\overline{\Lambda} : \Lambda_1, \varphi | \cdots | \Lambda_n, \varphi$ as Π_φ. An immediate generalization is the following.

Corollary 3.1. *By applying to (3.22) both (WL) and (WR) a sufficient number of times, we have*

$$\vdash_{\mathcal{N}^n} \overline{\Gamma} : \Delta_1, \varphi | \cdots | \Delta_n, \varphi \tag{3.24}$$

Theorem 3.2 (soundness and strong completeness). *For every* $\mathbf{\Pi}$ *and* Π*:*

$$\mathbf{\Pi} \models \Pi \text{ iff } \mathbf{\Pi} \vdash_{\mathcal{N}^n} \Pi \tag{3.25}$$

Proof:

soundness: Suppose that $\mathbf{\Pi} \vdash_{\mathcal{N}^n} \Pi$. The proof that $\mathbf{\Pi} \models \Pi$ is by induction on the derivation, by case analysis on the last rule applied. Let σ be an assignment s.t. $\models_\sigma \mathbf{\Pi}$.

initial poly-sequents: obvious.

shifting rules:

$(\overrightarrow{s}_i)$: Suppose σ satisfies the premise, $\models_\sigma \overline{\Gamma}_{\bar{i}} | \Gamma_i, \varphi : \overline{\Delta}$. We present only the case of non-vacuous satisfaction (see Definition 2.2).
There are two cases to consider.

$\sigma[\![\varphi]\!] = v_i$: In this case, σ a-supports $\overline{\Gamma}_{\bar{i}}|\Gamma_i, \varphi$. Therefore, σ c-supports $\overline{\Delta}_{\bar{j}}|\Delta_j, \varphi$ as well.

$\sigma[\![\varphi]\!] \neq v_i$: In this case, $\sigma[\![\varphi]\!] = v_j$ for some $j \in \bar{i}$. Hence, σ c-supports $\overline{\Delta}_{\bar{j}}|\Delta_j, \varphi$, and thereby satisfies $\overline{\Gamma} : \Delta_i|\overline{\Delta}_{\bar{i}}, \varphi$ itself.

In both cases, the conclusion of the rule is satisfied by σ.

Since σ is arbitrary, $\mathbf{\Pi}\models\Pi$.

$(\overleftarrow{s}_{i,j})$: Suppose σ satisfies the premise, $\models_\sigma \overline{\Gamma} : \overline{\Delta}_{\bar{i}}|\Delta_i, \varphi$. Again, we present only the case of non-vacuous satisfaction, and there are two cases to consider.

$\sigma[\![\varphi]\!] = v_i$: In this case, σ does not a-support $\overline{\Gamma}_{\bar{j}}|\Gamma_j, \varphi$, and the conclusion is satisfied vacuously by σ.

$\sigma[\![\varphi]\!] = v_k,\ k \neq i$: In this case, σ does not c-support $\overline{\Lambda}_{\bar{i}}|\Lambda_i, \varphi$; but the premise is satisfied by σ by assumption, hence σ c-supports $\overline{\Delta}_{\bar{k}}|\Delta_{k'}$.

operational rules: For simplicity, we consider the case where '$*$' is binary. Only the cases of non-vacuous satisfaction of the premises are presented.

$(*I_{i,j,k})$: Suppose σ satisfying the premises. Therefore, $\sigma[\![\varphi]\!] = v_i$ and $\sigma[\![\psi]\!] = v_j$. By assumption, the truth value of $\varphi * \psi$ in this case (by the truth-table for '$*$') is v_k. Hence, σ c-supports $\Delta_k, \varphi * \psi$, and thereby also satisfies the conclusion $\overline{\Gamma} : \overline{\Delta}_{\bar{k}}|\Delta_k, \varphi * \psi$.

$(*E_{I_k})$: Suppose σ satisfies the premises. Therefore, $\sigma[\![\varphi * \psi]\!] = v_k$. Suppose that $\sigma[\![\varphi]\!] = v_i$ and $\sigma[\![\psi]\!] = v_j$. Therefore, by the truth-table of '$*$', $\langle i, j\rangle \in I_k$. Therefore, σ also satisfies $\overline{\Gamma}_{\overline{i,j}}|\Gamma_i, \varphi|\Gamma_j, \psi$. By the satisfaction by σ of all the minor premises, σ satisfies $\overline{\Delta}$. Thereby, σ satisfies the conclusion $\overline{\Gamma} : \overline{\Delta}$.

completeness: We prove the contra-positive. Assume that $\mathbf{\Pi}\nvdash_{\mathcal{N}^n}\Pi$, and construct a counter model. We proceed in a number of stages.

Lemma 3.1. *For* $i \neq j$, $\vdash_{\mathcal{N}^n} \overline{\Lambda}_{\overline{\{i,j\}}}|\overline{\Lambda}_{\{i,j\}}, \varphi : \overline{\Lambda}$

The derivation is

$$\frac{\overline{\Lambda}_{\bar{i}}|\Lambda_i, \varphi : \overline{\Lambda}_{\bar{i}}|\Lambda_i, \varphi}{\overline{\Lambda}_{\overline{\{i,j\}}}|\overline{\Lambda}_{\{i,j\}}, \varphi : \overline{\Lambda}}\ (\overleftarrow{s}_{i,j}) \tag{3.26}$$

Lemma 3.2. *If* $\nvdash_{\mathcal{N}^n}\overline{\Gamma} : \overline{\Delta}$, *then for no* φ *and no* i, j *s.t.* $i \neq j$ *is* $\Gamma = \overline{\Gamma}_{\overline{\{i,j\}}}|\Gamma_i, \varphi|\Gamma_j, \varphi$.

Assume, towards a contradiction, the contrary

$$\Gamma = \overline{\Gamma}_{\overline{\{i,j\}}}|\Gamma_i, \varphi|\Gamma_j, \varphi \tag{3.27}$$

By Lemma 3.1, $\vdash_{\mathcal{N}^n} \overline{\Lambda}_{\overline{\{i,j\}}}|\overline{\Lambda}_{\{i,j\}}, \varphi : \overline{\Lambda}$. By a number of weakening applied to (3.27), we get that $\vdash_{\mathcal{N}^n} \overline{\Gamma} : \overline{\Delta}$, contradicting the assumption.

In view of the simplification proposition (Proposition 3.2) we may assume w.l.o.g both that all the poly-sequents in $\mathbf{\Pi}$ as well as Π itself are a-simple; namely, $\mathbf{\Pi} = \overline{\Lambda} : \overline{\Delta}^1, \cdots$ and $\Pi = \overline{\Lambda} : \overline{\Delta}$, respectively. Also, we may assume that the derivable poly-sequents Π_φ ((3.22), for all φ) belongs[7] to $\mathbf{\Pi}$.

Next, we construct a sequence $\overline{\Gamma}^i$, $i \geq 0$, s.t.:

- *(nded)* $\mathbf{\Pi} \nvdash_{\mathcal{N}^n} \overline{\Gamma}^i : \overline{\Delta}$.
- *(dis)* For every $1 \leq k \leq i$, and some j, $\Delta_j^k \cap \Gamma_j^i \neq \emptyset$.

Let $\overline{\Gamma}^0 = \overline{\Lambda}$. Then, *(nded)* holds by assumption and *(dis)* holds vacuously.

Assume $\overline{\Gamma}^i$ has been constructed.
Claim: For some j and some φ in $\overline{\Delta}_j^{i+1}$, $\mathbf{\Pi} \nvdash_{\mathcal{N}^n} \overline{\Gamma}_{\overline{j}}^i|\Gamma_j^i, \varphi : \overline{\Delta}$.

Otherwise, that is, for every j and every φ in Δ_j^{i+1}, $\mathbf{\Pi} \vdash_{\mathcal{N}^n} \overline{\Gamma}_{\overline{j}}^i|\Gamma_j^i, \varphi : \overline{\Delta}$, then by applying several *(cut)*s with $\overline{\Lambda} : \overline{\Delta}^{i+1}$, we get $\mathbf{\Pi} \vdash_{\mathcal{N}^n} \overline{\Gamma}^i : \overline{\Delta}$, contradicting *(nded)* for $\overline{\Gamma}^i$.

We can now define $\overline{\Gamma}^{i+1}$ by

$$\overline{\Gamma}^{i+1} =^{\text{df.}} \overline{\Gamma}_{\overline{j}}^i|\Gamma_j^i, \varphi \tag{3.28}$$

(for the φ and j from the claim).

Then, *(nded)* holds by the claim, and *(dis)* holds since it holds for $\overline{\Gamma}^i$ and $\varphi \in \overline{\Delta}_j^{i+1}$.

Put $\hat{\overline{\Gamma}} =^{\text{df.}} \bigcup_{i \geq 0} \overline{\Gamma}^i$ (where the union is compartment-wise).

We can now define the counter-model, an assignment $\hat{\sigma}$, by:

$$\hat{\sigma}[\![p]\!] = v_k \text{ iff } p \in \hat{\Gamma}_k \tag{3.29}$$

The assignment $\hat{\sigma}$ is well-defined by Lemma 3.2.

By Lemma 3.3 below, this property of $\hat{\sigma}$ extends to every φ.

[7] Therefore, $\mathbf{\Pi}$ is infinite.

Lemma 3.3.

$$\hat{\sigma}[\![\varphi]\!] = v_k \text{ iff } \varphi \in \hat{\Gamma}_k \tag{3.30}$$

The proof is by induction on the structure of φ. The basis is (3.29). For the induction step, suppose $\varphi = *(\varphi_1, \cdots, \varphi_p)$.

Let, for $1 \leq j \leq p$, $\hat{\sigma}[\![\varphi_j]\!] = v_{i_j}$.

By the induction hypothesis, for $1 \leq j \leq p$, $\varphi_j \in \hat{\Gamma}_{i_j}$.

1. Suppose $\hat{\sigma}[\![\varphi]\!] = v_k$, and assume, towards a contradiction, that for some $k' \neq k$ it holds that $\varphi \in \hat{\overline{\Gamma}}_{k'}$.
   Let m be such that $\Gamma^m_{k'}$ contains φ, and $\Gamma^m_{i_j}$ contains φ_j.
   From the initial sequents $\overline{\Gamma}_{\overline{i_j}}|\Gamma_{i_j}, \varphi_{i_j} : \overline{\Delta}_{\overline{i_j}}|\Delta_{i_j}, \varphi_{i_j}$, $1 \leq j \leq p$ we get by weakenings
   $$\mathbf{\Pi} \vdash_{\mathcal{N}^n} \overline{\Gamma}^m : \overline{\Delta}_{\overline{\{i_j, 1 \leq j \leq p\}}}|\Delta_{i_j}, \varphi_{i_j}, \qquad 1 \leq j \leq p \tag{3.31}$$
   By applying the rule $(*I_{i_1,\cdots,i_l,k})$, we get
   $$\mathbf{\Pi} \vdash_{\mathcal{N}^n} \overline{\Gamma}^m : \overline{\Delta}_{\overline{i_j}}|\Delta_k, \varphi \tag{3.32}$$
   But by applying the coordination rule $(c_{k,k'})$ to (3.32) and the initial poly-sequent $\overline{\Lambda}_{\overline{i}} \mid \Lambda_{k'}, \varphi \ : \ \overline{\Lambda}_{\overline{k'}} \mid \Lambda_{k'}, \varphi$, we get $\mathbf{\Pi} \vdash_{\mathcal{N}^n} \overline{\Gamma}^m : \overline{\Delta}$, contradicting $(nded)$ for m.
2. Suppose $\varphi \in \hat{\Gamma}_k$. Then, by part 1 of the proof, $\varphi \in \hat{\Gamma}_{k'}$ in contradiction with $\varphi \in \hat{\Gamma}_k$.
   We now show that indeed $\hat{\sigma}$ is a counter-model.
   First, we note that $\models_{\hat{\sigma}} \mathbf{\Pi}$, since by (dis), for some j, there is some $\varphi \in \Delta^i_j \cap \hat{\Gamma}_j$. By Lemma 3.3, $\hat{\sigma}[\![\varphi]\!] = v_j$, so $\models_{\hat{\sigma}} \overline{\Lambda} : \overline{\Delta}^i$.
   Assume, towards a contradiction, that for some $1 \leq j \leq n$ and some $\varphi \in \Delta_j$ it holds that $\hat{\sigma}[\![\varphi]\!] = v_j$. It follows by Lemma 3.3 that for some m, $\varphi \in \Gamma^m_j$. Hence, by applying a number of weakenings to the initial poly-sequent $\overline{\Lambda}_{\overline{j}}|\Lambda_j, \varphi : \overline{\Lambda}_{\overline{j}}|\Lambda_j, \varphi$, we get $\mathbf{\Pi} \vdash_{\mathcal{N}^n} \overline{\Gamma}^m : \overline{\Delta}$, contradicting $(nded)$ for m.

Corollary 3.2 (admissibility of the E-rules). *The E-rules of $\mathcal{N}^n$ are admissible.*

The corollary follows from the completeness proof, that makes no use of the E-rules. Still, they are useful both in derivations and in showing that other n-valued logics are retrievable from $\mathcal{N}^n$.

3.6 Harmony and stability of $\mathcal{N}^n$

According to the theory of meaning known as Proof-Theoretic Semantics (PTS) [10], the meaning of the connectives is not given by their truth-table, but is determined by the I/E-rules of a meaning-conferring ND-system. This approach to meaning has not been applied so far to multi-valued logics.

As is well known, not every ND-system qualifies as a meaning-conferring system. A major requirement for such qualification [6] is that there is a *balance* between the I/E rules, no group overpowers the other. This balance is often formalized by means of two properties, known as *local soundness* and *local completeness* [18, 5], explained below.

Definition 3.6 (maximal poly-sequent). *A* maximal[8] poly-sequent *in an $\mathcal{N}^n$-derivation is a poly-sequent that is both the consequence of an application of an I-rule (for some '$*$'), as well as a major premise of the application of an E rule (for the same '$*$').*

Note that a maximal poly-sequent cannot be c-simple.

Definition 3.7 (reduction). *A* reduction *of an $\mathcal{N}^n$-derivation $\mathcal{D}$ having an occurrence of a maximal poly-sequent is a transformation of $\mathcal{D}$ to an equivalent derivation $\mathcal{D}'$ (having the same open assumptions and conclusion) in which the above mentioned occurrence of a maximal poly-sequent is removed*[9].

Definition 3.8 (local soundness). *An ND-system is* locally sound *iff every derivation with an occurrence of a maximal (here: poly-sequent) is reducible.*

Violating local soundness means that the E-rules are too weak compared to the I-rules: there are conclusions obtainable only by means of introductions.

Theorem 3.3 (local soundness of $\mathcal{N}^n$). *$\mathcal{N}^n$ is locally sound.*

Proof: A derivation with a maximal occurrence of a (non c-simple) poly-sequent

[8] Traditionally ([22]), this notion is defined over formulas. Since here derivations are over poly-sequents, we use the natural adaptation of the definition.

[9] The absence of an infinite sequence of reductions is called (strong) *normalization*. Here, only a single reduction step is considered.

$\Pi = \overline{\Gamma} \;:\; \overline{\Delta}_{\overline{k}} \mid \Delta_k, \varphi * \psi$ has the following form[10].

$$\cfrac{\cfrac{\begin{matrix}\mathcal{D}_i \\ \overline{\Gamma} : \overline{\Delta}_{\overline{i}}|\Delta_i, \varphi\end{matrix} \quad \begin{matrix}\mathcal{D}_j \\ \overline{\Gamma} : \overline{\Delta}_{\overline{j}}|\Delta_j, \psi\end{matrix}}{\overline{\Gamma} : \overline{\Delta}_{\overline{k}}|\Delta_k, \varphi * \psi}\,(*I_{i,j,k}) \qquad \begin{matrix}\mathcal{D}_{i,j} \\ \{\overline{\Gamma}_{\overline{\{i,j\}}} \mid \Gamma_i, \varphi|\Gamma_j, \psi : \overline{\Delta},\;\; \langle i,j\rangle \in I_k\}\end{matrix}}{\overline{\Gamma} : \overline{\Delta}}\,(*E_{k,l}) \tag{3.33}$$

reduces to

$$\cfrac{\begin{matrix}\mathcal{D}_i \\ \overline{\Gamma} : \overline{\Delta}_{\overline{i}}|\Delta_i, \varphi\end{matrix} \qquad \cfrac{\begin{matrix}\mathcal{D}_j \\ \overline{\Gamma} : \overline{\Delta}_{\overline{j}}|\Delta_j, \psi\end{matrix} \quad \begin{matrix}\mathcal{D}_{i,j} \\ \overline{\Gamma}_{\overline{\{i,j\}}} \mid \Gamma_i, \varphi|\Gamma_j, \psi : \overline{\Delta}\end{matrix}}{\overline{\Gamma}_{\overline{\{i\}}}|\Gamma_i, \varphi : \overline{\Delta}}\,(cuts)}{\overline{\Gamma} : \overline{\Delta}}\,(cuts) \tag{3.34}$$

We now turn to the other direction of the balance requirement.

Definition 3.9 (expansion). *An* expansion *of a $\mathcal{N}^n$-derivation $\mathcal{D}$, say with a (non c-simple) conclusion $\Pi = \overline{\Gamma} \;:\; \overline{\Delta}_{\overline{k}} \mid \Delta_k, \varphi * \psi$, is a transformation to an equivalent derivation $\mathcal{D}'$, in which all the E-rules of '$*$' are applied, followed by applications of the I-rules of '$*$'.*

Definition 3.10 (local completeness). *An ND-system is locally complete iff every derivation with a non c-simple conclusion has an expansion.*

Violating local completeness means that that the I-rules are too weak compared to the E-rules. A conclusion can be decomposed by applications of the E-rules such that it cannot be reconstructed by applications of I-rules.

Theorem 3.4 (local completeness of $\mathcal{N}^n$). *$\mathcal{N}^n$ is locally complete.*

Proof: Consider any $\mathcal{N}^n$-derivation $\mathcal{D}$ with a (non c-simple) conclusion $\Pi = \overline{\Gamma} : \overline{\Delta}_{\overline{k}}|\Delta_k, \varphi * \psi$. Its expansion has the following form.
Let $\langle i,j\rangle \in I_k$:

$$\cfrac{\cfrac{\begin{matrix}\mathcal{D} \\ \overline{\Gamma} : \overline{\Delta}_{\overline{k}}|\Delta_k, \varphi * \psi\end{matrix} \quad \begin{matrix}\{\overline{\Gamma}_{\overline{i,j}}|\Gamma_i, \varphi|\Gamma_j, \psi : \\ \overline{\Delta}_i|\Delta_j, \psi,\;\; \langle i,j\rangle \in I_k\}\end{matrix}}{\overline{\Gamma} : \overline{\Delta}_{\overline{j}}|\Delta_j, \psi}\,(*E_{I_k}) \quad \cfrac{\begin{matrix}\mathcal{D} \\ \overline{\Gamma} : \overline{\Delta}_{\overline{k}}|\Delta_k, \varphi * \psi\end{matrix} \quad \begin{matrix}\{\overline{\Gamma}_{\overline{i,j}}|\Gamma_i, \varphi|\Gamma_j, \psi : \\ \overline{\Delta}_i|\Delta_i, \varphi,\;\; \langle i,j\rangle \in I_k\}\end{matrix}}{\overline{\Gamma} : \overline{\Delta}_{\overline{i}}|\Delta_i, \varphi}\,(*E_{I_k})}{\overline{\Gamma} : \overline{\Delta}_{\overline{k}}|\Delta_k, \varphi * \psi}\,(*I_{i,j,k}) \tag{3.35}$$

[10] To simplify the notation, we consider only the case of a binary '$*$'.

3.7 Designated truth-values

We next turn to the designated values $\mathcal{V}_d$ determining the consequence relation $\Theta \models_{\mathcal{L}} \varphi$ over formulas for the multi-valued logic $\mathcal{L}$ the truth-tables of which gave rise to $\mathcal{N}^n$. We show how this relation can be recovered from the consequence relation over poly-sequents.

Let D be the set of indices corresponding to $\mathcal{V}_d$, and ND the other indices (of non-designated truth-value). Our first step is to embed, under a given truth-value assignment σ, any formula φ as a poly-sequent Π_φ.

Definition 3.11 (formula embedding). *For φ an object-language formula, its D-embedding poly-sequent Π_φ is*

$$\Pi_\varphi =^{\text{df.}} \overline{\Lambda} : \overline{\Lambda}_{ND} | \overline{\Lambda}_D, \varphi$$

The definition directly implies the following proposition.

Proposition 3.4 (embedding). *For every truth-value assignment σ and every φ:*

$$\sigma[\![\varphi]\!] \in \mathcal{V}_d \text{ iff } \models_\sigma \Pi_\varphi \tag{3.36}$$

We next extend the embedding to finite sets of formulas. Consider $\Theta = \{\varphi_1, \cdots, \varphi_m, m \geq 1\}$ a finite collection of formulas. Let the D-embedding of Θ be $\mathbf{\Pi}_\Theta = \{\Pi_{\varphi_1}, \cdots, \Pi_{\varphi_m}\}$.

We now have the following theorem.

Theorem 3.5 (consequence).

$$\Theta \models_{\mathcal{L}} \varphi \text{ iff } \mathbf{\Pi}_\Theta \models \Pi_\varphi \tag{3.37}$$

4 Retrievability of multi-valued logics in $\mathcal{N}^n$

4.1 Introduction

In the literature on multi-valued logics, there were many proposals of *specific* such logics, defined over formulas (sometimes, signed formulas), based on various consequence relations. In this section, we consider the generality of the $\mathcal{N}^n$ calculi by showing how specific n-valued logics can be *retrieved* in $\mathcal{N}^n$. We present in detail two cases with $n = 3$:

1. Kleene's strong three-valued logic K_3.
2. Priest's paraconsistent logic LP.

We leave the full details for other ns to be presented elsewhere. These include:

2-valued poly-sequent logic: The following ND-systems are retrievable from $\mathcal{N}^2$.

- Gentzen's NK for classical logic.
- The axiomatic (Hilbert-like) system for a logic for *falsification* presented in [16].

4-valued poly-sequent logic: The following variants of the Belnap-Dunn logic for *first-degree entailment (FDE)* [3, 4, 7] are retrievable from $\mathcal{N}^4$.

- An ND-system for *relatedness to truth* from [31].
- A new ND-system for ETL (*exactly-true* logic) of [19]. Only a sequent-calculus for this logic is presented in the literature, in [33].
- An ND-system for a logic with a different negation, as in [15] or [25].

We refer here to the third truth-value mnemonically, in accordance to its role in the various three-valued logics assigning to it some specific behaviour.

Remark: While the system constructed by Baaz et.al. [2] is also capable of retrieving some multi-valued ND-systems, it can do so for logics the consequence relation of which is based on the preservation of *a single designated value*, our system can deal (as shown below) with *any number* of designated values.

4.2 Canonical translation

In a saturated logic $\mathcal{N}^n$ (where $n \geq 2$), one is interested in retrieving any *given ND-system* (over formulas) for an n-valued logic $\mathcal{L}_n$, corresponding to a consequence relation defined in terms of the preservation of *designated values*, a subset $\mathcal{V}_d$ of $\mathcal{V}$.

Let $\mathcal{V}_{nd}$ be the non-designated truth-values. Recall that D is the set of indices of the truth-values in $\mathcal{V}_d$ and ND – the indices of the truth-values in $\mathcal{V}_{nd}$. Also, let $d = |D|$.

What does it mean that a specific ND-system for an n-valued logic $\mathcal{L}_n$ is retrievable in $\mathcal{N}^n$? In order to define this notion of an ND-system retrieval, we introduce the following definition.

When a specific ND-system[11] for a logic $\mathcal{L}_n$ (over formulas) is to be retrieved, the rules of $\mathcal{N}_{\mathcal{L}_n}$ are interpreted as preserving designated truth-values: for every rule (ρ) of $\mathcal{N}_{\mathcal{L}_n}$, if each premise of (ρ) has a designated truth-value, so does the conclusion of (ρ).

Definition 4.12 (canonical rule translation). *Let*

$$\frac{\begin{matrix}[\psi_1] & & [\psi_m]\\ \vdots & & \vdots\\ \varphi_1 & \cdots & \varphi_m\end{matrix}}{\psi}\ (\rho)$$

be a rule in a natural deduction system $\mathcal{N}_{\mathcal{L}_n}$ for a given n-valued logic $\mathcal{L}_n$. For simplicity, we assume that each premise discharges at most one assumption. There are d^m canonical translations of each rule, since each ψ_j can be placed in any of the d designated truth-values assumption positions. The k'th canonical translation of (ρ) into $\mathcal{N}^n$ is given by

$$\frac{\Pi_{1,k} \quad \cdots \quad \Pi_{m,k}}{\Pi_\psi}\ (\mathcal{T}(\rho)_k) \tag{4.38}$$

where

$$\Pi_{j,k} =^{\text{df.}} \overline{\Gamma}_{D-\{\alpha_j\}}|\Gamma_{\alpha_j}, \psi_j|\overline{\Lambda}_{ND} : \overline{\Delta}_D, \varphi_j|\overline{\Lambda}_{ND},\ 1 \le j \le m,\ \alpha_j \in D$$

and[12]

$$\Pi_\psi =^{\text{df.}} \overline{\Gamma}_D|\overline{\Lambda}_{ND} : \overline{\Delta}_D, \psi|\overline{\Lambda}_{ND}$$

[11]Note the difference between $\mathcal{N}_{\mathcal{L}_n}$, a given ND-system for $\mathcal{L}_n$, and $\mathcal{N}^n$, our uniform poly-sequent ND-stem.

[12]Note that this notation was defined somewhat differently in Section 3.7, with $\overline{\lambda}$ instead of $\overline{\Gamma}$. The reason for this difference is that here we deal not merely with embedding a formula, but embedding *a premise*, that may depend on assumptions.

That is, the jth premise of (ρ) is converted into a poly-sequent Π_j in which:

- A discharged assumption is placed in *one* of the designated Γ-compartments (with empty non-designated Γ-compartments). If no assumption is discharged by the j'th premise, this step is ignored.
- An assumption φ_j is placed in *all* the designated Δ-compartments (with empty non-designated Δ-compartments).

Definition 4.13 (retrievability).

- *A rule (ρ) in an ND-system $\mathcal{N}_{\mathcal{L}_n}$ for an n-valued logic $\mathcal{L}_n$ is* retrievable *in $\mathcal{N}^n$ iff each of its canonical translations (as in (4.38)) satisfies*

$$\Pi_{1,k}, \cdots, \Pi_{m,k} \vdash_{\mathcal{N}^n} \Pi_\psi \tag{4.39}$$

- *An ND-system $\mathcal{N}_{\mathcal{L}_n}$ for an n-valued logic $\mathcal{L}_n$ is* retrievable *in $\mathcal{N}^n$ iff each of the rules of $\mathcal{N}_{\mathcal{L}_n}$ is* retrievable *in $\mathcal{N}^n$.*

The need for a canonical rule-translation arises from the need to translate formulas (from the object language of $\mathcal{L}_n$) serving as premises of a rule to poly-sequent. This translation has to take into account the definition of consequence relations based on the preservation (or propagation) of *designated values*, varying from one $\mathcal{N}_{\mathcal{L}_n}$ to another (for the same n and even the same object language), even when the truth-tables for the connectives remain the same.

We have the following theorem as a consequence from the definition of retrievability.

Theorem 4.6 (retrieval). *Let $\mathcal{N}_{\mathcal{L}_n}$ be a complete ND-system for $\mathcal{L}_n$.*

$$\varphi_1, \cdots, \varphi_m \vdash_{\mathcal{N}_{\mathcal{L}_n}} \psi \tag{4.40}$$

iff for each function $f : \{\varphi_1, \cdots, \varphi_m\} \to D$

$$\vdash_{\mathcal{N}^n} \overline{\Lambda}_{ND} | \overline{\Gamma^f_D} : \overline{\Lambda}_{ND} | \overline{\Lambda}_D, \psi \tag{4.41}$$

where, for $i \in D$, $\Gamma^f_i = f^{-1}(i)$.[13]

[13] That is, Γ^f_i consists of all assumptions which have the designated value v_i.

$\neg$		$\wedge$	t	n	f	$\vee$	t	n	f	$\supset$	t	n	f	
t	f	t	t	n	f	t	t	t	t	t	t	n	f	(4.42)
n	n	n	n	n	f	n	t	n	n	n	t	n	n	
f	t	f	f	f	f	f	t	n	f	f	t	t	t	

Figure 1: The three-valued truth-tables for K_3

Proof:

only if: By induction on the derivation, based on (4.39) for every rule application.

if: By the soundness of $\mathcal{N}^n$, the poly-sequent $\overline{\Lambda}_{ND}|\{\Lambda_j, \varphi_j \mid 1 \leq j \leq m,\ v_j \in D\} : \overline{\Lambda}_{ND}|\overline{\Lambda}_D, \psi$ is valid. Hence, if all the φ_js have the indicated designated truth-values, by (4.40) so does ψ. The result follows by the completeness of $\mathcal{N}_{\mathcal{L}_n}$.

4.3 Kleene's strong K_3

We start with Kleene's [14] strong three-valued logic K_3, in which the third truth value will be referred to as n, representing *neither true nor false.* The object language is over the classical operators, but having the truth-tables as presented in Figure 1 (see [23], p. 119).

4.3.1 The structure of $\mathcal{N}^3_{K_3}$

Poly-sequents have contexts with three compartments, corresponding to the three truth-values.

$$\Gamma_f|\Gamma_n|\Gamma_t : \Delta_f|\Delta_n|\Delta_t \tag{4.43}$$

Here it is more convenient to represent the exclusion of one of the compartments by indexing Γ and Δ with the remaining ones; for example, $\overline{\Gamma}_{f,t}$ excludes Γ_n, and $\overline{\Delta}_{f,n}$ excludes Δ_t.

$$\frac{\overline{\Gamma} : \overline{\Delta}_{f,n} | \Delta_t, \varphi}{\overline{\Gamma} : \overline{\Delta}_{n,t} | \Delta_f, \neg\varphi}\ (\neg I_f) \qquad \frac{\overline{\Gamma} : \overline{\Delta}_{n,t} | \Delta_f, \varphi}{\overline{\Gamma} : \overline{\Delta}_{f,n} | \Delta_t, \neg\varphi}\ (\neg I_t) \qquad \frac{\overline{\Gamma} : \overline{\Delta}_{f,t} | \Delta_n, \varphi}{\overline{\Gamma} : \overline{\Delta}_{f,t} | \Delta_n, \neg\varphi}\ (\neg I_n) \tag{4.47}$$

$$\frac{\overline{\Gamma} : \overline{\Delta}_{n,t} | \Delta_f, \neg\varphi}{\overline{\Gamma} : \overline{\Delta}_{f,n} | \Delta_t, \varphi}\ (\neg E_f) \qquad \frac{\overline{\Gamma} : \overline{\Delta}_{f,n} | \Delta_t, \neg\varphi}{\overline{\Gamma} : \overline{\Delta}_{n,t} | \Delta_f, \varphi}\ (\neg E_t) \qquad \frac{\overline{\Gamma} : \overline{\Delta}_{f,t} | \Delta_n, \neg\varphi}{\overline{\Gamma} : \overline{\Delta}_{f,t} | \Delta_n, \varphi}\ (\neg E_n) \tag{4.48}$$

Figure 2: $\mathcal{N}^3_{K_3}$: negation

The initial sequents are:

$$\frac{}{\Gamma_f, \varphi | \overline{\Gamma}_{n,t} : \Delta_f, \varphi | \overline{\Delta}_{n,t}}\ (Ax^f)$$

$$\frac{}{\Gamma_n, \varphi | \overline{\Gamma}_{f,t} : | \Delta_n, \varphi | \overline{\Delta}_{f,t}}\ (Ax^n) \tag{4.44}$$

$$\frac{}{\Gamma_t, \varphi | \overline{\Gamma}_{f,n} : \Delta_t, \varphi | \overline{\Delta}_{f,n}}\ (Ax^t)$$

The shifting rules are:

For every $i, j \in \{f, n, t\}$:

$$\frac{\overline{\Gamma}_{\bar{i}} | \Gamma_i, \varphi : \overline{\Delta}}{\overline{\Gamma} : \Delta_i | \overline{\Delta}_{\bar{i}}, \varphi}\ (\overrightarrow{s}_i) \qquad \frac{\overline{\Gamma} : \overline{\Delta}_{\bar{i}} | \Delta_i, \varphi}{\Gamma_{\bar{j}} | \Gamma_j, \varphi : \overline{\Delta}}\ (\overleftarrow{s}_{i,j}), j \neq i \tag{4.45}$$

The (admissible) coordination rules are the following.

$$\frac{\overline{\Gamma} : \overline{\Delta}_{\bar{i}} | \Delta_i, \varphi \quad \overline{\Gamma} : \overline{\Delta}_{\bar{j}} | \Delta_j, \varphi}{\overline{\Gamma} : \overline{\Delta}}\ (c_{i,j}), \quad i \neq j \tag{4.46}$$

The operational rules for the connectives are presented in figures 2 (negation), 3, 4, (conjunction), 5, 6 (disjunction) and 7, 8 (implication). Note that in the formulation of the rules, we used certain optimizations of the I/E-rules, easy to formulate for unary and binary operators, as is the case here. The optimization simplifies the rules in case a truth-table has a whole row (or a whole column) with *identical* entries; also, in case some truth-value *has a unique occurrence* in the table.

$$\frac{\overline{\Gamma} : \overline{\Delta}_{f,n}|\Delta_t, \varphi \quad \overline{\Gamma} : \overline{\Delta}_{f,n}|\Delta_t, \psi}{\overline{\Gamma} : \Delta_{f,n}|\Delta_t, \varphi\wedge\psi} (\wedge I_t) \tag{4.49}$$

$$\frac{\overline{\Gamma} : \Delta_f, \varphi|\overline{\Delta}_{n,t}}{\overline{\Gamma} : \Delta_f, \varphi\wedge\psi|\overline{\Delta}_{n,t}} (\wedge I_{f,1}) \qquad \frac{\overline{\Gamma} : \Delta_f, \psi|\overline{\Delta}_{n,t}}{\overline{\Gamma} : \Delta_f, \varphi\wedge\psi|\overline{\Delta}_{n,t}} (\wedge I_{f,2}) \tag{4.50}$$

$$\frac{\overline{\Gamma} : \overline{\Delta}_{f,t}|\Delta_n, \varphi \quad \overline{\Gamma} : \overline{\Delta}_{f,t}|\Delta_n, \psi}{\overline{\Gamma} : \overline{\Delta}_{f,t}|\Delta_n, \varphi\wedge\psi} (\wedge I_{n,1}) \qquad \frac{\overline{\Gamma} : \Delta_f, \varphi|\overline{\Delta}_{n,t} \quad \overline{\Gamma} : \overline{\Delta}_{f,t}|\Delta_n, \psi}{\overline{\Gamma} : \Delta_f|\Delta_n, \varphi\wedge\psi} (\wedge I_{n,2})$$

$$\frac{\overline{\Gamma} : \overline{\Delta}_{f,t}|\Delta_n, \varphi \quad \overline{\Gamma} : \Delta_f, \psi|\overline{\Delta}_{n,t}}{\overline{\Gamma} : \overline{\Delta}_{f,t}|\Delta_n, \varphi\wedge\psi} (\wedge I_{n,3}) \tag{4.51}$$

Figure 3: $\mathcal{N}^3_{K_3}$: conjunction introduction

$$\frac{\overline{\Gamma} : \Delta_t, \varphi\wedge\psi|\overline{\Delta}_{f,n} \quad \Gamma_{f,n}|\Gamma_t, \varphi : \overline{\Delta}}{\overline{\Gamma} : \overline{\Delta}} (\wedge E_{t,1}) \qquad \frac{\overline{\Gamma} : \Delta_t, \varphi\wedge\psi \mid \overline{\Delta}_{f,n} \quad \Gamma_{f,n}|\Gamma_t, \psi : \overline{\Delta}}{\overline{\Gamma} : \overline{\Delta}} (\wedge E_{t,2}) \tag{4.52}$$

$$\frac{\overline{\Gamma} : \Delta_n, \varphi\wedge\psi|\overline{\Delta}_{f,t} \quad \Gamma_f|\Gamma_t, \varphi|\Gamma_n, \psi : \overline{\Delta} \quad \overline{\Gamma}_{f,t}|\Gamma_n, \varphi, \psi : \overline{\Delta} \quad \Gamma_f|\Gamma_n, \varphi|\Gamma_t, \psi : \overline{\Delta}}{\overline{\Gamma} : \overline{\Delta}} (\wedge E_n) \tag{4.53}$$

$$\frac{\overline{\Gamma} \ : \ \Delta_f, \varphi\wedge\psi \mid \overline{\Delta}_{n,t} \quad \begin{array}{l} \Gamma_n \ |\Gamma_f, \varphi|\Gamma_t, \psi : \overline{\Delta} \\ \Gamma_t \ |\Gamma_f, \varphi|\Gamma_n, \psi : \overline{\Delta} \\ \Gamma_n|\Gamma_t, \varphi|\Gamma_f, \psi : \overline{\Delta} \\ \Gamma_t \ |\Gamma_n, \varphi|\Gamma_f, \psi : \overline{\Delta} \\ \Gamma_t \ |\Gamma_n|\Gamma_f, \varphi, \psi : \overline{\Delta} \end{array}}{\overline{\Gamma} : \overline{\Delta}} (\wedge E_f) \tag{4.54}$$

Figure 4: $\mathcal{N}^3_{K_3}$: conjunction elimination

$$\frac{\overline{\Gamma} : \Delta_t, \varphi | \overline{\Delta}_{f,n}}{\overline{\Gamma} : \Delta_t, \varphi \vee \psi | \overline{\Delta}_{f,n}} (\vee I_{t,1}) \qquad \frac{\overline{\Gamma} : \Delta_t, \psi | \overline{\Delta}_{f,n}}{\overline{\Gamma} : \Delta_t, \varphi \vee \psi | \Delta_{f,n}} (\vee I_{t,2}) \tag{4.55}$$

$$\frac{\overline{\Gamma} : \overline{\Delta}_{f,t} | \Delta_n, \varphi \quad \overline{\Gamma} : \overline{\Delta}_{f,t} | \Delta_n, \psi}{\overline{\Gamma} : \Delta_n, \varphi \vee \psi | \overline{\Delta}_{f,t}} (\vee I_{n,1})$$

$$\frac{\overline{\Gamma} : \overline{\Delta}_{n,t} | \Delta_f, \varphi \quad \overline{\Gamma} : \overline{\Delta}_{f,t} | \Delta_n, \psi}{\overline{\Gamma} : \Delta_n, \varphi \vee \psi | \overline{\Delta}_{f,t}} (\vee I_{n,2}) \quad \frac{\overline{\Gamma} : \overline{\Delta}_{f,t} | \Delta_n, \varphi \quad \overline{\Gamma} : \overline{\Delta}_{n,t} | \Delta_f, \psi}{\overline{\Gamma} : \Delta_n, \varphi \vee \psi | \overline{\Delta}_{f,t}} (\vee I_{n,3}) \tag{4.56}$$

$$\frac{\overline{\Gamma} : \overline{\Delta}_{n,t} | \Delta_f, \varphi \quad \overline{\Gamma} : \overline{\Delta}_{n,t} | \Delta_f, \psi}{\overline{\Gamma} : \Delta_f, \varphi \vee \psi | \overline{\Delta}_{n,t}} (\vee I_f) \tag{4.57}$$

Figure 5: $\mathcal{N}^3_{K_3}$: disjunction introduction

$$\frac{\overline{\Gamma} : \Delta_t, \varphi \vee \psi | \overline{\Delta}_{f,n} \quad \begin{array}{r} \Gamma_f | \Gamma_t, \varphi, \psi | \Gamma_n : \overline{\Delta} \\ \Gamma_f | \Gamma_t, \varphi | \Gamma_n, \psi : \overline{\Delta} \\ \Gamma_n | \Gamma_t, \varphi | \Gamma_f, \psi : \overline{\Delta} \\ \Gamma_f | \Gamma_n, \varphi | \Gamma_t, \psi : \overline{\Delta} \\ \Gamma_n | \ \Gamma_f, \varphi | \Gamma_t, \psi : \overline{\Delta} \end{array}}{\overline{\Gamma} : \overline{\Delta}} (\vee E_t) \tag{4.58}$$

$$\frac{\overline{\Gamma} : \Delta_t | \Delta_n, \varphi \vee \psi | \Delta_f \quad \begin{array}{r} \Gamma_t | \Gamma_n, \varphi, \psi | \Gamma_f : \overline{\Delta} \\ \Gamma_t | \Gamma_n, \varphi | \Gamma_f, \psi : \overline{\Delta} \\ \Gamma_t | \Gamma_n, \psi | \Gamma_f, \varphi : \overline{\Delta} \end{array}}{\overline{\Gamma} : \overline{\Delta}} (\vee E_n) \tag{4.59}$$

$$\frac{\overline{\Gamma} : \Delta_t | \Delta_n | \Delta_f, \varphi \vee \psi \quad \Gamma_t | \Gamma_n | \Gamma_f, \varphi, \psi : \overline{\Delta}}{\overline{\Gamma} : \overline{\Delta}} (\vee E_f) \tag{4.60}$$

Figure 6: $\mathcal{N}^3_{K_3}$: disjunction elimination

$$\frac{\overline{\Gamma} : \Delta_f, \varphi | \overline{\Delta}_{n,t}}{\overline{\Gamma} : \Delta_t, \varphi \supset \psi | \overline{\Delta}_{f,n}} (\supset I_{t,1}) \qquad \frac{\overline{\Gamma} : \Delta_t, \psi | \overline{\Delta}_{f,n}}{\overline{\Gamma} : \Delta_t, \varphi \supset \psi | \overline{\Delta}_{f,n}} (\supset I_{t,2}) \tag{4.61}$$

$$\frac{\overline{\Gamma} : \Delta_t, \varphi | \overline{\Delta}_{f,n} \quad \overline{\Gamma} : \Delta_n, \psi | \overline{\Delta}_{f,t}}{\overline{\Gamma} : \Delta_n, \varphi \supset \psi | \overline{\Delta}_{f,t}} (\supset I_{n,1}) \qquad \frac{\overline{\Gamma} : \Delta_n, \varphi | \overline{\Delta}_{f,t} \quad \overline{\Gamma} : \Delta_n, \psi | \overline{\Delta}_{f,t}}{\overline{\Gamma} : \Delta_n, \varphi \supset \psi | \overline{\Delta}_{f,t}} (\supset I_{n,2})$$

$$\frac{\overline{\Gamma} : \Delta_n, \varphi | \overline{\Delta}_{f,t} \quad \overline{\Gamma} : \Delta_f, \psi | \overline{\Delta}_{n,t}}{\overline{\Gamma} : \Delta_n, \varphi \supset \psi | \overline{\Delta}_{f,t}} (\supset I_{n,3}) \tag{4.62}$$

$$\frac{\overline{\Gamma} : \Delta_t, \varphi | \overline{\Delta}_{f,n} \quad \overline{\Gamma} : \Delta_f, \psi | \overline{\Delta}_{n,t}}{\overline{\Gamma} : \Delta_f, \varphi \supset \psi | \overline{\Delta}_{n,t}} (\supset I_f) \tag{4.63}$$

Figure 7: $\mathcal{N}^3_{K_3}$: implication introduction

$$\frac{\overline{\Gamma} : \Delta_t, \varphi \supset \psi | \overline{\Delta}_{f,n} \quad \begin{array}{c} \overline{\Gamma}_{f,n} | \Gamma_t, \varphi, \psi : \overline{\Delta} \\ \Gamma_f \, | \Gamma_n, \varphi | \Gamma_t, \psi : \overline{\Delta} \\ \Gamma_n | \Gamma_f, \varphi | \Gamma_t, \psi : \overline{\Delta} \\ \Gamma_n | \Gamma_f, \varphi | \Gamma_t, \psi : \overline{\Delta} \\ \Gamma_f | \Gamma_f, \varphi | \Gamma_n, \psi : \overline{\Delta} \\ \overline{\Gamma}_{n,t} | \Gamma_f, \varphi, \psi : \overline{\Delta} \end{array}}{\overline{\Gamma} : \overline{\Delta}} (\supset E_t) \tag{4.64}$$

$$\frac{\overline{\Gamma} : \Delta_n, \varphi \supset \psi | \overline{\Delta}_{f,t} \quad \begin{array}{c} \Gamma_f | \Gamma_t, \varphi | \Gamma_n, \psi : \overline{\Delta} \\ \overline{\Gamma}_{f,t} | \Gamma_n, \varphi, \psi : \overline{\Delta} \\ \Gamma_f | \Gamma_n, \varphi | \Gamma_t, \psi : \overline{\Delta} \end{array}}{\overline{\Gamma} : \overline{\Delta}} (\supset E_n) \tag{4.65}$$

$$\frac{\overline{\Gamma} : \Delta_{n,t} | \Delta_f, \varphi \supset \psi \quad \overline{\Gamma}_{n,t} | \Gamma_f, \varphi : \overline{\Delta}}{\overline{\Gamma} : \overline{\Delta}} (\supset E_{f,1}) \qquad \frac{\overline{\Gamma} : \Delta_{n,t} | \Delta_f, \varphi \supset \psi \quad \overline{\Gamma}_{f,n} | \Gamma_t, \psi : \overline{\Delta}}{\overline{\Gamma} : \overline{\Delta}} (\supset E_{f,2}) \tag{4.66}$$

Figure 8: $\mathcal{N}^3_{K_3}$: implication elimination

$$\frac{\varphi \quad \neg\varphi}{\psi}\ (EFQ) \qquad \frac{\varphi}{\neg\neg\varphi}\ (DN) \tag{4.67}$$

$$\frac{\varphi \quad \psi}{\varphi\wedge\psi}\ (\wedge I) \qquad \frac{\varphi\wedge\psi}{\varphi}\ (\wedge E_1) \qquad \frac{\varphi\wedge\psi}{\psi}\ (\wedge E_2) \tag{4.68}$$

$$\frac{\varphi}{\varphi\vee\psi}\ (\vee I_1) \qquad \frac{\psi}{\varphi\vee\psi}\ (\vee I_2) \qquad \frac{\varphi\vee\psi \quad \begin{matrix}[\varphi]_u\\ \vdots \\ \chi\end{matrix} \quad \begin{matrix}[\psi]_v\\ \vdots \\ \chi\end{matrix}}{\chi}\ (\vee E^{u,v}) \tag{4.69}$$

$$\frac{\neg(\varphi\vee\psi)}{\neg\varphi\wedge\neg\psi}\ (DeM_\vee) \qquad \frac{\neg(\varphi\wedge\psi)}{\neg\varphi\vee\neg\psi}\ (DeM_\wedge) \tag{4.70}$$

Figure 9: The rules of ND_{K_3}

4.3.2 The system ND_{K_3}

An ND system for K_3, called ND_{K_3} (see Figure 9), is presented in [32] and shown to be sound and complete w.r.t. the K_3 truth-tables, based on a consequence relation assuring the propagation of truth. Let $T = \{t\}$ and $UT = \{f, n\}$. A formula is a T-formula if its truth-value is in T, an UT-formula otherwise.

Definition 4.14 (K_3 consequence based on preserving truth). $\Gamma \models_{K_3} \varphi$ *iff whenever* $\psi \in T$ *for every* $\psi \in \Gamma$, $\varphi \in T$ *too.*

The logic considers implication as a defined connective, and contains rules for conjunction, disjunction and their negations, presented in Figure 9. Note that, strictly speaking, ND_{K_3} is *not* an ND-system. The DM-rules (de Morgan) are *not* I/E-rules as generally understood.

4.3.3 Retrieving ND_{K_3} in $\mathcal{N}^3_{K_3}$

Below we present the derivations of ND_{K_3}-rules within $\mathcal{N}^3_{K_3}$.

(DN): The canonical translation of (DN) is

$$\frac{\overline{\Lambda}_{f,n}|\Gamma_t : \overline{\Lambda}_{f,n}|\Lambda_t, \varphi}{\overline{\Lambda}_{f,n}|\Gamma_t : \overline{\Lambda}_{f,n}|\Lambda_t, \neg\neg\varphi} \ (\mathcal{T}(DN)) \tag{4.71}$$

The derivations of (DN) in $\mathcal{N}^3_{K_3}$ is:

$$\frac{\dfrac{\overline{\Lambda}_{f,n}|\Gamma_t : \overline{\Lambda}_{f,n}|\Lambda_t, \varphi}{\overline{\Lambda}_{f,n}|\Gamma_t : \Lambda_f, \neg\varphi|\overline{\Lambda}_{n,t}} \ (\neg I_f)}{\overline{\Lambda}_{f,n}|\Gamma_t : \overline{\Lambda}_{f,n}|\Lambda_t, \neg\neg\varphi} \ (\neg I_t) \qquad \frac{\dfrac{\overline{\Lambda}_{f,n}|\Gamma_t : \overline{\Lambda}_{f,n}|\Lambda_t, \neg\neg\varphi}{\overline{\Lambda}_{f,n}|\Gamma_t : \Lambda_f, \neg\varphi|\overline{\Lambda}_{n,t}} \ (\neg E_f)}{\overline{\Lambda}_{f,n}|\Gamma_t : \overline{\Lambda}_{f,n}|\Lambda_t, \varphi} \ (\neg E_t) \tag{4.72}$$

(EFQ): The canonical translation of (EFQ) is

$$\frac{\overline{\Lambda}_{f,n}|\Gamma_t : \overline{\Lambda}_{f,n}|\Lambda_t, \varphi \quad \overline{\Lambda}_{f,n}|\Gamma_t : \overline{\Lambda}_{f,n}|\Lambda_t, \neg\varphi}{\overline{\Lambda}_{f,n}|\Gamma_t : \overline{\Lambda}_{f,n}|\Lambda_t, \psi} \ (\mathcal{T}(EFQ)) \tag{4.73}$$

The derivation of (EFQ) in $\mathcal{N}^3_{K_3}$ is:

$$\frac{\dfrac{\overline{\Lambda}_{f,n}|\Gamma_t : \overline{\Lambda}_{f,n}|\Lambda_t, \varphi \quad \dfrac{\overline{\Lambda}_{f,n}|\Gamma_t : \overline{\Lambda}_{f,n}|\Lambda_t, \neg\varphi}{\overline{\Lambda}_{f,n}|\Gamma_t : \Lambda_f, \varphi|\overline{\Lambda}_{n,t}} \ (\neg E_t)}{\overline{\Lambda}_{f,n}|\Gamma_t : \overline{\Lambda}} \ (c_{t,f})}{\overline{\Lambda}_{f,n}|\Gamma_t : \overline{\Lambda}_{f,n}|\Lambda_t, \psi} \ (WR) \tag{4.74}$$

$(\wedge I)$: The canonical translation of $(\wedge I)$ is

$$\frac{\overline{\Lambda}_{f,n}|\Gamma_t : \overline{\Lambda}_{f,n}|\Lambda_t, \varphi \quad \overline{\Lambda}_{f,n}|\Gamma_t : \overline{\Lambda}_{f,n}|\Lambda_t, \psi}{\overline{\Lambda}_{f,n}|\Gamma_t : \overline{\Lambda}_{f,n}|\Lambda_t, \varphi\wedge\psi} \ (\mathcal{T}(\wedge I)) \tag{4.75}$$

The derivation is a direct application of $(\wedge I)$ of $\mathcal{N}^3$.

$(\wedge E)$: Skipped.

$(\vee I)$: Skipped.

$(\vee E_t)$: The canonical translation of $(\vee E_t)$ is:

$$\frac{\overline{\Lambda}_{f,n}|\Gamma_t : \overline{\Lambda}_{f,n}|\Lambda_t, \varphi\vee\psi \quad \overline{\Lambda}_{f,n}|\Gamma_t, \varphi : \overline{\Lambda}_{f,n}|\Lambda_t, \chi \quad \overline{\Lambda}_{f,n}|\Gamma_t, \psi : \overline{\Lambda}_{f,n}|\Lambda_t, \chi}{\overline{\Lambda}_{f,n}|\Gamma_t : \overline{\Lambda}_{f,n}|\Lambda_t, \chi} \ (\mathcal{T}(\vee E_t)) \tag{4.76}$$

The derivation of $(\vee E_t)$ in $\mathcal{N}_s^3$ is:

$$\dfrac{\overline{\Lambda}_{f,n}|\Gamma_t : \overline{\Lambda}_{f,n}|\Lambda_t, \varphi\vee\psi \quad \begin{array}{c} \dfrac{\overline{\Lambda}_{f,n}|\Gamma_t, \varphi : \overline{\Lambda}_{f,n}|\Lambda_t, \chi}{\overline{\Lambda}_{f,n}|\Gamma_t, \varphi, \psi : \overline{\Lambda}_{f,n}|\Lambda_t, \chi}(WL_t) \\[2ex] \dfrac{\overline{\Lambda}_{f,n}|\Gamma_t, \varphi : \overline{\Lambda}_{f,n}|\Lambda_t, \chi}{\Lambda_f|\Lambda_n, \psi|\Gamma_t, \varphi : \overline{\Lambda}_{f,n}|\Lambda_t, \chi}(WL_n) \\[2ex] \dfrac{\overline{\Lambda}_{f,n}|\Gamma_t, \varphi : \overline{\Lambda}_{f,n}|\Lambda_t, \chi}{\Lambda_f, \psi|\Lambda_n|\Gamma_t, \varphi : \overline{\Lambda}_{f,n}|\Lambda_t, \chi}(WL_f) \\[2ex] \dfrac{\overline{\Lambda}_{f,n}|\Gamma_t, \psi : \overline{\Lambda}_{f,n}|\Lambda_t, \chi}{\Lambda_f|\Lambda_n, \varphi|\Gamma_t, \psi : \overline{\Lambda}_{f,n}|\Lambda_t, \chi}(WL_n) \\[2ex] \dfrac{\overline{\Lambda}_{f,n}|\Gamma_t, \psi : \overline{\Lambda}_{f,n}|\Lambda_t, \chi}{\Lambda_f, \varphi|\Lambda_n|\Gamma_t, \psi : \overline{\Lambda}_{f,n}|\Lambda_t, \chi}(WL_f) \end{array}}{\overline{\Lambda}_{f,n}|\Gamma_t : \overline{\Lambda}_{f,n}|\Lambda_t, \chi}(\vee E_t) \tag{4.77}$$

4.3.4 The system ND_{LP}

The rules for ND_{LP} are again those in Figure 9, but with (EFQ) excluded. Let $NF = \{n, t\}$ (not false) and $F = \{f\}$.

Definition 4.15 (LP consequence based on preserving non-falsity). $\Gamma \models_{LP} \varphi$ *iff whenever* $\psi \in NF$ *for every* $\psi \in \Gamma$, $\varphi \in NF$ *too.*

4.3.5 Retrieving ND_{LP} within $\mathcal{N}_{K_3}^3$

Below we present some sample derivations of (ND_{LP})-rules from within $\mathcal{N}_{K_3}^3$. Note the difference between the derivation of (DN) and the corresponding derivation of (DN) as an $\mathcal{N}_{K_3}^3$-rule. The derivation of the other rules differ in the same way from the derivations of their ND_{K_3} counterparts and are skipped.

(DN) The canonical translation of (DN) (not discharging any assumptions) is

$$\frac{\overline{\Gamma}_{n,t}|\Lambda_f : \Lambda_f|\Lambda_n, \varphi|\Lambda_t, \varphi}{\overline{\Gamma}_{n,t}|\Lambda_f : \Lambda_f|\Lambda_n, \neg\neg\varphi|\Lambda_t, \neg\neg\varphi}\,(\mathcal{T}(DN)) \tag{4.78}$$

Compare this translation with that in (4.71), where in the latter φ is placed in Δ_t only.

The derivation of $(\mathcal{T}(DN))$ in $\mathcal{N}^3_{K_3}$ is:

$$\frac{\dfrac{\overline{\Gamma}_{n,t}|\Lambda_f : \Lambda_f|\Lambda_n, \varphi|\Lambda_t, \varphi}{\overline{\Gamma}_{n,t}|\Lambda_f : \Lambda_f, \neg\varphi|\Lambda_n, \neg\varphi|\Lambda_t}\,(\neg I_f, \neg I_n)}{\overline{\Gamma}_{n,t}|\Lambda_f : \Lambda_f|\Lambda_n, \neg\neg\varphi|\Lambda_t, \neg\neg\varphi}\,(\neg I_n, \neg I_t) \qquad \frac{\dfrac{\overline{\Gamma}_{n,t}|\Lambda_f : \Lambda_f|\Lambda_n, \neg\neg\varphi|\Lambda_t, \neg\neg\varphi}{\overline{\Gamma}_{n,t}|\Lambda_f : \Lambda_f, \neg\varphi|\Lambda_n\ \neg\varphi|\Lambda_t}\,(\neg E_t, \neg E_n)}{\overline{\Gamma}_{n,t}|\Lambda_f : \Lambda_f|\Lambda_n, \varphi|\Lambda_t, \varphi}\,(\neg E_f, \neg E_n) \tag{4.79}$$

$(DeM_\vee)$: The canonical translation of $(DeM_\vee)$ is

$$\frac{\overline{\Gamma}_{n,t}|\Lambda_f : \Lambda_f|\Lambda_n, \neg(\varphi\vee\psi)|\Lambda_t, \neg(\varphi\vee\psi)}{\overline{\Gamma}_{n,t}|\Lambda_f : \Lambda_f|\Lambda_n, \neg\varphi\wedge\neg\psi|\Lambda_t, \neg\varphi\wedge\neg\psi}\,(\mathcal{T}(DeM)) \tag{4.80}$$

The one derivation is shown; the other is similar and skipped.

$$\frac{\dfrac{\dfrac{\dfrac{\overline{\Gamma}_{n,t}|\Lambda_f : \Lambda_f|\Lambda_n, \neg(\varphi\vee\psi)|\Lambda_t, \neg(\varphi\vee\psi)}{\overline{\Gamma}_{n,t}|\Lambda_f : \Lambda_f, \varphi\vee\psi|\Lambda_n, \varphi\vee\psi|\Lambda_t}\,(\neg E_t, \neg E_n) \quad \overline{\Gamma}_{n,t}|\Lambda_f|\varphi, \psi : \varphi|\overline{\Lambda}_{n,t}}{\overline{\Gamma}_{n,t}|\Lambda_f : \Lambda_f, \varphi|\overline{\Lambda}_{n,t}}\,(\vee E_f)}{\overline{\Gamma}_{n,t}|\Lambda_f : \overline{\Lambda}_{f,n}|\Lambda_t, \neg\varphi}\,(\neg I_t) \quad \dfrac{\dfrac{\text{similar}}{\overline{\Gamma}_{n,t}|\Lambda_f : \Lambda_f, \psi|\overline{\Lambda}_{n,t}}}{\overline{\Gamma}_{n,t}|\Lambda_f : \overline{\Lambda}_{f,n}|\Lambda_t, \neg\psi}\,(\neg I_t)}{\overline{\Gamma}_{n,t}|\Lambda_f : \overline{\Lambda}_{f,n}|\Lambda_t, \neg\varphi\wedge\neg\psi}\,(\wedge I_t) \tag{4.81}$$

5 Conclusion

In this paper, we presented a uniform and *direct* construction of an ND-system for a multi-valued logic, given the truth-tables for its connectives. The construction is based on a formula structure called poly-sequents that provides a direct means for relating to the truth-value of a formula within a derivation. Such an ND-system allows both assumptions and conclusions of a rule to have any truth-value. The constructed ND-system is shown to be sound and strongly complete. Furthermore, it is shown how specific multi-valued logics, as found in the literature, can be retrieved in the uniformly constructed logic.

References

[1] A. Avron, The method of hypersequents in proof theory of propositional non-classical logics, in: W. Hodges, M. Hyland, C. Steinhorn, J. K. Truss (Eds.), *Logic: Foundations to Applications*, Oxford Science Publications, 1996, pp. 1–32.

[2] M. Baaz, C. G. Fermüler, R. Zach, Systematic construction of natural deduction systems for many-valued logics, in: *Proceedings of the 23rd international symposium on multiple valued logics*, Sacramento, CA, May 1993, IEEE press, Los Alamitos, 1993, pp. 208–213.

[3] N. D. Belnap, How a computer should think, in: G. Ryle (Ed.), *Contemporary Aspects of Philosophy*, Stocksfield:Oriel Press, 1976, pp. 30–56.

[4] N. D. Belnap, A useful four-valued logic, in: J. M. Dunn, G. Epstein (Eds.), *Modern Uses of Multiple-valued Logic*, Dordrecht:Reidel, 1977, pp. 8–37.

[5] R. Davies, F. Pfenning, A modal analysis of staged computation, *Journal of the ACM* 48 (3) (2001) 555–604.

[6] M. Dummett, *The Logical Basis of Metaphysics*, Harvard University Press, Cambridge, MA., 1993 (paperback), hard copy 1991.

[7] J. M. Dunn, Intuitive semantics for first-degree entailments and 'coupled trees', *Philosophical Studies* 29 (1976) 149–168.

[8] C. Englander, E. H. Haeusler, L. C. Pereira, Finitely many-valued logics and natural deduction, *Logic Journal of the IGPL* 22 (2) (2014) 333–354.

[9] N. Francez, Bilateralism, trilateralism, multilateralism and poly-sequents, *Journal of Philosophical Logic*, Doi 10.1007/s10992-018-9464-3.

[10] N. Francez, *Proof-theoretic Semantics*, College Publications, London, 2015.

[11] N. Francez, R. Dyckhoff, A note on harmony, *Journal of Philosophical Logic* 41 (3) (2012) 613–628.

[12] N. Francez, M. Kaminski, Transparent truth-value predicates in multi-valued logics, *Logique et Analyse*, 62 (245) (2019) 55-71.

[13] M. Kaminski, N. Francez, Calculi for multi-valued logics(In preparation).

[14] S. C. Kleene, *Introduction to Metamathematics*, North-Holland, Amsterdam, 1952.

[15] J. Lukasiewicz, A system of modal logic, *Journal of Computing Systems* 1 (1953) 111–149.

[16] C. G. Morgan, Sentential calculus for logical falsehood, *Notre Dame Journal of Formal Logic* XIV (3) (1973) 347–353.

[17] S. Negri, J. von Plato, *Structural Proof Theory*, Cambridge University Press, Cambridge, UK, 2001.

[18] F. Pfenning, R. Davies, A judgmental reconstruction of modal logic, *Mathematical Structures in Computer Science* 11 (2001) 511–540.

[19] A. Pietz, U. Rivieccio, Nothing but the truth, *Journal of Philosophical Logic* 42 (1) (2013) 125–135, doi:10.1007/s10992-011-9215-1.

[20] J. von Plato, Natural deduction with general elimination rules, *Archive for Mathematical Logic* 40 (2001) 541–567.

[21] G. Pottinger, Uniform, cut-free formulations of T, S4 and S5, *Journal of Symbolic Logic* 48 (1983) 900, (abstract only).

[22] D. Prawitz, *Natural Deduction: A Proof-Theoretical Study*, Almqvist and Wicksell, Stockholm, 1965, soft cover edition by Dover, 2006.

[23] G. Priest, *Non-Classical Logic*, Cambridge University Press, 2001.

[24] S. Read, General-elimination harmony and the meaning of the logical constants, *Journal of Philosophic Logic* 39 (2010) 557–576.

[25] N. Rescher, An intuitive interpretation of systems of four-valued logic, *Notre Dame Journal of Formal Logic* VI (2) (1965) 154–156.

[26] G. H. Rousseau, Sequents in many-valued logic I, *Fundamenta Mathematicae* 60 (1967) 23–33.

[27] G. H. Rousseau, Sequents in many-valued logic II, *Fundamenta Mathematicae* 67 (1) (1967) 125–131.

[28] P. Schroeder-Heister, A natural extension of natural deduction, *Journal of Symbolic Logic* 49 (1984) 1284–1300.

[29] I. Rumfitt, 'Yes' and 'No', *Mind* 169 (436) (2000) 781–823.

[30] P. Schroeder-Heister, On the notion of *assumption* in logical systems, in: R. Bluhm, C. Nimtz (Eds.), *Philosophy-Science-Scientific Philosophy*, Mentis, Paderborn, 2004, selected papers of the 5th int. congress of the society for Analytic Philosophy, Bielfield, September 2003.

[31] A. M. Tamminga, K. Tanaka, A natural deduction system for first degree entailment, *Notre Dame Journal of Formal Logic* 40 (2) (1999) 258–272.

[32] A. M. Tamminga, Correspondence analysis for strong three-valued logic, *Logical Investigations* 20 (2) (2014) 255–268.

[33] S. Wintein, R. Muskens, Nothing but the truth, *Journal of Philosophical Logic* 45 (4) (2016) 451–465.

 Received 16 January 2016

Logic of Bipartite Truth with Uncertainty Dimension

O. M. Grigoriev
Department of Logic, Faculty of Philosophy, Lomonosov Moscow State University, Russia.
grig@philos.msu.ru

Abstract

This research continues the studies of a system of two-dimensional truth values equipped with a pair of unary negation-like operations and their logics. The basic ideas of this approach historically were first presented in [11] and thoroughly investigated further in [18, 19].

According to the methodology of [11, 18, 19] a truth value consists of a pair of entities each representing a certain aspect of "being true" or "being false" property, eg. ontological and epistemic aspects, as assumed in papers cited above. The basic set of two-dimensional truth values constitutes a four-element diamond-shaped lattice endowed with the pair of unary operations which give rise to the corresponding propositional connectives, so called *semi-negations*, on syntactical level. Intuitively these semi-negation affects only one of the coordinates of a truth value, thus only partially transforming information which a particular truth value encodes. In this paper we extend the initial structure of the truth values adding uncertainty dimensions in each position of a pair thereby obtaining a nine-element distributive lattice. We present an axiomatization for the logic of this semantic structure along with correctness and completeness proofs. Then we abstract away from the finite semantic structures and explore relational semantics for the same logic but without distribution laws. We use an approach related to the methods of [16, 1, 3, 10].

Keywords: Generalized truth values, Non-classical logic, Relational semantics, Lattice based logics.

1 Introduction

The starting point of this research is the idea to consider a truth value as a complex object and distinguish some parts, dimensions or facets of it. In [11, 18, 19] there was proposed a system of truth values such that each value has two – *ontological* and *epistemic* – parts. These terms, though have a philosophical flavour to them, just reflect the fact of (in)dependence of a truth value assigned to a sentence from the attitudes of a rational agent or coalitions of them. Thus one may interpret the ontological part of a truth value as an abstract entity corresponding to the statements concerning to some general theory explaining a relevant part of the universe or to the items of a large knowledge database (ontology in a narrow sense), or even to an amount of truths about some closed system in quantum information theory. At the same time the term "epistemic dimension" refers to a level of some individual rational agent's knowledge (or common knowledge belonging to a collection of them) so that a valuation of sentences depends on agent's local supply of knowledge, its presuppositions and so on. The distinction between the two, objective and subjective, dimensions of a truth value seems to be one of the most fundamental, though of course not a single one. In the present paper we still utilize these two parts of a truth value without paying too much attention to their possible interpretations.

The informal model of compound truth values can be described mathematically via basic set-theoretical operations, power-set or Cartesian product construction, thus placing the intended research into the area of *generalized truth values* studies. The concept of generalized truth value was introduced in [15] and marked the whole line of exploration initiated in [6, 2].

As we will see in the next section, this new family of truth values is naturally partially ordered thereby defining *a four element* lattice, resembling the well known lattices of Dunn-Belnap's truth values. This new lattice appears especially interesting with the introduction of two unary complementation-like operations, each responsible for the operating only on its own part of a truth value (thus in a sense only "semi-complementing" the whole truth value). So, it seems natural to think of these operations as "semi-complementations".

Another reason to use the term "semi-complementation" comes from the striking fact that the composition of the two distinct operations behaves exactly like boolean complementation in the four element lattice. So, the boolean complementation in that lattice looks like something composed of two "halves". At the same time a single semi-complementation loses some properties of the boolean complementation, like contraposition, and saves some other, like introduction and elimination of double complementation. Paper [19] adopts the term *semi-boolean complementations* for the described operations to stress the boolean nature of their composition. In the same paper a logic of an underlying semantic structure is called $\mathbf{FDE}_{scl}$ (where the subscript *scl* means *super-classical logic*) because of close similarity with the famous relevant First Degree Entailment system.

The axiomatization of $\mathbf{FDE}_{scl}$ has been published in [17]. It follows the style

a	$\sim_t a$	$\sim_1 a$
$\langle t, 1\rangle$	$\langle f, 1\rangle$	$\langle t, 0\rangle$
$\langle t, 0\rangle$	$\langle f, 0\rangle$	$\langle t, 1\rangle$
$\langle f, 0\rangle$	$\langle t, 0\rangle$	$\langle f, 1\rangle$
$\langle f, 1\rangle$	$\langle t, 1\rangle$	$\langle f, 0\rangle$

Table 1: Definition of operations $\sim_t$ and $\sim_1$ in the lattice $\mathscr{L}\mathbf{4}^{oe}$

of well known axiomatization of First Degree Entailment as a binary symmetric consequence relation system (see [5]). It is not difficult to construct prefixed analytic-tableaux calculus, as has been done in [12].

In the next section we briefly sketch some technical details of the basic four valued approach to keep the exposition self-contained.

2 Four valued model

Let us choose two elementary bases, $\{t, f\}$ and $\{1, 0\}$, where the former is a set of ontological, while the latter is a set of epistemic truth values. Application of Cartesian product produces the set of four elements: $\langle t, 1\rangle$, $\langle t, 0\rangle$, $\langle f, 1\rangle$ and $\langle f, 0\rangle$. They constitute a collection of new generalized truth values with ontological and epistemic components. Let us refer to this set as $\mathbf{4}^{oe}$. Now we can define a natural partial order relation on $\mathbf{4}^{oe}$ provided that each of the bases has its own order, say $\leqslant_t$ and $\leqslant_1$, where, in particular, $f \leqslant_t t$, $0 \leqslant_1 1$. Then the partial order on $\mathbf{4}^{oe}$ is defined component-wise: for all $a, b \in \mathbf{4}^{oe}, a \leqslant b \Leftrightarrow a \leqslant_t b$ and $a \leqslant_1 b$. The resulting structure is a four element diamond shaped lattice with the top $\langle t, 1\rangle$ and the bottom $\langle f, 0\rangle$.

Alternatively it is possible of course to define meet and join on $\mathbf{4}^{oe}$ via classical disjunction and conjunction applied to the elements of $\mathbf{4}^{oe}$ component-wise. Left side of Figure 1 shows graphical representation of the resulting lattice, $\mathscr{L}\mathbf{4}^{oe}$.

Now we introduce a pair of unary operations (semi-complementations) on $\mathscr{L}\mathbf{4}^{oe}$, $\sim_t$ and $\sim_1$, each of them change a corresponding component of a pair leaving another unchanged. For the convenience we summarized the behavior of the operations in Table 1.

Some basic properties of semi-complementations are listed below:

$$\sim_t\sim_t a = \sim_1\sim_1 a = a, \quad \sim_t\sim_1 a = \sim_1\sim_t a,$$
$$\sim_t a \leqslant \sim_1 b \Leftrightarrow \sim_1 a \leqslant \sim_t b, \quad \sim_1 a \leqslant \sim_t b \Leftrightarrow \sim_t a \leqslant \sim_1 b,$$
$$a \leqslant b \Leftrightarrow \sim_t\sim_1 b \leqslant \sim_t\sim_1 a.$$

We are interested in a propositional logic determined by the semantic structure just described. It is natural to consider a signature consisting of $\wedge$ and $\vee$, corresponding to meet and join of the lattice and a pair of unary propositional connectives, *semi-negations* $\neg_t$ and $\neg_1$, corresponding to $\sim_t$ and $\sim_1$.

As we noted in Introduction, an axiomatization of logic of $\mathscr{L}\mathbf{4}^{oe}$ was proposed in [17]. Among other postulates it contains some schemata resembling paradoxes of classical logic. For example the scheme $\neg_t A \wedge \neg_1 A \vdash B$ which is an analog of classical consequence $A \wedge \neg A \vdash B$, is an axiom. Likewise in this logic we have $B \vdash \neg_t A \vee \neg_1 A$, analog of classical $B \vdash A \vee \neg A$. These consequences evidently witness some drawbacks of proposed semantics.

3 Uncertainty dimension

Historically generalized truth values were intended to deal with incomplete or overdetermined information. The evident insufficiency of the truth values system described in the previous section reveals itself in the presence of paradoxical consequences.

The straightforward way of resolving this difficulty within the complex truth-values approach is to use an additional sign for a "being indefinite" value, as it sometimes is done in three-valued logics. We denote this new components as u_t and u_1 replacing the old bases by $\{f, u_t, t\}$ and $\{0, u_1, 1\}$. The order relations within bases are modified correspondingly, satisfying in particular $f \leqslant_t u_t \leqslant_t t$ and $0 \leqslant_1 u_1 \leqslant_1 1$.

Applying Cartesian product again we obtain a nine-valued structure where elements ordered by relation, defined from $\leqslant_t$ and $\leqslant_1$ as before. Of course the old semi-complementations are also redefined. Table 2 shows their new definition.

Nine element lattice $\mathscr{L}\mathbf{9}^u$ is displayed on the right hand side of Figure 1.

4 Logic of $\mathscr{L}\mathbf{9}^u$

In this section we describe a propositional logic of the lattice $\mathscr{L}\mathbf{9}^u$. Let us firstly choose an appropriate propositional language. A collection of symbols consists of: 1) denumerable set PV of propositional variables $\{p_0, p_1, p_2, \ldots\}$; 2) logical connectives $\wedge, \vee, \neg_t, \neg_1$; 3) left and right parentheses (,).

a	$\sim_t a$	$\sim_1 a$	a	$\sim_t a$	$\sim_1 a$
$\langle t, 1\rangle$	$\langle f, 1\rangle$	$\langle t, 0\rangle$	$\langle t, u_1\rangle$	$\langle f, u_1\rangle$	$\langle t, u_1\rangle$
$\langle t, 0\rangle$	$\langle f, 0\rangle$	$\langle t, 1\rangle$	$\langle u_t, 1\rangle$	$\langle u_t, 1\rangle$	$\langle u_t, 0\rangle$
$\langle u_t, u_1\rangle$	$\langle u_t, u_1\rangle$	$\langle u_t, u_1\rangle$	$\langle u_t, 0\rangle$	$\langle u_t, 0\rangle$	$\langle u_t, 1\rangle$
$\langle f, 1\rangle$	$\langle t, 1\rangle$	$\langle f, 0\rangle$	$\langle f, u_1\rangle$	$\langle t, u_1\rangle$	$\langle f, u_1\rangle$
$\langle f, 0\rangle$	$\langle t, 0\rangle$	$\langle f, 1\rangle$			

Table 2: Definition of $\sim_t$ and $\sim_1$ in the lattice $\mathscr{L}\mathbf{9}^u$

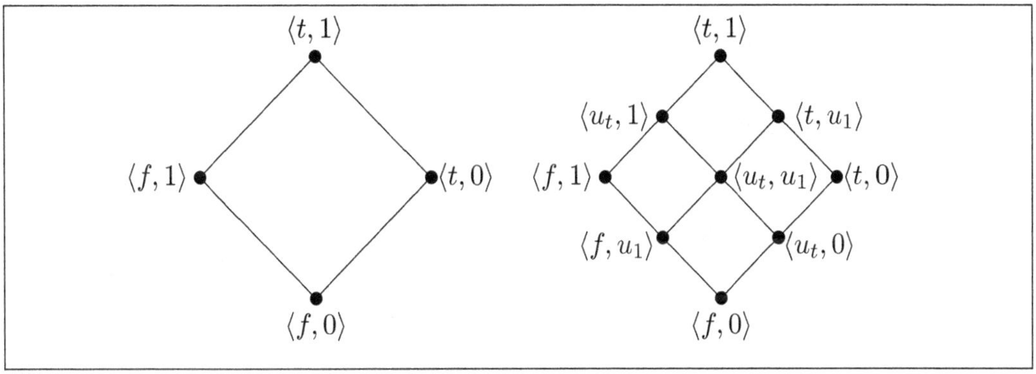

Figure 1: Lattices $\mathscr{L}\mathbf{4}^{oe}$ and $\mathscr{L}\mathbf{9}^u$

The definition of a formula is standard with the evident modifications for the cases of two semi-negations. We will refer to this language as $\mathcal{L}_{\text{LLSN}}$ in the sequel and will use the term "formula" in the same meaning as "formula in the language $\mathcal{L}_{\text{LLSN}}$".

Definition 1. *A* valuation function v *is a mapping* $PV \to \mathscr{L}\mathbf{9}^u$. *The following equations extend the valuation function to the set of all formulas:*

$$v(A \wedge B) = v(A) \cap v(B), \qquad v(A \vee B) = v(A) \cup v(B), \tag{1}$$

$$v(\neg_t A) = \sim_t v(A), \qquad v(\neg_1 A) = \sim_1 v(A). \tag{2}$$

We define a consequence relation via valuation function and order on $\mathscr{L}\mathbf{9}^u$.

Definition 2. $A \vDash B \Leftrightarrow v(A) \leqslant v(B)$.

Next simple lemma directly follows from the definition of $\sim_t$, $\sim_1$ and definition 1. To make things compact we display only the half concerning t component while the rest is just a "mirror image" obtained when t is replaced by 1 and f by 0.

Lemma 1. *For each formula A and every valuation v:*

$$t \in v(A) \Leftrightarrow t \in v(\neg_1 A) \Leftrightarrow t \in v(\neg_t \neg_t A)$$
$$f \in v(A) \Leftrightarrow f \in v(\neg_1 A) \Leftrightarrow f \in v(\neg_t \neg_t A)$$
$$t \notin v(\neg_t A) \Leftrightarrow (t \in v(A) \text{ or } u_t \in v(A)) \Leftrightarrow f \notin v(A)$$
$$u_t \in v(A) \Leftrightarrow u_t \in v(\neg_t A) \Leftrightarrow u_t \in v(\neg_1 A)$$
$$t \in v(A \wedge B) \Leftrightarrow t \in v(A) \text{ and } t \in v(B)$$
$$f \in v(A \wedge B) \Leftrightarrow f \in v(A) \text{ or } f \in v(B)$$
$$f \in v(A \vee B) \Leftrightarrow f \in v(A) \text{ and } f \in v(B)$$
$$t \in v(A \vee B) \Leftrightarrow t \in v(A) \text{ or } t \in v(B)$$
$$u_t \in v(A \wedge B) \Leftrightarrow (u_t \in v(A) \text{ and } t \in v(B)) \text{ or } (u_t \in v(B) \text{ and } t \in v(A)) \text{ or } (u_t \in v(A) \text{ and } u_t \in v(B))$$
$$u_t \in v(A \vee B) \Leftrightarrow (u_t \in v(A) \text{ and } f \in v(B)) \text{ or } (u_t \in v(B) \text{ and } f \in v(A)) \text{ or } (u_t \in v(A) \text{ and } u_t \in v(B))$$

Remark. The first line reflects the fact that epistemic negation is not sensible to the ontological part of a truth value and vice versa. The fourth line means that a semi-negation cannot 'break uncertainty'.

Note that the problematic entailments of logic $\mathscr{L}\mathbf{4}^{oe}$, $\neg_t A \wedge \neg_1 A \models B$ and $B \models \neg_t A \vee \neg_1 A$, are no longer valid in $\mathscr{L}\mathbf{9}^u$ semantics. It is worth noting that unlike $\mathscr{L}\mathbf{4}^{oe}$ case, not for all $a \in \mathscr{L}\mathbf{9}^u$, $a \cup \sim_t\sim_1 a$ equals to the top element of lattice $\mathscr{L}\mathbf{9}^u$, likewise not for all $a \in \mathscr{L}\mathbf{9}^u$, $a \cap \sim_t\sim_1 a$ equals to the bottom element. Thus $\sim_t\sim_1$ do not produce a complement for each element of $\mathscr{L}\mathbf{9}^u$. Nevertheless we will continue to use the term "semi-negation" in the sequel for the corresponding logical connectives.

5 Axiomatizing the consequence relation

We propose an axiomatization of logic determined by $\mathscr{L}\mathbf{9}^u$ following the style of so called *symmetric consequence system* (the term used e. g. in [14][1]).

Axioms (1–12) and rules of inference (R1-R5) for symmetric consequence system **LLSN**[2] axiomatizing logic of $\mathscr{L}\mathbf{9}^u$ is listed in figure 2.

[1]See also Chapter 6 of [9] where different representations of logics are discussed. In particular the representation we follow in the present paper is called *binary implicational system* there. Binary implicational system in turn is a particular case of symmetric consequence system.

[2]Lattice Logic with Semi-Negations.

1. $A \wedge B \vdash A$	7. $A \dashv\vdash \neg_t\neg_t A$
2. $A \wedge B \vdash B$	8. $\neg_1\neg_t A \dashv\vdash \neg_t\neg_1 A$
3. $A \vdash A \vee B$	9. $\neg_t A \wedge \neg_t B \vdash \neg_t(A \vee B)$
4. $B \vdash A \vee B$	10. $\neg_1 A \wedge \neg_1 B \vdash \neg_1(A \vee B)$
5. $A \wedge (B \vee C) \vdash (A \wedge B) \vee C$	11. $\neg_t(A \wedge B) \vdash \neg_t A \vee \neg_t B$
6. $\neg_1\neg_1 A \dashv\vdash A$	12. $\neg_1(A \wedge B) \vdash \neg_1 A \vee \neg_1 B$
R1. $A \vdash B, A \vdash C/\ A \vdash B \wedge C$	R4. $\neg_t A \vdash \neg_1 B/\ \neg_t B \vdash \neg_1 A$
R2. $A \vdash C, B \vdash C/\ A \vee B \vdash C$	
R3. $A \vdash B, B \vdash C/\ A \vdash C$	R5. $\neg_1 A \vdash \neg_t B/\ \neg_1 B \vdash \neg_t A$

Figure 2: Symmetric consequence system **LLSN**.

Definition 3. *A* derivation *of a sequent $A \vdash B$ in* **LLSN** *is a finite sequence of sequents σ where each element is either an axiom or obtained from earlier elements of σ using some rule from the list R1–R5 and the last element of σ is $A \vdash B$. A sequent $A \vdash B$ is* provable *in* **LLSN** *if there is a derivation of it in* **LLSN**.

In the sequel we often abuse notation and write $A \vdash B$ to mean that the sequent $A \vdash B$ is provable in **LLSN** (or its non-distributive version in the context of the next section), wheras $A \nvdash B$ means that $A \vdash B$ is not provable.

5.1 Soundness

The soundness proof consists of a routine check that all schemata and rules of inference preserve consequence relation. We explore only few of them.

Theorem 2. *For all formulas A and B, $A \vdash B \Rightarrow A \models B$.*

Proof. For the proof $v(A) \leqslant v(B)$ it is enough to check that if $t \in v(A)$, then $t \in v(B)$, if $1 \in v(A)$, then $1 \in v(B)$, if $f \in v(B)$, then $f \in v(A)$ and if $0 \in v(B)$, then $0 \in v(A)$.

1. Let us take as an example axiom scheme 9. What we need is to show that $v(\neg_t A \wedge \neg_t B) \leqslant v(\neg_t(A \vee B))$. Lemma 1 is crucial here, we use it tacitly throughout the proof.

Suppose $t \in v(\neg_t A \wedge \neg_t B)$. Then $t \in v(\neg_t A) \cap v(\neg_t B)$, by definition 1, so $t \in v(\neg_t A)$ and $t \in v(\neg_t B)$. From $t \in v(\neg_t A)$ we get $t \notin v(A)$ and $u_t \notin v(A)$; $t \in v(\neg_t B)$ implies $t \notin v(B)$ and $u_t \notin v(B)$. Now $t \notin v(A)$ and $t \notin v(B)$ imply $t \notin (v(A) \cup v(B))$ and, by definition 1, $t \notin v(A \vee B)$; likewise $u_t \notin v(A)$ and $u_t \notin v(B)$ leads to $u_t \notin v(A \vee B)$. Finally, $t \notin v(A \vee B)$ and $u_t \notin v(A \vee B)$ imply $t \in v(\neg_t(A \vee B))$.

Next suppose that $1 \in v(\neg_t A \wedge \neg_t B)$. It follows that $1 \in v(\neg_t A) \cap v(\neg_t B)$; $1 \in v(\neg_t A)$ and $1 \in v(\neg_t B)$, therefore $1 \in v(A)$ and $1 \in v(B)$. Evidently then $1 \in v(A) \cup v(B)$, hence $1 \in v(A \vee B)$, by definition 1, so $1 \in v(\neg_t(A \vee B))$.

Let $f \in v(\neg_t(A \vee B))$. This implies $t \in v(A \vee B)$, so $t \in v(A) \cup v(B)$, hence $t \in v(A)$ or $t \in v(B)$ which means $f \in v(\neg_t A)$ or $f \in v(\neg_t B)$. The latter assertion implies $f \in v(\neg_t A \wedge \neg_t B)$.

Finally, $0 \in v(\neg_t(A \vee B))$ implies $0 \in v(A \vee B)$, so $0 \in v(A) \cup v(B)$, $0 \in v(A)$ or $0 \in v(B)$, $0 \in v(\neg_t A)$ or $0 \in v(\neg_t B)$, thus $0 \in v(\neg_t A \wedge \neg_t B)$.

2. Let us also verify one of the rules, say R4. Suppose $\neg_t A \models \neg_1 B$ but $\neg_t B \not\models \neg_1 A$. The latter assumption means that some of the condition which consequence relation must satisfy are violated.

Suppose $t \in v(\neg_t B)$, but $t \notin v(\neg_1 A)$. Thus $f \in v(B)$, hence $f \in v(\neg_1 B)$, $t \notin v(A)$, so $t \notin v(\neg_t \neg_t A)$. The latter means that $f \notin v(\neg_t A)$, thus $\neg_t A \not\models \neg_1 B$, a contradiction.

Next suppose $f \in v(\neg_1 A)$, but $f \notin v(\neg_t B)$. First, $f \in v(\neg_1 A)$ implies $f \in v(A)$, so $t \in v(\neg_t A)$. Second, from $f \notin v(\neg_t B)$ it follows that $t \notin v(\neg_t \neg_t B)$, $t \notin v(B)$, so $t \notin v(\neg_1 B)$, contradicting to the assumption.

The other cases are similar. □

5.2 Completeness

For the completeness proof we use a variant of well known structures called *theories* and *counter-theories*.

Definition 4. *A set $\mathcal{T}$ of formulas is a* theory *if for all formulas A and B:*

1. *if $A \in \mathcal{T}$ and $A \vdash B$, then $B \in \mathcal{T}$;*

2. *if $A, B \in \mathcal{T}$ then $A \wedge B \in \mathcal{T}$.*

A theory is $\mathcal{T}$ is prime *if $A \vee B \in \mathcal{T}$ implies $A \in \mathcal{T}$ or $B \in \mathcal{T}$.*

A set $\mathcal{T}^\circ$ of formulas is a counter-theory *if for all formulas A and B:*

1. *if $B \in \mathcal{T}^\circ$ and $A \vdash B$, then $A \in \mathcal{T}^\circ$;*

2. *if $A, B \in \mathcal{T}^\circ$ then $A \vee B \in \mathcal{T}^\circ$.*

A counter-theory $\mathcal{T}^{\circ}$ *is* prime *if* $A \wedge B \in \mathcal{T}^{\circ}$ *implies* $A \in \mathcal{T}^{\circ}$ *or* $B \in \mathcal{T}^{\circ}$.

One of the most important properties of theories and counter-theories for the subsequent constructions is their closure under one of the semi-negations. Formally we define a $\neg_t$- or $\neg_1$-closed set of formulas in the following way.

Definition 5. *A set of formulas* Γ *is* $\neg_t$*-closed (*$\neg_1$*-closed) if for each formula* A*, if* $A \in \Gamma$*, then* $\neg_t A \in \Gamma$ *(*$\neg_1 A \in \Gamma$*).*

Let us denote by $\mathcal{T}_t$ (by $\mathcal{T}_1$) a $\neg_t$-*closed* ($\neg_1$-*closed*) *theory* and by $\mathcal{T}_t^{\circ}$ (by $\mathcal{T}_1^{\circ}$) a $\neg_t$-*closed* ($\neg_1$-*closed*) *counter-theory*. We introduce the following notation. For a formula A the expression $A \neq \neg_1 B$ means that for no formula B, A is of the form $\neg_1 B$. In other words A has not $\neg_1$ as its main connective. The same reading is supposed for $A \neq \neg_t B$. Let Γ be a set of formulas. Then $\neg_t \Gamma = \{\neg_t A\colon A \in \Gamma \textit{ and } A \neq \neg_t B\} \cup \{A\colon \neg_t A \in \Gamma\}$ and similarly $\neg_1 \Gamma = \{\neg_1 A\colon A \in \Gamma \textit{ and } A \neq \neg_1 B\} \cup \{A\colon \neg_1 A \in \Gamma\}$.

The following simple fact is useful: if $\neg_t A \in \mathcal{T}_t$ then $A \in \mathcal{T}_t$. Indeed, $\neg_t A \in \mathcal{T}_t$ implies $\neg_t \neg_t A \in \mathcal{T}_t$ (by the closure of $\mathcal{T}_t$ under $\neg_t$) and then $\neg_t \neg_t A \vdash A$ gives $A \in \mathcal{T}_t$ (by the closure of $\mathcal{T}_t$ under $\vdash$). Similarly the following implications are also true: if $\neg_1 A \in \mathcal{T}_1$, then $A \in \mathcal{T}_1$, if $\neg_t A \in \mathcal{T}_t^{\circ}$, then $A \in \mathcal{T}_t^{\circ}$, if $\neg_1 A \in \mathcal{T}_1^{\circ}$, then $A \in \mathcal{T}_1^{\circ}$.

Now we prove an important fact about the relationship between theories and counter-theories in the next lemma.

Lemma 3. *For each theory* $\mathcal{T}_t$ *(counter-theory* $\mathcal{T}_t^{\circ}$*), the set* $\neg_1 \mathcal{T}_t$ *(*$\neg_1 \mathcal{T}_t^{\circ}$*) is a* $\neg_t$*-closed counter-theory (a* $\neg_t$*-closed theory). For each theory* $\mathcal{T}_1$ *(counter-theory* $\mathcal{T}_1^{\circ}$*), the set* $\neg_t \mathcal{T}_1$ *(*$\neg_t \mathcal{T}_1^{\circ}$*) is a* $\neg_1$*-closed counter-theory (a* $\neg_1$*-closed theory).*

Proof. We prove the first assertion (the proofs for the other statements are analogues). For some theory $\mathcal{T}_t$ consider a set $\neg_1 \mathcal{T}_t$. First of all we prove $\neg_t$-closure of $\neg_1 \mathcal{T}_t$. Suppose $A \in \neg_1 \mathcal{T}_t$ for some formula A.

1. $\neg_1 A \in \mathcal{T}_t$. Thus $\neg_t \neg_1 A \in \mathcal{T}_t$ by $\neg_t$-closure of $\mathcal{T}_t$. Using axiom 8 we then obtain $\neg_1 \neg_t A \in \mathcal{T}_t$, hence $\neg_t A \in \neg_1 \mathcal{T}_t$ by the definition of $\neg_1 \mathcal{T}_t$.

2. $A = \neg_1 B$ for some $B \in \mathcal{T}_t$. Then we have $\neg_1 \neg_1 B \in \mathcal{T}_t$ applying axiom 6, $\neg_t \neg_1 \neg_1 B \in \mathcal{T}_t$ by $\neg_t$-closure of $\mathcal{T}_t$, $\neg_1 \neg_t \neg_1 B \in \mathcal{T}_t$ by axiom 8, $\neg_t \neg_1 B = \neg_t A \in \neg_1 \mathcal{T}_t$ by the definition of $\neg_1 \mathcal{T}_t$.

Next step is to show the backward closure of $\neg_1 \mathcal{T}_t$ under $\vdash$ relation. Assume $A \in \neg_1 \mathcal{T}_t$ and $C \vdash A$. We need to show that $C \in \neg_1 \mathcal{T}_t$. To this aim we consider two cases again.

1. $\neg_1 A \in \mathcal{T}_t$. Using the axioms 6 and 7 we get $\neg_t\neg_t C \vdash \neg_1\neg_1 A$ from $C \vdash A$. Applying R4 we deduce $\neg_t\neg_1 A \vdash \neg_1\neg_t C$. Note that $\neg_t\neg_1 A \in \mathcal{T}_t$ by $\neg_t$-closure of $\mathcal{T}_t$. Thus $\neg_1\neg_t C \in \mathcal{T}_t$ since $\mathcal{T}_t$ is also $\vdash$-closed. Therefore $\neg_t C \in \neg_1\mathcal{T}_t$, so $C \in \neg_1\mathcal{T}_t$. Here we are using the fact of $\neg_t$-closure of $\neg_1\mathcal{T}_t$ proved above.

2. $A = \neg_1 B$ for some $B \in \mathcal{T}_t$. Applying axiom 7 we get $\neg_t\neg_t C \vdash \neg_1 B$ which implies $\neg_t B \vdash \neg_1\neg_t C$ by R4. Since $\neg_t B \in \mathcal{T}_t$, we have $\neg_1\neg_t C \in \mathcal{T}_t$. The latter implies $\neg_t C \in \neg_1\mathcal{T}_t$, so $C \in \neg_1\mathcal{T}_t$.

To show the closure of $\neg_1\mathcal{T}_t$ under disjunction assume $A, B \in \neg_1\mathcal{T}_t$.

1. Suppose $\neg_1 A \in \mathcal{T}_t$ and $\neg_1 B \in \mathcal{T}_t$. Using $\neg_t$-closure of $\mathcal{T}_t$ under conjunction we get $\neg_1 A \wedge \neg_1 B \in \mathcal{T}_t$, hence $\neg_1(A \vee B) \in \mathcal{T}_t$ in virtue of axiom 10, so $A \vee B \in \neg_1\mathcal{T}_t$.

2. Assume $A = \neg_1 C$, $B = \neg_1 D$ for some $C, D \in \mathcal{T}_t$. Applying axiom 6 and the closure of $\mathcal{T}_t$ under conjunction we obtain $\neg_1\neg_1 C \wedge \neg_1\neg_1 D \in \mathcal{T}_t$. Now axiom 10 and $\vdash$-closure of $\mathcal{T}_t$ imply $\neg_1(\neg_1 C \vee \neg_1 D) \in \mathcal{T}_t$, thus $\neg_1 C \vee \neg_1 D = A \vee B \in \neg_1\mathcal{T}_t$.

3. Now assume $\neg_1 A \in \mathcal{T}_t$, $B = \neg_1 C$. Since $C \in \mathcal{T}_t$, so $\neg_1\neg_1 C \in \mathcal{T}_t$ in virtue of axiom 6 and $\vdash$-closure of $\mathcal{T}_t$. Thus $\neg_1 A \wedge \neg_1\neg_1 C \in \mathcal{T}_t$ by the conjunctive closure of a theory; $\neg_1(A \vee \neg_1 C) \in \mathcal{T}_t$ by axiom 10. Therefore $A \vee \neg_1 C = A \vee B \in \neg_1\mathcal{T}_t$.

4. The last subcase is similar to the previous one.

Thus $\neg_1\mathcal{T}_t$ is a $\neg_t$-closed counter-theory. □

Theories and counter-theories are intended to play the same role as Lindenbaum sets in completeness proofs for systems of classical logic. We will use them to define a canonical valuation of formulas. Another purpose of theories and counter-theories is a *separation* tool for the relation $\vdash$. Namely, if $A \nvdash B$ then there is a theory $\mathcal{T}$ such that $A \in \mathcal{T}$ but $B \notin \mathcal{T}$ or a counter-theory $\mathcal{T}^\circ$ such that $B \in \mathcal{T}^\circ$ but $A \notin \mathcal{T}^\circ$. The existence of these separating sets is a subject of the next lemma. It is well known how to construct theories in the case of "prototypic" logics of generalized truth values, see e.g. [5, lemma 8]. But now we should provide an appropriate procedure which builds a theory *or* a counter-theory closed under a semi-negation and separating some specific pair of formulas.

Lemma 4. *If $A \nvdash B$ then there exists a prime theory $\mathcal{T}_i$ such that $A \in \mathcal{T}_i$ and $B \notin \mathcal{T}_i$, or there exists a prime counter-theory $\mathcal{T}_i^\circ$ such that $B \in \mathcal{T}_i^\circ$ and $A \notin \mathcal{T}_i^\circ$, where $i \in \{t, 1\}$.*

Proof. Assume $A \nvdash B$. Then we can construct a sequence of theories which ultimately gives rise to a "large" prime theory maximal with respect to containing A and not containing B or construct a sequence of counter-theories and then generate a counter-theory maximal with respect to containing B, but not containing A.

1. Let $S_1 = \{D \wedge \neg_t D \colon A \vdash D\}$, $S_2 = \{D \wedge \neg_1 D \colon A \vdash D\}$. We can choose an S_i for which $B \notin S_i$, $i \in \{1, 2\}$. Let us assume that it is the S_1. Note that S_1 is not a theory yet. We have to generate the closure of S_1 under $\wedge$ and $\vdash$. Thereby we get the theory, let us denote it as $\mathcal{T}_{t0}$.

Still one problem remains: what if $B \in \mathcal{T}_{t0}$? It may be the case, indeed. For example, there may be some D', $A \vdash D'$, $D' \wedge \neg_t D' \vdash B$. Evidently $A \nvdash D' \wedge \neg_t D'$.

To cope with this problem, we construct the set $S_1^\circ = \{E \colon E \vdash B\}$ and close it under $\vee$ and $\vdash$ (backward direction), thus obtaining counter-theory $\mathcal{T}_{t0}^\circ$. It is not difficult to see that $A \notin \mathcal{T}_{t0}^\circ$.

So, if $B \notin \mathcal{T}_{t0}$, then we take $\mathcal{T}_{t0}$ as a first element of a sequence we are about to construct, else put $\mathcal{T}_{t0}^\circ$ as its first element.

2. Let $C_0, C_1, C_2, \ldots$ be some enumeration of all $\mathcal{L}_{\mathrm{LLSN}}$-formulas. For definiteness' sake assume $A \in \mathcal{T}_{t0}$, $B \notin \mathcal{T}_{t0}$. Now suppose $\mathcal{T}_{tn}$ is already constructed. If $\mathcal{T}_{tn} \cup \{C_n \wedge \neg_t C_n\}$ being closed under $\wedge$ and $\vdash$ do not contain B, then $\mathcal{T}_{tn+1} = \mathcal{T}_{tn} \cup \{C_n \wedge \neg_t C_n\}$ else $\mathcal{T}_{tn} = \mathcal{T}_{tn+1}$. An analogues procedure works when starting theory is $\mathcal{T}_1$ or counter-theory. In the latter case we need closure under $\vee$ and backward closure under $\vdash$.

3. Finally we take the union of all theories (or counter-theories) and obtain a prime theory (or prime counter-theory) maximal with respect to containing A and not containing B (respectively containing B and not containing A).

4. The primeness of obtained theory (or counter-theory) can be shown in a standard way using distribution laws, see [5, lemma 8]. □

Now we are ready to define a canonical valuation v_c with the help of prime theories and counter-theories. Note that the canonical valuation should be coherent in a sense that it captures all possible assignments to a formula A. That is each case, $t \in v_c(A)$, $u_t \in v_c(A)$, $f \in v_c(A)$, etc. must be covered.

Definition 6 (Canonical valuation). *Let $\mathcal{T}_i$ be a $\neg_i$-closed prime theory ($\mathcal{T}_i^\circ$ be a $\neg_i$-closed counter-theory), $i \in \{t, 1\}$. For each propositional variable p, a canonical valuation v_c to be defined as follows:*

$$p \in \mathcal{T}_t \Leftrightarrow 1 \in v_c(p) \text{ or } u_1 \in v_c(p), \qquad p \in \mathcal{T}_t^\circ \Leftrightarrow 0 \in v_c(p) \text{ or } u_1 \in v_c(p),$$
$$p \in \mathcal{T}_1 \Leftrightarrow t \in v_c(p) \text{ or } u_t \in v_c(p), \qquad p \in \mathcal{T}_1^\circ \Leftrightarrow f \in v_c(p) \text{ or } u_t \in v_c(p).$$

Now we would like to have an extended valuation for all formulas defined in a standard way. This is possible according to the next lemma.

Lemma 5. *The canonical valuation v_c can be extended to the set of all formulas.*

Proof. By induction of a formula construction. To accomplish the proof we need to consider four cases, depending on the main connective of a formula, each case then has four subcases, according to the definition 6. We provide a careful inspection of the proofs for the typical subcases while the remaining parts can be easily restored using the similar patterns of reasoning.

Case $\neg_t A$.

Subcase 1. ($\Rightarrow$) Assume $\neg_t A \in \mathcal{T}_t$. Then $A \in \mathcal{T}_t$ by $\neg_t$-closure of $\mathcal{T}_t$. Thus $1 \in v_c(A)$ or $u_1 \in v_c(A)$ by induction hypothesis (IH in the sequel). Each of the disjuncts implies $1 \in v_c(\neg_t A)$ or $u_1 \in v_c(\neg_t A)$ by lemma 1. ($\Leftarrow$) Assume $1 \in v_c(\neg_t A)$ or $u_1 \in v_c(\neg_t A)$. To apply reasoning by cases we take firstly $1 \in v_c(\neg_t A)$ which gives $1 \in v_c(A)$ by lemma 1, hence $1 \in v_c(A)$ or $u_1 \in v_c(A)$; then use the IH and $\neg_t$-closure of $\mathcal{T}_t$. Similarly work with the second disjunct.

Subcase 2. ($\Rightarrow$) Let $\neg_t A \in \mathcal{T}_1$. By lemma 3, $\neg_t \mathcal{T}_1$ is a counter-theory; $A \in \neg_t \mathcal{T}_1$ by the construction of $\neg_t \mathcal{T}_1$. Then IH gives $f \in v_c(A)$ or $u_t \in v_c(A)$. Again, each of the disjuncts implies $t \in v_c(\neg_t A)$ or $u_t \in v_c(\neg_t A)$ by lemma 1. ($\Leftarrow$) Similar to subcase 1, the result is obtained via reasoning by cases and lemma 1, along with IH.

The other subcases and the case $\neg_1 A$ present no difficulties and their proof uses essentially the same technique.

Case $A \wedge B$.

Subcase 1. ($\Rightarrow$) Suppose first $A \wedge B \in \mathcal{T}_t$. What we have to show is that $1 \in v_c(A \wedge B)$ or $u_1 \in v_c(A \wedge B)$. Axioms of $\wedge$-elimination imply $A \in \mathcal{T}_t$ and $B \in \mathcal{T}_t$. Then IH implies [$1 \in v_c(A)$ or $u_1 \in v_c(A)$] and [$1 \in v_c(B)$ or $u_1 \in v_c(B)$]. Elementary classical transformations give thus the following four disjuncts: [$1 \in v_c(A)$ and $1 \in v_c(B)$], [$1 \in v_c(A)$ and $u_1 \in v_c(B)$], [$u_1 \in v_c(A)$ and $1 \in v_c(B)$] and [$u_1 \in v_c(A)$ and $u_1 \in v_c(B)$]. It is easy to see that in each disjunctive case lemma 1 yields the result.

($\Leftarrow$) Now suppose $1 \in v_c(A \wedge B)$ or $u_1 \in v_c(A \wedge B)$. Then we apply reasoning by cases. First $1 \in v_c(A \wedge B)$ implies $1 \in v_c(A)$ and $1 \in v_c(B)$ by lemma 1. Then from the first conjunct we have $1 \in v_c(A)$ or $u_1 \in v_c(A)$ along with $1 \in v_c(B)$ or $u_1 \in v_c(B)$ from the second one. Thus $A \in \mathcal{T}_t$ and $B \in \mathcal{T}_t$ by IH and hence $A \wedge B \in \mathcal{T}_t$ by $\wedge$-closure of $\mathcal{T}_t$. Second case, $u_1 \in v_c(A \wedge B)$, is a bit more tedious because it produces three clauses: [$u_1 \in v_c(A)$ and $1 \in v_c(B)$], [$u_1 \in v_c(B)$ and $1 \in v_c(A)$], [$u_1 \in v_c(A)$ and $u_1 \in v_c(B)$] according to lemma 1. But simple reasoning shows that each of the disjuncts with the help of IH and $\wedge$-closure of $\mathcal{T}_t$ implies $A \wedge B \in \mathcal{T}_t$.

Subcase 2. ($\Rightarrow$) Assume that $A \wedge B \in \mathcal{T}_1^\circ$. From the primeness of $\mathcal{T}_1^\circ$ it follows that $A \in \mathcal{T}_1^\circ$ or $B \in \mathcal{T}_1^\circ$ and, by IH, [$f \in v_c(A)$ or $u_t \in v_c(A)$] or [$f \in v_c(B)$ or $u_t \in v_c(B)$]. The presence of f in the assigned truth value immediately yields $f \in v_c(A \wedge$

B) so [$f \in v_c(A \wedge B)$ or $u_t \in v_c(A \wedge B)$] – that is what we need to show. If u_t is an element of a formula value, say A, then t cannot be in $v_c(A \wedge B)$, so, again, [$f \in v_c(A \wedge B)$ or $u_t \in v_c(A \wedge B)$].

($\Leftarrow$) Now suppose [$f \in v_c(A \wedge B)$ or $u_t \in v_c(A \wedge B)$]. This subcase proof includes lots of routine reasoning by cases arguments. We just sketch some of them. First, suppose $f \in v_c(A \wedge B)$. Then $f \in v_c(A)$ or $f \in v_c(B)$ by lemma 1. Suppose, further, $f \in v_c(A)$. Immediately $f \in v_c(A)$ or $u_t \in v_c(A)$. It means that $A \in \mathcal{T}_1^\circ$ by IH. But $A \wedge B \vdash A$ and $\mathcal{T}_1^\circ$ backwardly closed under $\vdash$, hence $A \wedge B \in \mathcal{T}_1^\circ$. Similar argument applied when $f \in v_c(B)$. Next assume $u_t \in v_c(A \wedge B)$. According to lemma 1, we have three clauses [$u_t \in v_c(A)$ and $u_t \in v_c(B)$], [$t \in v_c(A)$ and $u_t \in v_c(B)$], [$u_t \in v_c(A)$ and $t \in v_c(B)$], but each of them with the help of IH and $\vdash$-closure of $\mathcal{T}_1^\circ$ yields the result. For instance $u_t \in v_c(A)$ follows from the first clause, then $u_t \in v_c(A)$ or $f \in v_c(A)$ implies $A \in \mathcal{T}_1^\circ$ by IH and hence $A \wedge B \in \mathcal{T}_1^\circ$.

Case $A \vee B$.

Subcase 1. ($\Rightarrow$) Let $A \vee B \in \mathcal{T}_1$. Then, by the primeness of $\mathcal{T}_1$, $A \in \mathcal{T}_1$ or $B \in \mathcal{T}_1$.

Thus we have four disjuncts: $t \in v_c(A)$, $t \in v_c(B)$, $u_t \in v_c(A)$ and $u_t \in v_c(B)$ using IH. Again the presence of t in a truth value assigned to A or in a truth value assigned to B immediately gives $t \in v_c(A \vee B)$ and so [$t \in v_c(A \vee B)$ or $u_t \in v_c(A \vee B)$], while u_t guarantees [$u_t \in v_c(A \vee B)$ or $t \in v_c(A \vee B)$] as required.

($\Leftarrow$) Now assume $t \in v_c(A \vee B)$ or $u_t \in v_c(A \vee B)$. The first disjunct leads to the result almost straightforwardly, since $t \in v_c(A)$ or $t \in v_c(B)$ and IH give $A \in \mathcal{T}_1$ or $B \in \mathcal{T}_1$, so $A \vee B \in \mathcal{T}_1$ follows from each of disjuncts using the introduction of $\vee$ axioms. Next assume $u_t \in v_c(A \vee B)$. Lemma 1 then gives three possible situations: [$f \in v_c(A)$ and $u_t \in v_c(B)$], [$u_t \in v_c(A)$ and $f \in v_c(B)$] or [$u_t \in v_c(A)$ and $u_t \in v_c(B)$]. Each of them along with IH implies $A \vee B \in \mathcal{T}_1$.

Subcase 2. ($\Rightarrow$) Suppose $A \vee B \in \mathcal{T}_1^\circ$. Then, by the backward $\vdash$-closure of $\mathcal{T}_1^\circ$ and the $\vee$-introduction axioms, $A \in \mathcal{T}_1^\circ$ and $B \in \mathcal{T}_1^\circ$. IH leads to the following two assertions: [$f \in v_c(A)$ or $u_t \in v_c(A)$] and [$f \in v_c(B)$ or $u_t \in v_c(B)$]. Again we transform these into four clauses: [$f \in v_c(A)$ and $f \in v_c(B)$], [$f \in v_c(A)$ and $u_t \in v_c(B)$], [$f \in v_c(B)$ and $u_t \in v_c(A)$], [$u_t \in v_c(A)$ and $u_t \in v_c(B)$]. Evidently in each case we obtain $f \in v_c(A \vee B)$ or $u_t \in v_c(A \vee B)$.

($\Leftarrow$) Assume $f \in v_c(A \vee B)$ or $u_t \in v_c(A \vee B)$. The result easily follows from each of the disjuncts. Assuming $f \in v_c(A \vee B)$ we derive $f \in v(A)$ and $f \in v_c(B)$ in virtue of lemma 1. IH then gives $A \in \mathcal{T}_1^\circ$, $B \in \mathcal{T}_1^\circ$ which implies $A \vee B \in \mathcal{T}_1^\circ$ since $\mathcal{T}_1^\circ$ is closed under $\vee$. Case for $u_t \in v_c(A \vee B)$ is a bit more involved as it provides three disjuncts according to lemma 1. Each of them is easy to explore. For instance, let $u_t \in v_c(A)$ and $f \in v_c(A)$. $u_t \in v_c(A)$ implies $u_t \in v_c(A)$ or $f \in v_c(A)$, so, by IH $A \in \mathcal{T}_1^\circ$. Likewise $B \in \mathcal{T}_1^\circ$ from the second conjunct. Finally, by $\vee$-closure of $\mathcal{T}_1^\circ$, $A \vee B \in \mathcal{T}_1^\circ$. □

Theorem 6 (Completeness). *For all formulas A, B: if $A \models B$ then $A \vdash B$.*

Proof. Let $A \nvdash B$. Then, using lemma 4, we construct a prime theory $\mathcal{T}_i$ such that $A \in \mathcal{T}_i$ and $B \notin \mathcal{T}_i$ or a prime counter-theory $\mathcal{T}_i^\circ$, such that $B \in \mathcal{T}_i^\circ$ and $A \notin \mathcal{T}_i^\circ$, $i \in \{1, t\}$. Let us assume that the prime theory $\mathcal{T}_1$ has been constructed and $A \in \mathcal{T}_1$. Then due to lemma 5 we have $t \in v_c(A)$ or $u_t \in v_c(A)$ but $t, u_t \notin v_c(B)$. If $t \in v_c(A)$ then clearly $A \nvDash B$, according to the definition 12. Assume $u_t \in v_c(A)$. It's easy to see from truth table 2 that $f \in v_c(B)$ but $f \notin v_c(A)$, so, again $A \nvDash B$. The counter-theory case is analogous. □

6 Relational Semantics for Non-distributive LLSN

In this section we change the direction of thought and consider some different semantics for the calculus based on the old **LLSN** system. More specifically, we omit the distribution scheme from the set of axiomatic schemata and escape from the realm of finite lattice-ordered sets of generalized truth values.

Unlike classical logic, not all non-classical logical systems presuppose distributivity of conjunction over disjunction and vise versa. Moreover, the distribution laws are even not considered as representing the inherent properties of conjunction and disjunction at all. As stated in [4], "nondistributive logics are perfectly natural". The reason for this claim is presumably stems from the fact that the very basic properties of meet and join in a lattice (not necessarily distributive in general), represented by an appropriate family of logical deductive rules, do not entail the distribution laws.

From the semantical perspective non-distributive logics are determined by a different class of structures and demand essentially novel insights in completeness proof techniques. In the literature one can find examples of semantics for positive non-distributive propositional logics as well as approaches aimed to handle different negation-like operations in the context of non-distributivity, see, eg. [4, 14].

Closely related research area is a pure lattice-theoretic studies in representation theory of lattices with negations [10, 3, 13, 14]. Some of these sources treat negation as modal operator.

Non-distributive version of **LLSN** is obtained from original **LLSN** by dropping scheme 5 which implies the distribution laws for $\wedge$ and $\vee$.

For our purposes we will modify the approach to construction of a relational semantics for Linear logic from §6 of [1][3], where the mathematical apparatus of representation theory developed in preceding sections transforms into the customary logical framework for describing relationship between syntax and semantics. Linear

[3]As one of the reviewers pointed out this article actually deals with MALL and some related logics.

logic has rich signature including De Morgan negation along with $\wedge$ and $\vee$ which do not enjoy the distribution laws. Semantically this means that we do not have a sufficient supply of prime theories and counter-theories to separate formulas. From the relational semantics point of view, a world of a frame can be an element of some theory, or counter-theory, or do not belong to any of them. Thus the valuation of propositional variables relative to possible worlds appears to be three-valued, which is close to our understanding of generalized truth values with uncertainty component.

The basic underlying semantic structure is a set equipped with a family of quasi-orders.

Definition 7. *A* frame *is a structure* $(X, (\leqslant_i, \preccurlyeq_i)_{i\in\{1,2\}})$, *where* X *is a non-empty set and each of* $\leqslant_i$ *and* $\preccurlyeq_i$, $(i \in \{1,2\})$ *is a quasi-order on* X *satisfying the following conditions for all* $x, y \in X$*:*

$$x \leqslant_i y \,\&\, x \preccurlyeq_i y \Rightarrow x = y, \; i \in \{1,2\}. \tag{3}$$

The elements of X are usually called 'worlds' or 'states' as is customary for relational structures. The relations on X induce two couples of functions: r_1, l_1 and r_2, l_2 of the type $\mathcal{P}(X) \to \mathcal{P}(X)$, defined via the following equations:

$$l_i(A) = \{x \in X : \forall y \in X(x \leqslant_i y \Rightarrow y \notin A)\},$$
$$r_i(A) = \{x \in X : \forall y \in X(x \preccurlyeq_i y \Rightarrow y \notin A)\},$$

where $i \in \{1,2\}$.

The key fact, proved in [16], is that these two mappings constitute a Galois connection between $\leqslant_i$-increasing and $\preccurlyeq_i$-increasing subsets of X. Therefore the compositions of two mappings, $r_i l_i$ and $l_i r_i$, $(i \in \{1,2\})$, are closure operators on the subsets of X. Slightly modifying the corresponding notions from [16] we call a set $A \subseteq X$ *l_i-stable* if $A = l_i(r_i(A))$; A is *r_i-stable* if $A = r_i(l_i(A))$, $i \in \{1,2\}$.

l- and r-stable subsets are key components of semantic constructions studied in the literature on the related topics (e.g. [10, 3]).

The presence of semi-negations in the signature of logic forces the appearance of their semantic counterparts. We adopt here the use of so called generalized star operations (following the lines of [1, 10]) and denote them as N_t and N_1 to keep clear the correspondence with the semi-negations. We then call a doubly ordered set endowed with N_t and N_1 a star frame, using the term from [7] (although we have two star-like operations here and do not actually designating them as stars).

Definition 8. *A tuple* $F = (X, (\leqslant_i, \preccurlyeq_i)_{i\in\{1,2\}}, N_t, N_1)$ *is called* a star frame *if* $(X, (\leqslant_i, \preccurlyeq_i)_{i\in\{1,2\}})$ *is a frame,* N_t *and* N_1 *are the unary operations on* X *such that for all* $x, y \in X$, $i \in \{1,2\}$*:*

1. $N_t(N_t(x)) = x$,
2. $N_1(N_1(x)) = x$,
3. $N_t(N_1(x)) = N_1(N_t(x))$,
4. $N_t(x) \leqslant_i N_1(y) \Leftrightarrow N_1(x) \preccurlyeq_i N_t(y)$,
5. $x \leqslant_1 y \Leftrightarrow N_t(x) \preccurlyeq_1 N_t(y)$,
6. $x \leqslant_2 y \Leftrightarrow N_1(x) \preccurlyeq_2 N_1(y)$.

Note that we also have the following expressions derivable from the above definition:

$$x \preccurlyeq_1 y \Leftrightarrow N_t(x) \leqslant_1 N_t(y), \tag{4}$$
$$x \preccurlyeq_2 y \Leftrightarrow N_1(x) \leqslant_2 N_1(y), \tag{5}$$
$$N_t(x) \preccurlyeq_1 y \Leftrightarrow x \leqslant_1 N_t(y), \tag{6}$$
$$N_1(x) \preccurlyeq_2 y \Leftrightarrow x \leqslant_2 N_1(y), \tag{7}$$
$$N_t(x) \preccurlyeq_i N_1(y) \Leftrightarrow N_1(x) \leqslant_i N_t(y). \tag{8}$$

Indeed, $x \preccurlyeq_1 y \Rightarrow N_t(N_t(x)) \preccurlyeq_1 N_t(N_t(y)) \Rightarrow N_t(x) \leqslant_1 N_t(y)$, using items 1 and 5 of definition 8. Now we have the following derivation:

$$\begin{aligned}
N_t(x) \preccurlyeq_1 N_1(y) &\Leftrightarrow N_t(N_t(x)) \leqslant_1 N_t(N_1(x)) && \text{def. } 8-1\\
&\Leftrightarrow N_t(N_t(x)) \leqslant_1 N_1(N_t(x)) && \text{def. } 8-3\\
&\Leftrightarrow N_1(N_t(x)) \preccurlyeq_1 N_t(N_t(x)) && \text{def. } 8-4\\
&\Leftrightarrow N_t(N_1(x)) \preccurlyeq_1 N_t(N_t(x)) && \text{def. } 8-3\\
&\Leftrightarrow N_1(x)) \leqslant_1 N_t(x) && \text{def. } 8-5
\end{aligned}$$

To define a valuation of propositional variables on a star frame we use two valuation functions, v_t and v_1, each responsible for a valuation in its own basic truth values set. This choice is not essential but makes our definitions slightly less cumbersome.

Definition 9. *Let $F = (X, (\leqslant_i, \preccurlyeq_i)_{i \in \{1,2\}}, N_t, N_1)$ be a star frame. The valuation functions v_t, v_1 are the mappings $v_t \colon PV \times X \to \{t, u_t, f\}$ and $v_1 \colon PV \times X \to \{1, u_1, 0\}$ such that for each $p \in PV$, $x \in X$:*

1. *If $P^p_t = \{x \colon v_t(p, x) = t\}$ and $P^p_f = \{x \colon v_t(p, x) = f\}$ then $r_1(P^p_t) = P^p_f$ and $l_1(P^p_f) = P^p_t$;*

2. *If* $P_1^p = \{x\colon v_1(p,x) = 1\}$ *and* $P_0^p = \{x\colon v_1(p,x) = 0\}$ *then* $r_2(P_t^p) = P_f^p$ *and* $l_2(P_f^p) = P_t^p$.

The correspondence between the truth and falsity sets of a propositional variable p (that is between components in the pairs (P_t^p, P_f^p) and (P_1^p, P_0^p)) postulated in the above definition shows that these sets are stable with respect to the corresponding composition of l_i and r_i, $i \in \{1,2\}$. For example for a propositional variable p, $l_1(r_1(P_t^p)) = P_t^p$, so P_t^p is an l_1-stable set.

Next we introduce the satisfaction relations between elements of X and formulas, relativized with respect to the value of a valuation function. On the level of propositional variables this correspondence provided by the valuation functions and described by the following expressions, where p ranges over PV:

Definition 10.

$$x \vDash_t p \Leftrightarrow v_t(x,p) = t \qquad x \vDash_1 p \Leftrightarrow v_1(x,p) = 1 \tag{9}$$

$$x \vDash_{u_t} p \Leftrightarrow v_t(x,p) = u_t \qquad x \vDash_{u_1} p \Leftrightarrow v_1(x,p) = u_1 \tag{10}$$

$$x \vDash_f p \Leftrightarrow v_t(x,p) = f \qquad x \vDash_0 p \Leftrightarrow v_1(x,p) = 0 \tag{11}$$

Having defined relativized satisfaction relations we can model our old truth values from $\mathscr{L}\mathbf{9}^u$ in terms of these relations. For example we will write $x \vDash_{t1} p$ to mean that $x \vDash_t p$ and $x \vDash_1 p$ simultaneously hold in some world x. Before extending the satisfaction relation to the complex formulas we introduce another notational convention. An expression $x \vDash_{-\mathrm{v}} A$ means that a formula A is in some satisfaction relation with x (under the valuations v_t and v_1) except of the relation with particular value v. For instance $x \vDash_{-0} A$ is understood as $x \vDash_{t1} A$, $x \vDash_{fu_1} A$, etc., that is encodes a collection of conjunctions of valuations for a formula A except of those containing conjunct $x \vDash_0 A$. For the complex formulas we put $x \vDash_{u_t} A \Leftrightarrow \nvDash_t A$ and $\nvDash_f A$; $x \vDash_{u_1} A \Leftrightarrow \nvDash_1 A$ and $\nvDash_0 A$. Now we are ready to spell out the definitions of the satisfaction relations for $\wedge$, $\vee$ and the semi-negations.

Definition 11.

$$x \vDash_t A \wedge B \Leftrightarrow x \vDash_t A \text{ and } x \vDash_t B, \quad x \vDash_f A \vee B \Leftrightarrow x \vDash_f A \text{ and } x \vDash_f B \tag{12}$$
$$x \vDash_1 A \wedge B \Leftrightarrow x \vDash_1 A \text{ and } x \vDash_1 B, \quad x \vDash_0 A \vee B \Leftrightarrow x \vDash_0 A \text{ and } x \vDash_0 B \tag{13}$$
$$x \vDash_t A \vee B \Leftrightarrow \forall y(x \leqslant_1 y \Rightarrow (y \vDash_{-f} A \text{ or } y \vDash_{-f} B)), \tag{14}$$
$$x \vDash_1 A \vee B \Leftrightarrow \forall y(x \leqslant_2 y \Rightarrow (y \vDash_{-0} A \text{ or } y \vDash_{-0} B)), \tag{15}$$
$$x \vDash_f A \wedge B \Leftrightarrow \forall y(x \preccurlyeq_1 y \Rightarrow (y \vDash_{-t} A \text{ or } y \vDash_{-t} B)), \tag{16}$$
$$x \vDash_0 A \wedge B \Leftrightarrow \forall y(x \preccurlyeq_2 y \Rightarrow (y \vDash_{-1} A \text{ or } y \vDash_{-1} B)), \tag{17}$$
$$x \vDash_t \neg_t A \Leftrightarrow N_t(x) \vDash_f A, \quad x \vDash_t \neg_1 A \Leftrightarrow x \vDash_t A, \tag{18}$$
$$x \vDash_1 \neg_1 A \Leftrightarrow N_1(x) \vDash_0 A, \quad x \vDash_1 \neg_t A \Leftrightarrow x \vDash_1 A, \tag{19}$$
$$x \vDash_f \neg_t A \Leftrightarrow \forall y(x \preccurlyeq_1 y \Rightarrow N_t(y) \vDash_{-f} A), \tag{20}$$
$$x \vDash_0 \neg_1 A \Leftrightarrow \forall y(x \preccurlyeq_2 y \Rightarrow N_1(y) \vDash_{-0} A), \tag{21}$$
$$x \vDash_0 \neg_t A \Leftrightarrow \forall y(x \preccurlyeq_2 y \Rightarrow y \vDash_{-1} A), \tag{22}$$
$$x \vDash_f \neg_1 A \Leftrightarrow \forall y(x \preccurlyeq_1 y \Rightarrow y \vDash_{-t} A). \tag{23}$$

Remark. We would like to extend the definitions of truth and falsity sets of propositional variables given in definition 9 to arbitrary formulas. Thus, given a formula A, $P_i^A = \{x \colon x \vDash_i A\}$, where $i \in \{1, t, 0, f\}$. The key observation here is that $r_1(P_t^A) = P_f^A$, $r_2(P_1^A) = P_0^A$, $l_1(P_f^A) = P_t^A$ and $l_2(P_0^A) = P_1^A$. We will prove this fact in the next lemma.

Lemma 7. *For each formula A, star frame F and $x \in F$: if $P_t^A = \{x \colon x \vDash_t A\}$ and $P_f^A = \{x \colon x \vDash_f A\}$, then $r_1(P_t^A) = P_f^A$ and $l_1(P_f^A) = P_t^A$; if $P_1^A = \{x \colon x \vDash_1 A\}$ and $P_0^A = \{x \colon x \vDash_0 A\}$, then $r_2(P_1^A) = P_0^A$ and $l_2(P_0^A) = P_1^A$.*

Proof. 0. For the propositional variables the assertions of lemma hold by definition of canonical valuations and definition 16.

1. Assume $x \in r_1(P_t^{A \wedge B})$. Then $\forall y(x \preccurlyeq_1 y \Rightarrow y \notin P_t^{A \wedge B})$, that is $\forall y(x \preccurlyeq_1 y \Rightarrow y \vDash_{-t} A \wedge B)$ and, by definition 11, $\forall y(x \preccurlyeq_1 y \Rightarrow (y \vDash_{-t} A$ or $y \vDash_{-t} B))$. Therefore $x \vDash_f A \wedge B$ and $x \in P_f^{A \wedge B}$. For the other direction just move backward.

2. Next suppose $x \in r_1(P_t^{A \vee B})$. Then we have $\forall y(x \preccurlyeq_1 y \Rightarrow y \notin P_t^{A \vee B})$, which means that $\forall y(x \preccurlyeq_1 y \Rightarrow y \vDash_{-t} (A \vee B))$, so $\forall y(x \preccurlyeq_1 y \Rightarrow \exists z(y \leqslant_1 z$ and $(z \nvDash_{-f} A$ and $z \nvDash_{-f} B)))$ which gives

$$\forall y(x \preccurlyeq_1 y \Rightarrow \exists z(y \leqslant_1 z \text{ and } z \vDash_f A)) \text{ and} \tag{24}$$
$$\forall y(x \preccurlyeq_1 y \Rightarrow \exists z(y \leqslant_1 z \text{ and } z \vDash_f B)). \tag{25}$$

The last two expressions are just unwinding of $x \in r_1(l_1(P_f^A))$ and $x \in r_1(l_1(P_f^B))$. By IH we know that $r_1(l_1(P_f^A))$ and $r_1(l_1(P_f^B)$ are both r_1-stable, hence $x \in P_f^A$ and $x \in P_f^B$ and, finally, $x \in P_f^{A \vee B}$.

Let $x \in P_f^{A \vee B}$. Then we evidently have $x \in P_f^A$ and $x \in P_f^B$. Again, IH yields $x \in r_1(l_1(P_f^A))$ and $x \in r_1(l_1(P_f^B))$. These conjuncts can be rewritten in the form of (24) and (25) correspondingly. Those expressions imply $x \in r_1(P_t^{A \vee B})$.

3. Cases for the semi-negations are almost straightforward. Let us check for $\neg_t A$. So, suppose $x \in r_1(P_t^{\neg_t A})$. Then $\forall y(x \preccurlyeq_1 y \Rightarrow y \notin P_t^{\neg_t A})$, hence $\forall y(x \preccurlyeq_1 y \Rightarrow y \vDash_{-t} \neg_t A))$, so $\forall y(x \leqslant_1 y \Rightarrow N_t(y) \vDash_{-f} A)$. The last expression exactly means $x \vDash_f \neg_t A$. Moving backward we obtain the converse.

4. Cases for l_2 and r_2 are shown analogously. □

As a simple implication from the above lemma we have the following statements fixing the facts about stabilities of truth and falsity sets of an arbitrary formula A:

Corollary 8. *For each formula A, $P_t^A = l_1(r_1(P_t^A))$, $P_1(A) = l_2(r_2(P_1^A))$, $P_f^A = r_1(l_1(P_f^A))$, $P_0^A = r_2(l_2(P_0^A))$.*

The following definition introduces the central semantic concept, namely the *consequence relation.*

Definition 12. *For all A, B, $A \vDash B$ if for all $\mathcal{F}$ and all $x \in \mathcal{F}$ the following statements hold:*

$$x \vDash_t A \Rightarrow x \vDash_t B, \qquad x \vDash_1 A \Rightarrow x \vDash_1 B, \tag{26}$$

$$x \vDash_f B \Rightarrow x \vDash_f A, \qquad x \vDash_0 B \Rightarrow x \vDash_0 A. \tag{27}$$

6.1 Soundness

Let us stipulate and prove the soundness result.

Theorem 9. *For all formulas A and B, $A \vdash B \Rightarrow A \vDash B$.*

As usual the soundness proof supposes the routine check for all the axiom schemata and the rules of inference, but we take for the illustration purposes only scheme 11 and rule R5 cases, probably the most cumbersome, though.

Proof. 1. Let us take an arbitrary star frame F, $x \in F$ and assume that $x \vDash_t \neg_t(A \wedge B)$. Then, according to definition 11, $N_t(x) \vDash_f (A \wedge B)$, so $\forall y(N_t(x) \preccurlyeq_1 y \Rightarrow (y \vDash_{-t} A \text{ or } y \vDash_{-t} B))$. Next we use the fact that $N_t(x) \preccurlyeq_1 y$ is implied by $x \leqslant_1 N_t(y)$ for all $x, y \in F$. Consequently, $\forall y(x \leqslant_1 N_t(y) \Rightarrow (N_t(y) \models_{-f} \neg_t A$ or $N_t(y) \models_{-f} \neg_t B))$ which means, again by definition 11, that $x \models_t (\neg_t A \vee \neg_t B)$.

Further assume that $x \models_1 \neg_t(A \wedge B)$. From definition 11 we have $x \models_1 (A \wedge B)$, that is $x \in P_1(A \wedge B)$, hence $x \in l_2(r_2(A \wedge B))$ (see Remark after definition 11). Now assume $x \not\models_1 (\neg_t A \vee \neg_t B)$. Unwinding of this assumption is the expression $\exists y(x \preccurlyeq_2 y$ and $\forall z(y \leqslant_2 z \Rightarrow z \models_{-1} A$ and $z \models_{-1} B))$. But this exactly means that $x \notin l_2(r_2(A \wedge B))$ as it is the case when $\exists y(x \preccurlyeq_2 y$ and $\forall z(y \leqslant_2 z \Rightarrow z \models_{-1} A$ or $z \models_{-1} B))$.

Let $x \models_f \neg_t A \vee \neg_t B$. By definition 11 this implies $x \models_f \neg_t A$ and $x \models_f \neg_t B$. Let us take $x \models_f \neg_t A$. Applying the same definition again we get $\forall y(x \preccurlyeq_1 y \Rightarrow N_t(y) \models_{-f} A)$. Now consider consequent $N_t(y) \models_{-f} A$. It can be rewritten as $N_t(y) \notin P_f(A)$, so, by r_1-stability of $P_f(A)$, $N_t(y) \notin r_t(l_1(P_f(A)))$. The latter expression means $\exists z(N_t(y) \preccurlyeq_1 z$ and $z \in l_1(P_f(A)))$. Note that $l_1(P_f(A)) = P_t(A)$, so $z \models_t (A)$. The same reasoning gives $z \models_t B$ from $x \models_f \neg_t B$. But it means that $\forall y(x \preccurlyeq_1 y \Rightarrow \exists z(N_t(y) \preccurlyeq_1 z$ and $z \models_t A \wedge B))$, that is $\forall y(x \preccurlyeq_1 y \Rightarrow N_t(y) \models_{-f} A \wedge B)$ and, finally, $x \models_f \neg_t(A \wedge B)$.

Suppose $x \models_0 \neg_t A \vee \neg_t B$. Therefore $x \models_0 \neg_t A$ and $x \models_0 \neg_t B$, which is $\forall y(x \preccurlyeq_2 y \Rightarrow y \models_{-1} A$ and $y \models_{-1} B)$. But it implies $x \models_0 \neg_t(A \wedge B)$ by definition 11.

Now we check the rule $\neg_1 A \vdash \neg_t B / \neg_1 B \vdash \neg_t A$. Assume $\neg_1 B \not\models \neg_t A$. Let for some x $x \models_t \neg_1 B$, but $x \models_{-t} \neg_t A$. Then $x \models_t B$, hence $N_t(x) \models_{-f} \neg_1 B$. From $x \models_{-t} \neg_t A$ we have $N_t(x) \models_{-f} A$. It remains to note $N_t(x) \notin P_f(A)$ and from r_1-stability of $P_f(A)$ we infer $\exists z(N_t(x) \preccurlyeq_1 z$ and $z \models_t A)$, so $N_t(x) \models_{-f} \neg_1 A$. Consequently $\neg_1 A \not\models \neg_t B$. □

Next lemma is proved with the help of simple usage of soundness theorem.

Lemma 10. *For each formula A, canonical frame $\mathcal{F}$ and $x \in \mathcal{F}$:*

$$x \models_t A \Leftrightarrow N_t(x) \models_f \neg_t A \qquad x \models_f A \Leftrightarrow N_t(x) \models_t \neg_t A \tag{28}$$

$$x \models_1 A \Leftrightarrow N_1(x) \models_0 \neg_1 A \qquad x \models_0 A \Leftrightarrow N_1(x) \models_1 \neg_1 A \tag{29}$$

Proof. The equivalences on the left side are easy to see. For instance, $x \models_t A \Leftrightarrow N_t(N_t(x)) \models_t \neg_t \neg_t A \Leftrightarrow N_t(x) \models_f \neg_t A$. As for the equivalences on the right side, right-to-left direction is clear: $N_t(x) \models_t \neg_t A \Rightarrow N_t(N_t(x)) \models_f \neg_t \neg_t A \Rightarrow x \models_f A$. For the converse we need the following sequence of metalinguistic implications:

$$
\begin{array}{lr}
x \vDash_f A \Rightarrow x \vDash_f \neg_t\neg_t A & \text{axiom 7} \\
\quad \Rightarrow \forall y(x \preccurlyeq_1 y \Rightarrow N_t(y) \vDash_{-f} \neg_t A) & \text{def. 11} \\
\quad \Rightarrow (x \preccurlyeq_1 N_t(y) \Rightarrow N_t(N_t(y)) \vDash_{-f} \neg_t A) & \forall_{el} \\
\quad \Rightarrow (N_t(x) \leqslant_1 y \Rightarrow y \vDash_{-f} \neg_t A) & \text{def 8} - 1, 5, (4) \\
\quad \Rightarrow \forall y(N_t(x) \leqslant_1 y \Rightarrow y \vDash_{-f} \neg_t A) & \forall_{in}
\end{array}
$$

The last line and lemma 7 imply that $N_t(x) \in l_1(r_1(P_t^{\neg_t A})) = P_t^{\neg_t A}$, that is $N_t(x) \vDash_t \neg_t A$. □

6.2 Completeness

For the completeness proof we will make use of the notions of $\neg_t$-closed and $\neg_1$-closed theories and counter-theories (see the definitions 4, 6) again. To construct a canonical frame we need a modification of commonly used (*theory*, *counter-theory*) pair based approach (see e.g. [16, 10][4]) due to the distinction of $\neg_t$- and $\neg_1$-closed theories. Namely a *theory* component splits now into *a pair* of two theories $(\mathcal{T}_t, \mathcal{T}_1)$, $\neg_t$- and $\neg_1$-closed correspondingly. Likewise a *counter-theory* component will be replaced by $(\mathcal{T}_t^\circ, \mathcal{T}_1^\circ)$, a pair of $\neg_t$- and $\neg_1$-closed counter-theories. So, a world x of a canonical frame may be thought as a quadruple of the form $(\mathcal{T}_t, \mathcal{T}_1, \mathcal{T}_t^\circ, \mathcal{T}_1^\circ)$ for some theories $\mathcal{T}_t, \mathcal{T}_1$ and counter-theories $\mathcal{T}_t^\circ, \mathcal{T}_1^\circ$ closed according to their lower indices. For notational convenience we will represent quadruples in the form of pairs and call them "q-pairs". If $x = ((\mathcal{T}_t, \mathcal{T}_1), (\mathcal{T}_t^\circ, \mathcal{T}_1^\circ))$, we will refer to its pairs' components by adding x to the their superscript positions. For example $\mathcal{T}_1^x$ is the second element of the first pair of x, while $\mathcal{T}_1^{\circ x}$ is the second element of the second pair of x. To define the star-like operations N_t and N_1 in a canonical frame we will use the notations $\neg_1\mathcal{T}_t$, $\neg_1\mathcal{T}_t^\circ$, $\neg_t\mathcal{T}_1$ and $\neg_t\mathcal{T}_1^\circ$ in the same sense as before (see the explanations precluding lemma 3).

Definition 13. *A q-pair x is* maximal disjoint *if $\mathcal{T}_t^x$ is maximal in the set of $\neg_t$-closed theories disjoint from $\mathcal{T}_t^{\circ x}$ and $\mathcal{T}_t^{\circ x}$ is maximal in the set of $\neg_t$-closed theories disjoint from $\mathcal{T}_t^x$, while $\mathcal{T}_1^x$ is maximal in the set of $\neg_1$-closed theories disjoint from $\mathcal{T}_1^{\circ x}$ and $\mathcal{T}_1^{\circ x}$ is maximal in the set of $\neg_1$-closed counter-theories disjoint from $\mathcal{T}_1^x$.*

Definition 14. *A canonical frame is a structure $\mathcal{F} = (W, (\leqslant_i, \preccurlyeq_i)_{i\in\{1,2\}}, N_t, N_1)$, where W is set of maximal disjoint q-pairs such that for all $x, y \in W$,*

[4]To be more precise the approach developed in [16] utilizes the notion of disjoint filter-ideal pair.

1. $x \leqslant_1 y \Leftrightarrow \mathcal{T}_1^x \subseteq \mathcal{T}_1^y$, $x \leqslant_2 y \Leftrightarrow \mathcal{T}_t^x \subseteq \mathcal{T}_t^y$,

2. $x \preccurlyeq_1 y \Leftrightarrow \mathcal{T}_1^{\circ x} \subseteq \mathcal{T}_1^{\circ y}$, $x \preccurlyeq_2 y \Leftrightarrow \mathcal{T}_t^{\circ x} \subseteq \mathcal{T}_t^{\circ y}$,

3. $N_t(x) = ((\mathcal{T}_t^x, \neg_t \mathcal{T}_1^{\circ x}), (\mathcal{T}_t^{\circ x}, \neg_t \mathcal{T}_1^x))$,

4. $N_1(x) = ((\neg_1 \mathcal{T}_t^{\circ x}, \mathcal{T}_1^x), (\neg_1 \mathcal{T}_t^x, \mathcal{T}_1^{\circ x}))$.

We will write $x \leqslant y$, $x, y \in \mathcal{F}$, to mean $x \leqslant_i y$ and $x \leqslant_i y$, $i \in \{1, 2\}$. We also often use the notations $\mathcal{T}_t^{N_i(x)}$, $\mathcal{T}_1^{N_i(x)}$, $\mathcal{T}_t^{\circ N_i(x)}$ and $\mathcal{T}_1^{\circ N_i(x)}$ for the elements from $N_i(x)$, $i \in \{t, 1\}$.

We have an analogue of lemma 4 but now without assuming primeness of theories and counter-theories as this property requires the distribution laws absent from the set of axiom schemata.

Lemma 11. *If $A \nvdash B$ then there exists a theory $\mathcal{T}_i$ such that $A \in \mathcal{T}_i$ and $B \notin \mathcal{T}_i$ or there exists a counter-theory $\mathcal{T}_i^\circ$ such that $B \in \mathcal{T}_i^\circ$ and $A \notin \mathcal{T}_i^\circ$, where $i \in \{1, t\}$.*

Proof. Is essentially the same as in given for lemma 4, but without the last point concerning the primness as we do not have it this time. □

Lemma 12. *If $A \nvdash B$ then there exists a maximal disjoint q-pair x such that $A \in \mathcal{T}_i^x$ and $B \notin \mathcal{T}_i^x$ or $B \in \mathcal{T}_i^{\circ x}$ and $A \notin \mathcal{T}_i^{\circ x}$, $i \in \{t, 1\}$.*

Proof. By the previous lemma we know how to construct a theory or counter-theory separating A from B. Assume without loss of generality that $A \in \mathcal{T}_t$. Thus to compose a required q-pair we first need arbitrary counter-theory $\mathcal{T}_1^\circ$, disjoint from $\mathcal{T}_1$ and a disjoint pair $(\mathcal{T}_t, \mathcal{T}_t^\circ)$ which always can be found. Then extend this q-pair to the maximal disjoint x using standard techniques exploiting Zorn lemma. □

Next simple lemma is useful for the technical inferences below.

Lemma 13. *Let $\mathcal{F}$ be a canonical frame and $x \in \mathcal{F}$. Then*

1. $A \in \mathcal{T}_1^{\circ N_t(x)} \Leftrightarrow \neg_t A \in \mathcal{T}_1^x$ *or* $A = \neg_t A'$ *for some* $A' \in \mathcal{T}_1^x$ *and* $A' \neq \neg_t B$,

2. $A \in \mathcal{T}_1^{N_t(x)} \Leftrightarrow \neg_t A \in \mathcal{T}_1^{\circ x}$ *or* $A = \neg_t A'$ *for some* $A' \in \mathcal{T}_1^{\circ x}$ *and* $A' \neq \neg_t B$,

3. $A \in \mathcal{T}_t^{\circ N_1(x)} \Leftrightarrow \neg_1 A \in \mathcal{T}_t^x$ *or* $A = \neg_1 A'$ *for some* $A' \in \mathcal{T}_t^x$ *and* $A' \neq \neg_1 B$,

4. $A \in \mathcal{T}_1^{N_1(x)} \Leftrightarrow \neg_1 A \in \mathcal{T}_1^{\circ x}$ *or* $A = \neg_1 A'$ *for some* $A' \in \mathcal{T}_t^{\circ x}$ *and* $A' \neq \neg_1 B$.

Proof. Directly follows from the definitions of operations N_t and N_1 in a canonical frame. □

Lemma 14. *Let $\mathcal{F}$ be a canonical frame. Then $\mathcal{F}$ is a star frame.*

Proof. We have to check the properties 1-6 of definition 8 with respect to the unary operations of a canonical frame.

1. To prove equation 1 we have to show that for each theory $\mathcal{T}_1$, $\mathcal{T}_1 = \neg_t\neg_t\mathcal{T}_1$ and for each counter-theory $\mathcal{T}_1^\circ$, $\mathcal{T}_1^\circ = \neg_t\neg_t\mathcal{T}_1^\circ$. Let us check the equation $\mathcal{T}_1 = \neg_t\neg_t\mathcal{T}_1$. For a formula A we have to consider three sub-cases.

   (a) $A \neq \neg_t B$. Then we have $A \in \mathcal{T}_1 \Leftrightarrow \neg_t A \in \neg_t\mathcal{T}_1 \Leftrightarrow A \in \neg_t\neg_t\mathcal{T}_1$.

   (b) $A = \neg_t B$ and $B \neq \neg_t C$. Here we have $A \in \mathcal{T}_1 \Leftrightarrow B \in \neg_t\mathcal{T}_1 \Leftrightarrow \neg_t B = A \in \neg_t\neg_t\mathcal{T}_1$.

   (c) $A = \neg_t B$ and $B = \neg_t C$. In this case we have $A \in \mathcal{T}_1 \Leftrightarrow B \in \neg_t\mathcal{T}_1 \Leftrightarrow C \in \neg_t\neg_t\mathcal{T}_1$. Since $\neg_t\neg_t\mathcal{T}$ is a theory according to lemma 3, $\neg_t\neg_t C = A \in \neg_t\neg_t\mathcal{T}_1$.

   Similar reasoning shows that $\mathcal{T}_1^\circ = \neg_t\neg_t\mathcal{T}_1^\circ$.

2. For equation 3 it is enough to notice that for all $x \in \mathcal{F}$, $N_t(N_1(x)) = ((\neg_1\mathcal{T}_t^{\circ x}, \neg_t\mathcal{T}_1^{\circ x}), (\neg_1\mathcal{T}_t^{x}, \neg_t\mathcal{T}_1^{x})) = N_1(N_t(x))$.

3. For equation 4 assume $N_t(x) \leqslant_1 N_1(y)$. This means that inclusion $\mathcal{T}_1^{N_t(x)} \subseteq \mathcal{T}_1^{N_1(y)}$ holds. Note that $\mathcal{T}_1^{N_1(y)} = \mathcal{T}_1^{y}$ (as N_1 does not affect the $\mathcal{T}_1$ when changing from y to $N_1(y)$), so the inclusion can be rewritten as $\mathcal{T}_1^{N_t(x)} \subseteq \mathcal{T}_1^{y}$. The last expression implies $\neg_t\mathcal{T}_1^{N_t(x)} \subseteq \neg_t\mathcal{T}_1^{y}$ which is the same as $\mathcal{T}_1^{\circ N_1(x)} \subseteq \mathcal{T}_1^{\circ N_t(y)}$ (because $\neg_t\mathcal{T}_1^{N_t(x)} = \mathcal{T}_1^{\circ x} = \mathcal{T}_1^{\circ N_1(x)}$; $\neg_t\mathcal{T}_1^{y} = \mathcal{T}_1^{\circ N_t(y)}$). We conclude that $N_1(x) \preccurlyeq_1 N_t(y)$. For the other direction: $N_1(x) \preccurlyeq_1 N_t(y) \Rightarrow \mathcal{T}_1^{\circ N_1(x)} \subseteq \mathcal{T}_1^{\circ N_t(y)} \Rightarrow \neg_t\mathcal{T}_1^{\circ N_1(x)} \subseteq \neg_t\mathcal{T}_1^{\circ N_t(y)} \Rightarrow \mathcal{T}_1^{N_t(x)} \subseteq \mathcal{T}_1^{N_1(y)} \Rightarrow N_t(x) \leqslant_1 N_t(y)$.

4. The equations 5 and 6 are straightforward.

Now we need to make sure that for a maximal disjoint q-pair x, $N_t(x)$ and $N_1(x)$ are also maximal disjoint q-pairs. First of all lemma 3 ensures that operations N_t and N_1 transform theories to contertheories and vice versa. Disjointness is also straightforward: without loss of generality suppose for some $x \in \mathcal{F}$, $\neg_t\mathcal{T}_1^{x} \cap \neg_t\mathcal{T}_1^{\circ x} \neq \emptyset$. Thus there is some $A \in \neg_t\mathcal{T}_1^{x} \cap \neg_t\mathcal{T}_1^{\circ x}$. But then $\neg_t A \in \mathcal{T}_1^{x} \cap \mathcal{T}_1^{\circ x}$ or some $B \in \mathcal{T}_1^{x} \cap \mathcal{T}_1^{\circ x}$ such that $A = \neg_t B$, a contradiction. For the maximality suppose there is some $y \in \mathcal{F}$ such that $N_t(x) \leqslant y$. So, $\mathcal{T}_i^x \leqslant_i \mathcal{T}_i^y$, $\mathcal{T}_i^{\circ x} \preccurlyeq_i \mathcal{T}_i^{\circ y}$ $(i \in \{1, t\})$. First note that $\mathcal{T}_t^x = \mathcal{T}_t^{N_t(x)}$ and $\mathcal{T}_t^{\circ x} = \mathcal{T}_t^{\circ N_t(x)}$, so $\mathcal{T}_t^x = \mathcal{T}_t^y$ and $\mathcal{T}_t^{\circ x} = \mathcal{T}_t^{\circ y}$. Next we have $\neg_t \mathcal{T}_1^{\circ x} \subseteq \mathcal{T}_1^y$ and $\neg_t \mathcal{T}_1^x \subseteq \mathcal{T}_1^{\circ y}$. From the first conjunct we derive $\neg_t \neg_t \mathcal{T}_1^{\circ x} \subseteq \neg_t \mathcal{T}_1^y$, so $\mathcal{T}_1^{\circ x} \subseteq \neg_t \mathcal{T}_1^y$. But $N_t(y)$ is disjoint while x is maximal, so $\mathcal{T}_1^{\circ x} = \neg_t \mathcal{T}_1^x = \mathcal{T}_1^{\circ y}$. Likewise from the second conjunct we obtain $\mathcal{T}_1^x = \neg_t \mathcal{T}_1^{\circ x} = \mathcal{T}_1^y$. Thus $N_t(x)$ is equal to y and so is maximal. In the same way maximality of $N_1(x)$ can be shown. □

Let W be a carrier set of a canonical structure $\mathcal{F}$. Now we define the valuations of propositional variables and satisfaction relation in $\mathcal{F}$.

Definition 15. *Let $\mathcal{F} = (W, (\leqslant_i, \preccurlyeq_i)_{i \in \{1,2\}}, N_t, N_1)$ be a canonical frame. The functions $v_t^c\colon PV \times W \to \{t, u_t, f\}$ and $v_1^c\colon PV \times W \to \{1, u_1, 0\}$ are called* canonical valuation functions *iff conditions 1 and 2 of the definition 9 hold along with the following:*

$$
\begin{aligned}
&v_t^c(p,x) = t \Leftrightarrow p \in \mathcal{T}_1^x, && v_1^c(p,x) = 1 \Leftrightarrow p \in \mathcal{T}_t^x,\\
&v_t^c(p,x) = f \Leftrightarrow p \in \mathcal{T}_1^{\circ x}, && v_1^c(p,x) = 0 \Leftrightarrow p \in \mathcal{T}_f^{\circ x},\\
&v_t^c(p,x) = u_t \Leftrightarrow p \notin \mathcal{T}_1^x \text{ and } p \notin \mathcal{T}_1^{\circ x}, && v_1^c(p,x) = u_1 \Leftrightarrow p \notin \mathcal{T}_t^x \text{ and } p \notin \mathcal{T}_t^{\circ x}.
\end{aligned}
$$

Next definition reuses $\vDash$ symbol for a canonical versions of the satisfaction relation.

Definition 16. *Let $\mathcal{F}$ be a canonical structure. For $x \in \mathcal{F}$, $p \in PV$:*

$$
\begin{aligned}
&x \vDash_t p \Leftrightarrow v_t^c(p,x) = t \Leftrightarrow p \in \mathcal{T}_1^x, && x \vDash_1 p \Leftrightarrow v_1^c(p,x) = 1 \Leftrightarrow p \in \mathcal{T}_t^x, && (30)\\
&x \vDash_f p \Leftrightarrow v_t^c(p,x) = f \Leftrightarrow p \in \mathcal{T}_1^{\circ x}, && x \vDash_0 p \Leftrightarrow v_1^c(p,x) = 0 \Leftrightarrow p \in \mathcal{T}_t^{\circ x}, && (31)\\
&x \vDash_{u_t} p \Leftrightarrow x \nvDash_t p \text{ and } x \nvDash_f p && x \vDash_{u_1} p \Leftrightarrow x \nvDash_1 p \text{ and } x \nvDash_0 p. && (32)
\end{aligned}
$$

Now we establish the fact that satisfaction relations can be extended to the set of all formulas according to definition 11 in a way that the conditions 1 and 2 from definition 9 hold.

Lemma 15. *The satisfaction relation from definition 16 can be extended to the whole set of formulas in the language $\mathcal{L}_{\mathrm{LLSN}}$.*

Proof. 1. Let us assume $x \vDash_t A \vee B$ for an $x \in W$. Then, according to (14), we have for all y such that $x \leqslant_1 y$, $y \vDash_{-f} A$ or $y \vDash_{-f} B$. The latter disjunction can be conveniently rewritten as $\forall y(x \leqslant_1 y \Rightarrow (y \notin r_1(P_t^A)$ or $y \notin r_1(P_t^B)))$ which is enough to see that $x \in l_1(r_1(P_t^A))$ or $x \in l_1(r_1(P_t^B))$. By corollary 8 we then have

$x \in P_t^A$ or $x \in P_t^B$, hence $x \vDash_t A$ or $x \vDash_t B$. Now by IH we obtain $A \in \mathcal{T}_1^x$ or $B \in \mathcal{T}_1^x$. Each of disjuncts imply then $A \vee B \in \mathcal{T}_1^x$.

Conversely, suppose $A \vee B \in \mathcal{T}_1^x$. To prove $x \vDash_t A \vee B$ we have to show that for each y such that $x \leqslant_1 y$, $y \vDash_{-f} A$ or $y \vDash_{-f} B$. Assume it is not the case. Thus there is a z, $x \leqslant_1 z$, and both $z \vDash_f A$ and $z \vDash_f B$ that is, by IH we have $A \in \mathcal{T}_1^{\circ z}$, $B \in \mathcal{T}_1^{\circ z}$. The latter assumption means $A \vee B \in \mathcal{T}_1^{\circ z}$ because $\mathcal{T}_1^{\circ z}$ is a counter-theory. At the same time we have $A \vee B \in \mathcal{T}_1^z$, because $x \leqslant_1 z$, that is $\mathcal{T}_1^x \subseteq \mathcal{T}_1^z$. But this violates the disjointness of $\mathcal{T}_1^z$ and $\mathcal{T}_1^{\circ z}$.

2. Suppose $x \vDash_t \neg_t A$. We have to show $\neg_t A \in \mathcal{T}_1^x$. First we have $N_t(x) \vDash_f A$ and hence, by IH, $A \in \mathcal{T}_1^{\circ N_t(x)}$. Next, by lemma 13, we obtain the result. Indeed, $\neg_t A \in \mathcal{T}_1^x$ or $A = \neg_t A'$ for some $A' \in \mathcal{T}_1^x$ and $A' \neq \neg_t B$, then $\neg_t \neg_t A' = \neg_t A \in \mathcal{T}_1^x$.

Conversely assume $\neg_t A \in \mathcal{T}_1^x$. Then by lemma 13 it follows that $A \in \mathcal{T}_1^{\circ N_t(x)}$ and by IH $N_t(x) \vDash_f A$. The last statement gives $x \vDash_t \neg_t A$ from definition 16.

3. Suppose $x \vDash_t \neg_1 A$, then $x \vDash_t A$ and, by IH, $A \in \mathcal{T}_1^x$. Since $\mathcal{T}_1^x$ is $\neg_1$-closed, $\neg_1 A \in \mathcal{T}_1^x$. For the other direction suppose $\neg_1 A \in \mathcal{T}_1^x$, then $A \in \mathcal{T}_1^x$ by $\neg_1$-closure of $\mathcal{T}_1^x$, and then IH gives $x \vDash_t A$. Finally, (18) gives $x \vDash_t \neg_1 A$.

4. Now suppose for an $x \in W$, $x \vDash_f \neg_t A$. Then, by (20), for all y, $x \preccurlyeq_1 y$, we have $N_t(y) \notin P_f^A$, that is $N_t(y) \notin r_1(P_t^A)$. Eliminating universal quantifier we infer $x \preccurlyeq_1 N_t(y) \Rightarrow N_t(N_t(y)) \notin r_1(P_t^A)$. Now using the properties 1, (4) and (6) of the star-frames we get $N_t(x) \leqslant_1 y \Rightarrow y \notin r_t(P_t^A)$, so $\forall y(N_t(x) \leqslant_1 y \Rightarrow y \notin r_t(P_t^A))$ which is $N_t(x) \in l_t(r_t(P_t^A)) = P_t^A$. IH then gives $A \in \mathcal{T}_1^{N_t(x)}$. Lastly, $\neg_t A \in \mathcal{T}_1^{\circ x}$ by lemma 13.

For the converse let $\neg_t A \in \mathcal{T}_1^{\circ x}$. Then, by lemma 13, $A \in \mathcal{T}_1^{\circ N_t(x)}$. IH implies $N_t(x) \vDash_f A$, so applying lemma 10 we conclude that $x \vDash_f \neg_t A$. □

Before we formulate and prove completeness theorem it should be noted that the consequence relation remains the same as in the definition 12.

Theorem 16 (Completeness). *For all formulas A, B: if $A \vDash B$ then $A \vdash B$.*

Proof. Suppose $A \nvdash B$. Then, according to lemma 12 we can find a q-pair x such that one of its components provides a separation. Suppose, first, that $A \in \mathcal{T}_1^x$ and $B \notin \mathcal{T}_1^x$. So, by lemma 15, $x \vDash_t A$ and $x \vDash_{-t} B$. Evidently then $A \nvDash B$ and the result follows by contraposition. For a different situation assume $B \in \mathcal{T}_t^{\circ x}$, $A \notin \mathcal{T}_t^{\circ x}$. Then $x \vDash_0 B$ and $x \vDash_{-0} A$. Again according to the definition of consequence relation $A \nvDash B$. □

7 Concluding remarks

In this paper we have used the most straightforward approach to study the uncertainty phenomenon in the context of generalized truth values, namely to incorporate an intermediate value between truth and false in a basic set of values and render it as an uncertainty of a specific kind. It seems that a logic assuming uncertainty could be fruitfully studied from the probabilistic perspective. We would like to mention here paper [8] (see also references therein) which, in particular, suggests a probabilistic view on the systems of Dunn-Belnap's truth values which obtains new structure and reveals the definitions of probabilistic consequence relations.

References

[1] Gerard Allwein and J. Michael Dunn. Kripke models for linear logic. *J. Symbolic Logic*, 58(2):514–545, 06 1993.

[2] Nuel D. Belnap. *A Useful Four-Valued Logic*, pages 5–37. Springer Netherlands, Dordrecht, 1977.

[3] Katalin Bimbó. Functorial duality for ortholattices and de morgan lattices. *Logica Universalis*, 1(2):311–333, Oct 2007.

[4] Katalin Bimbó and J. Michael Dunn. Four-valued logic. *Notre Dame J. Formal Logic*, 42(3):171–192, 07 2001.

[5] J. M. Dunn. Partiality and its dual. *Studia Logica*, 65:5–40, 2000.

[6] J. Michael Dunn. Intuitive semantics for first-degree entailments and 'coupled trees'. *Philosophical Studies*, 29(3):149–168, Mar 1976.

[7] J. Michael Dunn. Star and perp: Two treatments of negation. *Philosophical Perspectives*, 7:331–357, 1993.

[8] J. Michael Dunn. Contradictory information: Too much of a good thing. *Journal of Philosophical Logic*, 39(4):425–452, Aug 2010.

[9] Michael Dunn and Gary Hardegree. *Algebraic Methods in Philosophical Logic*. Number 41 in Oxford Logic Guides. OUP, 2001.

[10] Wojciech Dzik, Ewa Orlowska, and Clint van Alten. *Relational Representation Theorems for General Lattices with Negations*, pages 162–176. Springer Berlin Heidelberg, Berlin, Heidelberg, 2006.

[11] Oleg Grigoriev. Bipartite truth and semi-negations. In *Proceedings of 7-th International conference 'Smirnov readings in logic', June 22–24, Moscow*, 2011.

[12] Oleg Grigoriev. A tableau calculus for a logic of two-component truth. In *Proceedings of 8-th International conference 'Smirnov readings in logic', June 19–21, Moscow*, 2013.

[13] Chrysafis Hartonas. *Reasoning with Incomplete Information in Generalized Galois Logics Without Distribution: The Case of Negation and Modal Operators*, pages 279–312. Springer International Publishing, Cham, 2016.

[14] Chrysafis Hartonas. Order-dual relational semantics for non-distributive propositional logics. *Logic Journal of the IGPL*, 25(2):145–182, 2017.

[15] Y. Shramko, J. M. Dunn, and T. Takenaka. The trilattice of constructive truth values. *Journal of Logic and Computation*, 11(6):761–788, Dec 2001.

[16] Alasdair Urquhart. A topological representation theory for lattices. *algebra universalis*, 8(1):45–58, Dec 1978.

[17] Dmitry Zaitsev. Logics of generalized classical truth values. In P. Arazim and M. Peliš, editors, *The Logica Yearbook 2014*, pages 331–341. College Publications London, 2015.

[18] Dmitry Zaitsev and Oleg Grigoriev. Two kinds of truth – one logic. In *Logical Investigations*, pages 121–139. Moscow, 2011.

[19] Dmitry Zaitsev and Yaroslav Shramko. Bi-facial truth: a case for generalized truth values. *Studia Logica*, 101(6):1299–1318, Dec 2013.

Received 28 February 2018

Relations between Assumption-Based Approaches in Non-Monotonic Logic and Formal Argumentation: From Structured Argumentation to Adaptive Logics

Jesse Heyninck
Ruhr-University Bochum
jesse.heyninck@rub.de

Abstract

This paper investigates the relation between two prominent frameworks for the formal explication of defeasible reasoning: assumption-based argumentation and adaptive logics. Assumption-based argumentation is a formalism that allows to make inferences from a strict rule base and a set of defeasible assumptions. Adaptive logics are a paradigmatic case of preferential reasoning, based on the idea that for defeasible reasoning, it is often sufficient to consider only a subset of the models of a premise set. In this paper, I study a translation from assumption-based argumentation into adaptive logic in order to explicate the exact relationship between these two approaches.

1 Introduction

This paper makes a contribution to the unification of formal models of defeasible reasoning. In particular, I will investigate the relation between two prominent frameworks for the formal explication of defeasible reasoning: assumption-based argumentation (in short, ABA) and adaptive logics. The reason for considering these two systems is that they are members of two different families of models for defeasible reasoning. Consequently, translations between these two systems will lead to insights into the relations between the two families they belong to. *Assumption-based argumentation* is an instance of what can be called the rule-based approach to defeasible reasoning: in this approach a set of (domain specific) Horn-rules is the main engine behind inferences. Other members of this family include default logic [38], logic programming [13] and other forms of structured argumentation [8] such as

ASPIC [1, 37, 31] and DeLP [24]. Assumption-based argumentation can be distinguished within this class of formal models by the fact that it interprets all the rules as strict and furthermore assumes a set of *defeasible assumptions* that are accepted until and unless a feasible counter-argument occurs. *Adaptive logics* (also studied as *formula-preferential systems* [2]), on the other hand, can be seen as a paradigmatic subclass of *preferential reasoning* as studied by Kraus, Lehman and Magidor [32, 41]. Preferential reasoning can be studied in a model-theoretic or semantic way and revolves around the idea that for defeasible reasoning, it is sufficient to look at a subset of all the models of a premise set. Adaptive logics base the selection of this subset on the satisfaction of a set of defeasible assumptions, also called *normality assumptions*. Furthermore, adaptive logics come equiped with a *dynamic proof theory*. So even though they are part of different families of *representational formats*, they are both based on the idea of delineating a set of defeasible assumptions that are assumed to be true as much as possible. Adaptive logics give formal substance to the phrase "as much as possible" by comparing models of the premise set with respect to the defeasible assumptions they satisfy, while ABA uses concepts from formal argumentation to formally substantiate this idea. This paper can thus be seen as an exploration in the similarities and differences between the way defeasible assumptions are handled in these two formalisms.

Outline of the paper: In Section 2 I present assumption-based argumentation and in Section 3 adaptive logics are introduced: first the standard format of adaptive logics is explained and in Section 3.1 I expose sequential combinations of adaptive logics. In Section 4, the main section of this paper, the translation of assumption-based argumentation into adaptive logics is presented. First, I set out criteria for such a translation to be adequate. Then, I motivate the translation by presenting some limitations a previous translation in [30] suffered from. Thereafter I present the translation and give adequacy results. Finally, related work is discussed and directions for future work are set out.

2 Assumption-Based Argumentation

ABA, thoroughly described in [12], is a formal model that allows one to use a set of plausible assumptions "to extend a given theory" [12, p.70] unless and until there are good arguments for not using such assumptions. In more detail, sets of defeasible assumptions can be in conflict with one another. To represent and resolve such conflicts, a formal argumentation framework is constructed on the basis of the strict rules and defeasible assumptions. In particular, argumentative attacks represent conflicts between sets of assumptions. Argumentation semantics from formal argumentation theory are then used to select sets of assumptions that can be upheld in an argumentative dialogue based on the argumentation framework under consideration. The interested reader can find helpful tutorials on ABA in [48, 18].

Inferences are implemented in ABA by means of *a deductive system* [12] consisting of a language and rules formulated over this language:

Definition 1 (Deductive System). *A* deductive system *is a pair* $(\mathcal{L}, \mathcal{R})$ *such that:*

- $\mathcal{L}$ *is a countable set of sentences;*
- $\mathcal{R}$ *is a set of inference rules of the form* $A_1, \ldots, A_n \to A$ *and* $\to A$*, where* $A, A_1 \ldots, A_n \in \mathcal{L}$*.*

Definition 2 ($\mathcal{R}$-deduction). *Where* $m > 0$*, an* $\mathcal{R}$-deduction *from* $\Gamma \subseteq \mathcal{L}$ *is a sequence* $B_1, \ldots, B_m$ *such that for all* $i = 1, \ldots, m$*:* $B_i \in \Gamma$ *or there exists an* $A_1, \ldots, A_n \to B_i \in \mathcal{R}$ *such that* $A_1, \ldots, A_n \in \{B_1, \ldots, B_{i-1}\}$*. I will write* $\Gamma \vdash_{\mathcal{R}} A$ *if there is an* $\mathcal{R}$*-deduction from* Γ *whose last element is* A*.*

I now introduce defeasible assumptions and a contrariness operator to express argumentative attacks. Given a rule system, an *assumption-based framework* [12] is defined as follows:

Definition 3 (Assumption-based framework). *An* assumption-based framework *is a tuple* $\mathtt{ABF} = ((\mathcal{L}, \mathcal{R}), \Gamma, \Lambda, \overline{})$ *where:*

- $(\mathcal{L}, \mathcal{R})$ *is a deductive system;*
- $\Gamma \subseteq \mathcal{L}$ *is a set of strict premisses;*
- $\emptyset \neq \Lambda \subseteq \mathcal{L}$ *is a finite set of candidate assumptions;*[1]
- $\overline{} : \Lambda \to \mathcal{L} \setminus \Lambda$ *is a contrariness operator.*[2]

I will restrict attention to so-called *flat* ABFs, i.e. assumption-based frameworks that contain no rules $A_1, \ldots, A_n \to A$ such that $A \in \Lambda$ (this restriction is also made in e.g. [12, 48]). Furthermore, restricting the image of $\overline{}$ to $\mathcal{L} \setminus \Lambda$ means

[1]That Λ is finite will be crucial in order to be able to employ *sequential combinations of adaptive logics* (which has semantics that are sound and complete for finite sets of abnormalities, see Section 3). The generalization of this translation to ABFs with infinite sets of assumptions is left for future work.

[2]Note that $\overline{}$ does *not* denote the set theoretic complement. I will sometimes abuse notation and assume that $\overline{A} \in \mathcal{L}$ to avoid clutter. Furthermore, in this paper, I will assume that $\overline{A} \in \mathcal{L}$ for any $A \in \Lambda$. This assumption, made for simplicity, does not result in any loss of generality (since I assume normal ABFs) and will allow to avoid clutter.

that attention is restricted so called *normal* ABFs [16]. It was proven that any flat non-normal ABF can be transformed into a flat normal ABF which is equivalent to the original framework. Consequently, this assumption does not result in any loss of generality.

In most structured accounts of argumentation, attacks are defined between arguments which are deductions in a given deductive or defeasible system (e.g., in ASPIC [37] or defeasible logic programming [24]) or sequents $\Gamma \vdash_{\mathbf{L}} A$ where $\mathbf{L}$ is an underlying core logic ([44, 9]). In contrast, ABA operates at a higher level of abstraction, since attacks are defined directly on the level of sets of assumptions instead of on the level of $\mathcal{R}$-deductions [12].[3] ABA can thus be viewed as operating on the level of equivalence classes consisting of arguments generated using the same assumptions.

Definition 4 (Attacks). *Given an* ABF $= ((\mathcal{L}, \mathcal{R}), \Gamma, \Lambda, \overline{})$*:*

- *a set of assumptions* $\Delta \subseteq \Lambda$ attacks *an assumption* $A \in \Lambda$ *iff* $\Gamma \cup \Delta \vdash_{\mathcal{R}} \overline{A}$.
- *a set of assumptions* $\Delta \subseteq \Lambda$ attacks *a set of assumptions* $\Delta' \subseteq \Lambda$ *iff* $\Gamma \cup \Delta \vdash_{\mathcal{R}} \overline{A}$ *for some* $A \in \Delta'$.

On the basis of argumentative attacks, *assumption labellings* [40] stipulate which assumptions are jointly acceptable. An assumption labelling assigns every assumption a label determining the status of the assumption. In more formal details, an assumption labelling $\mathbb{L}$ is a function that assigns to every assumption one of the labels `in`, `undec` or `out` (standing for accepted, undecided and rejected respectively). The more interesting assumption labellings are, of course, those that take into account rational criteria of acceptability, such as the requirement that a given set of assumptions should not attack itself (*conflict-freeness*), or it should be able to defend itself against attacks by other sets of assumptions (*admissibility*).

Definition 5 (Assumption Labellings). *Given an* ABF $= ((\mathcal{L}, \mathcal{R}), \Gamma, \Lambda, \overline{})$, *an assumption labelling is a function* $\mathbb{L} : \Lambda \to \{\texttt{in}, \texttt{undec}, \texttt{out}\}$.

Where $\texttt{X} \in \{\texttt{in}, \texttt{undec}, \texttt{out}\}$, I will write $\mathbb{L}(\Delta) = \texttt{X}$ iff $\mathbb{L}(A) = \texttt{X}$ for every $A \in \Delta$ and let $\texttt{X}(\mathbb{L}) := \{A \in \Lambda \mid \mathbb{L}(A) = \texttt{X}\}$.

Definition 6 (Conflict-free Labellings). *Given an* ABF $= ((\mathcal{L}, \mathcal{R}), \Gamma, \Lambda, \overline{})$, *an assumption labelling* $\mathbb{L}$ *is* conflict-free *iff: for no* $\Delta \subseteq \Lambda$ *that attacks* A, $\mathbb{L}(\Delta) = \texttt{in}$.[4]

[3] There are some formulations of ABA that define attacks on the level of individual arguments. However, since attacks are only possible 'on' assumptions, these formulations are equivalent (cf. also [48]).

[4] To the best of my knowledge, conflict-free (and naive assumption labellings, see Definition

For a labelling to be admissible, an assumption can be labelled `in` only if it can be defended from every attacker: for every attacking set Δ, there is at least one member of Δ that is labelled `out`. Furthermore, in an admissible labelling, an assumption A can be labelled `out` only if there is a set of accepted assumptions that attacks A. Finally, an admissible labelling should label every argument that is attacked by a set of accepted assumptions `out`, which means that assumptions can be labelled `undec` only if there is no attacking set that is labelled `in`. Notice that an admissible labelling is conflict-free, but that conflict-free labellings are not necessarily admissible.

Definition 7 (Admissible Labellings). *Given an* ABF $= ((\mathcal{L}, \mathcal{R}), \Gamma, \Lambda, \bar{\ })$, *an assumption labelling* $\mathbb{L}$ *is* admissible *iff:*

- *if* $\mathbb{L}(A) =$ `in` *then for each* $\Delta \subseteq \Lambda$ *that attacks* A*, there is a* $B \in \Delta$ *s.t.* $\mathbb{L}(B) =$ `out`.
- *if* $\mathbb{L}(A) =$ `out` *then there is a* $\Delta \subseteq \Lambda$ *that attacks* A *s.t.* $\mathbb{L}(\Delta) =$ `in`.
- *if* $\mathbb{L}(A) =$ `undec` *then for each* $\Delta \subseteq \Lambda$ *that attacks* A*, there is a* $B \in \Delta$ *s.t.* $\mathbb{L}(B) \neq$ `in`.

Complete labellings are admissible labellings that assign `in` to every member they defend. Consequently, in contrast to admissible labellings, defended assumptions are not allowed to be labelled `undec`, i.e. any assumption labelled `undec` has to be attacked by a set that contains no member which is labelled `out`.

Definition 8 (Complete Labellings). *Given an* ABF $= ((\mathcal{L}, \mathcal{R}), \Gamma, \Lambda, \bar{\ })$, *an assumption labelling* $\mathbb{L}$ *is* complete *iff it is admissible and if* $\mathbb{L}(A) =$ `undec` *for some* $A \in \Lambda$ *then for some* $\Delta \subseteq \Lambda$ *that attacks* A*,* $\mathbb{L}(\Delta) \neq$ `out`.

Based on conflict-free and complete labellings, other labellings can be defined by requiring maximality or minimality among either `in`($\mathbb{L}$) or `undec`($\mathbb{L}$).

Definition 9 (Naive, Grounded, Preferred, Semi-Stable Assumption Labellings). *Given an* ABF $= ((\mathcal{L}, \mathcal{R}), \Gamma, \Lambda, \bar{\ })$, *an assumption labelling* $\mathbb{L}$ *is:*

- naive *iff* `in`($\mathbb{L}$) *is maximal (with respect to set inclusion) among all conflict-free labellings of* ABF.

9) have not been presented before. In Appendix A I show soundness and completeness for these labellings w.r.t. the more conventional extension-based semantics as orginally introduced by [12] (see Appendix A for more details on these extension-based semantics).

- preferred *iff* $\mathtt{in}(\mathbb{L})$ *is maximal (with respect to set inclusion) among all complete labellings of* ABF.
- grounded *iff* $\mathtt{in}(\mathbb{L})$ *is minimal (with respect to set inclusion) among all complete labellings of* ABF.
- semi-stable *iff* $\mathtt{undec}(\mathbb{L})$ *is minimal (with respect to set inclusion) among all complete labellings of* ABF.

I will use $\mathcal{A}(\mathtt{ABF})$, $\mathcal{C}(\mathtt{ABF})$, $\mathcal{N}(\mathtt{ABF})$, $\mathcal{P}(\mathtt{ABF})$, $\mathcal{G}(\mathtt{ABF})$ and $\mathcal{S}(\mathtt{ABF})$ to denote the set of all admissible, complete, naive, preferred, grounded respectively semi-stable labellings of ABF.[5] Where ABF is clear and unambiguous from the context I will denote these sets simply by $\mathcal{A}$, $\mathcal{C}$, $\mathcal{N}$, $\mathcal{P}$, $\mathcal{G}$ and $\mathcal{S}$.

The following characterizations of preferred and semi-stable labellings will allow several simplifications in the translations provided below:

Theorem 1 ([40, Prop.7]). *Given an assumption-based framework* ABF, *an assumption labelling* $\mathbb{L}$ *is:*

- *preferred iff* $\mathtt{in}(\mathbb{L})$ *is maximal (with respect to set inclusion) among all admissible labellings of* ABF.
- semi-stable *iff* $\mathtt{undec}(\mathbb{L})$ *is minimal (with respect to set inclusion) among all admissible labellings of* ABF.

The concepts defined above are illustrated by two examples.

Example 1. *Let* $\mathtt{ABF} = ((\mathcal{L}, \mathcal{R}), \Gamma, \Lambda, \overline{})$ *with:*

- $\Lambda = \{p, q\}$
- $\mathcal{L} = \Lambda \cup \{\overline{q}\}$
- $\mathcal{R} = \{p \to \overline{q}\}$
- $\Gamma = \emptyset$.

The following are all the labellings of ABF*:*

[5]Another popular labelling is the *stable* labelling. In Subsection 4.5 I explain why this assumption labelling is not considered for the translation in this paper.

i	$\mathbb{L}_i(p)$	$\mathbb{L}_i(q)$	i	$\mathbb{L}_i(p)$	$\mathbb{L}_i(q)$
1	undec	undec	6	out	in
2	undec	out	7	in	undec
3	out	undec	8	in	out
4	out	out	9	in	in
5	undec	in			

Every labelling except $\mathbb{L}_9$ is conflict-free. $\mathbb{L}_5, \mathbb{L}_6, \mathbb{L}_7$ and $\mathbb{L}_8$ are naive. Only $\mathbb{L}_8$ is admissible, complete, preferred, semi-stable and grounded.

Example 2. *Let* $\texttt{ABF} = ((\mathcal{L}, \mathcal{R}), \Gamma, \Lambda, \overline{\ \ })$ *with:*

- $\Lambda = \{p, q, r, s, t\}$
- $\mathcal{L} = \Lambda \cup \{\overline{A} \mid A \in \Lambda\} \cup \{u\}$
- $\mathcal{R} = \begin{Bmatrix} p \rightarrow \overline{q}; & q \rightarrow \overline{p}; & q \rightarrow \overline{r}; & r \rightarrow \overline{r}; \\ s \rightarrow \overline{t}; & p \rightarrow u; & q \rightarrow u & \end{Bmatrix}$
- $\Gamma = \emptyset$.

The graph in Figure 1 conveys the attacks between assumptions.

The following are all the admissible labellings of ABF*:*

i	$\mathbb{L}_i(p)$	$\mathbb{L}_i(q)$	$\mathbb{L}_i(r)$	$\mathbb{L}_i(s)$	$\mathbb{L}_i(t)$
1	undec	undec	undec	undec	undec
2	undec	undec	undec	in	out
3	out	in	out	undec	undec
4	out	in	out	in	out
5	in	out	undec	undec	undec
6	in	out	undec	in	out

Among these labellings, $\mathbb{L}_2$, $\mathbb{L}_4$ and $\mathbb{L}_6$ are complete. $\mathbb{L}_4$ and $\mathbb{L}_6$ are preferred and $\mathbb{L}_4$ is semi-stable whereas $\mathbb{L}_2$ is grounded.

Based on the various kinds of assumption-based labellings defined above, several consequence relations for ABA can be defined:

Definition 10 (ABA-Consequence Relations). *Given an* $\texttt{ABF} = ((\mathcal{L}, \mathcal{R}), \Gamma, \Lambda, \overline{\ \ })$ *and* $\mathsf{Sem} \in \{\mathcal{N}, \mathcal{P}, \mathcal{G}, \mathcal{S}\}$*:*

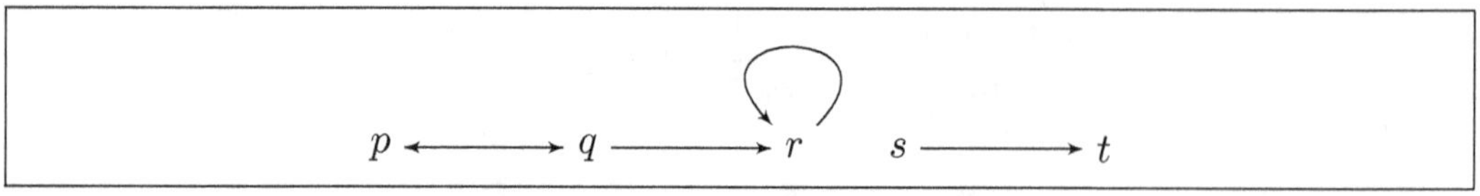

Figure 1: Attack diagram for Example 2. An arrow from A to B means A attacks B. The attack diagram is restricted to singletons and set brackets are omitted to avoid clutter.

- $\mathtt{ABF} \mid\!\sim^{\cup}_{\mathsf{Sem}} A$ *iff* $\mathtt{in}(\mathbb{L}) \vdash_{\mathcal{R}} A$ *for some* $\mathbb{L} \in \mathsf{Sem}(\mathtt{ABF})$;[6]
- $\mathtt{ABF} \mid\!\sim^{\cap}_{\mathsf{Sem}} A$ *iff* $\mathtt{in}(\mathbb{L}) \vdash_{\mathcal{R}} A$ *for every* $\mathbb{L} \in \mathsf{Sem}(\mathtt{ABF})$;
- $\mathtt{ABF} \mid\!\sim^{\Cap}_{\mathsf{Sem}} A$ *iff* $\bigcap_{\mathbb{L}.\in \mathsf{Sem}(\mathtt{ABF})} \mathtt{in}(\mathbb{L}) \vdash_{\mathcal{R}} A$.

Example 3. *[Example 2 continued] In Example 2, there are two preferred labellings,* $\mathbb{L}_4$ *and* $\mathbb{L}_6$. *Since* $\mathbb{L}_6(p) = \mathtt{in}$ *and* $\mathbb{L}_4(q) = \mathtt{in}$, $\mathtt{ABF} \mid\!\sim^{\cup}_{\mathcal{P}} p$ *and* $\mathtt{ABF} \mid\!\sim^{\cup}_{\mathcal{P}} q$ *whereas* $\mathtt{ABF} \not\mid\!\sim^{\cap}_{\mathcal{P}} p$ *or* $\mathtt{ABF} \not\mid\!\sim^{\cap}_{\mathcal{P}} q$ *(since* $\mathbb{L}_6(q) = \mathtt{out}$ *and* $\mathbb{L}_4(p) = \mathtt{out}$*). Furthermore, observe that* $\mathtt{ABF} \mid\!\sim^{\cap}_{\mathcal{P}} u$ *whereas* $\mathtt{ABF} \not\mid\!\sim^{\Cap}_{\mathcal{P}} u$.

3 Preferential Semantics

Adaptive logics [6, 5, 42] are a general framework for the formal explication of defeasible reasoning. They have been applied to a multitude of defeasible reasoning forms, such as non-monotonic forms of reasoning with inconsistent information, causal discovery, inductive generalisations, abductive hypothesis generation and normative reasoning (see [42, p.86]).

The semantics of ALs is based on the idea that to draw defeasible inferences from a premise set Γ, only a subset of the models of Γ, namely the *most normal* models of Γ, should be considered. In ALs, normality is defined with respect to a set of formulas called *abnormalities*: a model M is more normal than another model M' if the abnormalities M satisfies are a subset of the abnormalities satisfied by M'. More specifically, fix a compact Tarksi logic[7] $\mathbf{L}$ (the *core* or *lower limit logic*) in a formal

[6]Since for flat ABFs, there exists a unique grounded labelling (see [40, Theorem 5] and [18, Theorem 2.20]), all of the three consequence relations defined above coincide for $\mathsf{Sem} = \mathcal{G}$, i.e. $\mathtt{ABF} \mid\!\sim^{\cup}_{\mathcal{G}} A = \mathtt{ABF} \mid\!\sim^{\cap}_{\mathcal{G}} A = \mathtt{ABF} \mid\!\sim^{\Cap}_{\mathcal{G}} A$.

[7]Recall: a derivability relation $\vdash_{\mathbf{L}}$ characterizes a Tarski logic $\mathbf{L}$ iff $\vdash_{\mathbf{L}}$ is reflexive, transitive and monotonic. Furthermore, I say that $\mathbf{L}$ is compact if $\Gamma \vdash_{\mathbf{L}} A$ implies that $\Gamma' \vdash_{\mathbf{L}} A$ for some finite $\Gamma' \subseteq \Gamma$.

language $\mathcal{L}$ and with the derivability relation $\vdash_{\mathbf{L}}$ and fix a set of *abnormalities*[8] $\Omega \subseteq \mathcal{L}$. I will assume that the core logic **L** comes with an adequate model-theoretic semantics and an associated semantic consequence relation $\Vdash_{\mathbf{L}}$.[9] I write $\mathcal{M}(\Gamma)$ for the set of all models of a premise set Γ. Furthermore, where $M \in \mathcal{M}(\Gamma)$, $\Omega(M) = \{A \in \Omega \mid M \models A\}$. A model $M \in \mathcal{M}(\Gamma)$ is *minimally abnormal (w.r.t. Ω)* iff there is no $M' \in \mathcal{M}(\Gamma)$ for which $\Omega(M') \subset \Omega(M)$. I will denote the set of all models of Γ that are minimally abnormal (w.r.t. Ω) by $\min_{\Omega}(\Gamma)$. Based on this set of minimally abnormal models, several consequence relations can be defined (I refer to [42] for more detailed explanations).[10] The consequence relation $\Vdash_{\cap}^{\Omega,\mathbf{L}}$ is perhaps the most straightfoward consequence relation to explain. The basic idea is to say that a formula A is a consequence iff A is validated in every minimally abnormal model. A second consequence relation $\Vdash_{\Cap}^{\Omega,\mathbf{L}}$ is a bit more cautious: instead of just looking at the minimally abnormal models, it looks at all the so-called *reliable models* (which are in fact a superset of the minimally abnormal models). In more detail, an abnormality is unreliable if it is validated by at least one minimally abnormal model. Accordingly, reliable models are those models that validate no reliable abnormality, which means that a model is reliable iff it verifies only those abnormalities that are verified by some minimally abnormal model. More formally, a model M is reliable if $\Omega(M) \subseteq \bigcup_{M' \in \min_{\Omega}(\Gamma)} \Omega(M')$. The third consequence relation, $\Vdash_{\cup}^{\Omega,\mathbf{L}}$, is a so-called credulous consequence relation: instead of requiring that A is derivable from every minimally abnormal model, it suffices that A is derivable from every model in a set of minimally abnormal models that validate the same abnormalities. In other words, this consequence relation looks at equivalence classes of minimally abnormal models that validate the same set of abnormalities.

Definition 11 (AL-Consequence Relations)**.** *Where* **L** *is a compact Tarski logic and* $\Gamma \cup \Omega \cup \{A\} \subseteq \mathcal{L}$*:*

- $\Gamma \Vdash_{\cap}^{\Omega,\mathbf{L}} A$ *iff* $M \models A$ *for every* $M \in \min_{\Omega}(\Gamma)$.
- $\Gamma \Vdash_{\Cap}^{\Omega,\mathbf{L}} A$ *iff* $M \models A$ *for every* $M \in \mathcal{M}(\Gamma)$ *such that every member of* $\Omega(M)$ *is verified in some minimally abnormal model* $M' \in \mathcal{M}(\Gamma)$.

[8]It is usual to assume that abnormalities are characterized by some logical form [6, 42, 5], but it is shown in [2] that this assumption can be given up by showing stopperedness (also know as strong reassurance) for adaptive logics where abnormalities are made up of any set of formulas. In [51] one finds a similar generalization.

[9]As usual I will write $M \models A$ iff $v_M(A) = 1$ where v_M is a valuation function associated with the model M.

[10]In the orthodox nomenclature of adaptive logics, the variations in consequence relations are said to be determined by *strategies*, a term which stems from the dynamic proof theory of adaptive logics. $\Vdash_{\cap}^{\Omega,\mathbf{L}}$ is called the *minimally abnormality strategy*, $\Vdash_{\Cap}^{\Omega,\mathbf{L}}$ the *reliability strategy* and $\Vdash_{\cup}^{\Omega,\mathbf{L}}$ the *normal selections strategy* [6, 42, 5].

- $\Gamma \Vdash_{\Cup}^{\Omega,\mathbf{L}} A$ *iff there is a* $M \in \min_{\Omega}(\Gamma)$ *such that for all* $M' \in \mathcal{M}(\Gamma)$ *for which* $\Omega(M) = \Omega(M')$, $M' \models A$.

Adaptive logical consequences are defined in the usual way (for $\dagger \in \{\Cap, \Cap\!\!\!\!\cap, \Cup\}$): $Cn_{\dagger}^{\Omega,\mathbf{L}}(\Gamma) = \{A \mid \Gamma \Vdash_{\dagger}^{\Omega,\mathbf{L}} A\}$. I will also say that A is a consequence of the premise set Γ under the adaptive logic $\mathbf{L}_{\dagger}^{\Omega}$ if $A \in Cn_{\dagger}^{\Omega,\mathbf{L}}(\Gamma)$. It is clear that the semantics for ALs is a proper but rich subclass of the well known preferential semantics as defined in [32, 41].

Example 4. *I now give an example of a very simple adaptive logic. Suppose that* $\Omega = \{\neg p, \neg q, \neg s\}$, $\Gamma = \{p \supset \neg q\}$, $\mathcal{L}$ *is the closure of* $\{p, q, s\}$ *under the connectives* $\neg, \vee, \wedge$ *and the core logic is classical propositional logic. The following table characterizes the relevant parts of the* **CL**-*models of* Γ*:*

i	$v_{M_i}(p)$	$v_{M_i}(q)$	$v_{M_i}(s)$	$\Omega(M_i)$
1	0	0	0	$\neg p, \neg q, \neg s$
2	0	0	1	$\neg p, \neg q$
3	0	1	0	$\neg p, \neg s$
4	0	1	1	$\neg p$
5	1	0	0	$\neg q, \neg s$
6	1	0	1	$\neg q$

The set of minimally abnormal models of Γ *is* $\min_{\Omega}(\Gamma) = \{M_4, M_6\}$. *Consequently,* $\Gamma \Vdash_{\Cap}^{\Omega,\mathbf{CL}} p \vee q$. *Since* $M_4 \models \neg p$ *and* $M_6 \models \neg q$, *all abnormalities verified by* M_2 *are verified by some minimally abnormal model. Since* $M_2 \not\models p \vee q$, $\Gamma \nVdash_{\Cap\!\!\!\!\cap}^{\Omega,\mathbf{CL}} p \vee q$. *Note that for example,* $\Gamma \Vdash_{\Cap}^{\Omega,\mathbf{CL}} p \vee q$. *Finally, observe that* $\Gamma \Vdash_{\Cup}^{\Omega,\mathbf{CL}} p$ *and* $\Gamma \Vdash_{\Cup}^{\Omega,\mathbf{CL}} q$ *(but* $\Gamma \nVdash_{\Cup}^{\Omega,\mathbf{CL}} p \wedge q$*).*

3.1 Sequential Combinations of Adaptive Logics

For the translations presented in this paper, it will be necessary to combine various adaptive logics (for example to capture consequence relations based on preferred assumption labellings, it will prove necessary to first select all the models that correspond to an admissible labelling and then select those models that validate as many assumptions as possible). This will be done by combining adaptive logics in a *sequential* way.[11] A sequential combination of some adaptive logics $\mathbf{L1}_{\dagger_1}^{\Omega_1}, \ldots, \mathbf{Ln}_{\dagger_n}^{\Omega_n}$ applied to a premise set Γ amounts to first applying $\mathbf{L1}_{\dagger_1}^{\Omega_1}$ to Γ, then applying $\mathbf{L2}_{\dagger_2}^{\Omega_2}$

[11] In [52] one finds a comparative study of sequential combinations and various other ways to combine adaptive logics.

to the $\mathbf{L^1}^{\Omega_1}_{\dagger_1}$-consequence set of Γ, ... and finally applying $\mathbf{L^n}^{\Omega_n}_{\dagger_n}$ to the $\mathbf{L^{n-1}}^{\Omega_{n-1}}_{\dagger_{n-1}}$-consequence set of ... of the $\mathbf{AL}_1$-consequences of Γ.

Definition 12. *Where $\mathbf{L}_1, \ldots, \mathbf{L}_n$ are compact Tarski logics in the respective languages $\mathcal{L}_1, \ldots, \mathcal{L}_n$, $\Gamma \cup \Omega_1 \subseteq \mathcal{L}_1$, $\Omega_i \subseteq \mathcal{L}_i$ for every $1 < i \leqslant n$ and $\dagger_i \in \{\cup, \Cup, \cap\}$ for every $1 \leqslant i \leqslant n$,* The consequences of a sequential application of $\mathbf{L^1}^{\Omega_1}_{\dagger_1}, \ldots, \mathbf{L^n}^{\Omega_n}_{\dagger_n}$ to a premise set Γ *are:*

$$Cn^{\Omega_n, \mathbf{L}_n}_{\dagger_n}(\ldots Cn^{\Omega_1, \mathbf{L}_1}_{\dagger_1}(\Gamma)\ldots) \tag{1}$$

In what follows I will only consider sequential combinations of adaptive logics where every but possibly the outermost (i.e. the nth) adaptive logic uses the minimally abnormality-strategy (i.e. for every $i < n$, $\dagger_i = \cup$). Furthermore all adaptive logics in sequential combinations will make use of the same lower limit logic, i.e. $\mathbf{L}_1 = \ldots = \mathbf{L}_n$. $\mathbf{L}$ will be used to denote this lower limit logic. In the following discussion, I will implicitly make use of these assumptions.[12]

The semantics of such sequential combinations of adaptive logics work rather straightforwardly: one starts by selecting models in $\mathcal{M}(\Gamma)$ that are minimally abnormal according to Ω_1, i.e. one takes $\min_{\Omega_1}(\Gamma)$. As a second step, one takes all models *in* $\min_{\Omega_1}(\Gamma)$ that are minimally abnormal with *respect to* Ω_2 (denoted by $\min_{\Omega_2}(\min_{\Omega_1}(\Gamma))$, ... and finally one takes the models that are minimally abnormal with respect to Ω_n in $\min_{\Omega_{n-1}}(\ldots(\min_{\Omega_1}(\Gamma))\ldots)$.

More formally, first the definition of minimally abnormal models is generalized as to range over any set of models:

Definition 13 (Minimally Abnormal Models). *Where $\mathcal{M}$ is a set of models and $\Omega \subseteq \mathcal{L}$: $\min_\Omega(\mathcal{M})$ is the set of all models $M \in \mathcal{M}$ s.t. for no $M' \in \mathcal{M}$, $\Omega(M') \supset \Omega(M)$.*

The semantics for sequential combinations of adaptive logics described above (and formally stated in Theorem 2) are sound and complete with respect to the syntactic description of Definition 12:

Theorem 2 ([42]). *Where $\Omega_1 \cup \ldots \cup \Omega_n \cup \Gamma \subseteq \mathcal{L}$ are finite:*

- *$A \in Cn^{\Omega_n, \mathbf{L}}_{\cap}(\ldots(Cn^{\Omega_1, \mathbf{L}}_{\cap}(\Gamma))\ldots)$ iff $M \models A$ for every $M \in \min_{\Omega_n}(\ldots\min_{\Omega_1}(\Gamma)\ldots)$*
- *$A \in Cn^{\Omega_n, \mathbf{L}}_{\Cap}(\ldots(Cn^{\Omega_1, \mathbf{L}}_{\cap}(\Gamma))\ldots)$ iff $M \models A$ for every $M \in \min_{\Omega_{n-1}}(\ldots\min_{\Omega_1}(\Gamma)\ldots)$ s.t. every member of $\Omega_n(M)$ is verified by some $M' \in \min_{\Omega_n}(\ldots\min_{\Omega_1}(\Gamma)\ldots)$.*

[12]These restrictions are common in the literature and are motivated in [42, 50].

- $A \in Cn_{\Cap}^{\Omega_n,\mathbf{L}}(\ldots(Cn_{\cap}^{\Omega_1,\mathbf{L}}(\Gamma))\ldots)$ *iff there is a* $M \in \min_{\Omega_n}(\ldots \min_{\Omega_1}(\Gamma)\ldots)$ *s.t.* $M' \models A$ *for every* $M' \in \min_{\Omega_n}(\ldots \min_{\Omega_1}(\Gamma)\ldots)$ *s.t.* $\Omega_n(M) = \Omega_n(M')$.

4 Translating Assumption-Based Argumentation in Adaptive Logic

This section provides the main technical contribution of the paper. Here, the translation of assumption-based argumentation into adaptive logics is presented. First, I set out criteria for such a translation to be adequate. Then, I motivate the translation by presenting some limitations of a previous translation [30]. Finally I present the translation and give adequacy results.

4.1 Motivation and Goal.

The goal of this paper is to provide a translation τ such that for any ABF (and some $\dagger \in \{\cup, \cap, \Cap\}$), there is a sequential combination of adaptive logics $\mathbf{L}_{\mathtt{ABF}}{}^{\Omega_1}_{\dagger_1} \ldots, \mathbf{L}_{\mathtt{ABF}}{}^{\Omega_n}_{\dagger_n}$ that gives the same consequences for $\tau(\mathtt{ABF})$ as ABF:

$$\mathtt{ABF} \mid\!\sim^{\dagger}_{\mathsf{Sem}} A \text{ iff } A \in Cn_{\dagger_n}^{\Omega_n,\mathbf{L}_n}\left(\ldots Cn_{\dagger_1}^{\Omega_1,\mathbf{L}_1}(\tau(\mathtt{ABF}))\ldots\right)$$

Since both adaptive logics and assumption-based argumentation support so-called cautious (e.g. the $\cap$-and $\Cap$-based consequence relations) and credulous (the $\cup$-based consequence relations) reasoning strategies, the parameters of the adaptive logic consequence relations that are open for manipulation in terms of the ABF under consideration are:

- the lower limit logic **L**
- the set of abnormalities Ω.

L will only depend on the ABF in the sense that the language of ABF will determine the language $\mathcal{L}_{\mathtt{ABF}}$ of the lower limit logic **L**. Ω will be determined solely by Λ and Sem.

The translation of this paper will be required to be *faithful on an extensional level*, meaning that for every labelling $\mathbb{L}$ one can find a corresponding minimally abnormal model M of the translated ABF and vice versa. This constraint has been adapted from the context of translations between default logic and autoepistemic logic [25]. In more detail, I say that a sequential selection $\min_{\Omega_n}(\ldots(\min_{\Omega_1}(.))\ldots)$ (where $n > 0$) is *extensionally faithful* for $\tau(\mathtt{ABF})$ and a set of labellings $\mathcal{B}$ iff:

1. for every $M \in \min_{\Omega_n}(\ldots(\min_{\Omega_1}(.))\ldots)$ there is a labelling $\mathbb{L} \in \mathcal{B}$ such that for every $A \in Ab$:
   - $M \models A$ iff $\mathbb{L}(A) = \texttt{in}$, and
   - $M \models \overline{A}$ iff $\mathbb{L}(A) = \texttt{out}$.
2. for every labelling $\mathbb{L} \in \mathcal{B}$ there is an $M \in \min_{\Omega_n}(\ldots(\min_{\Omega_1}(.))\ldots)$ such that for all $A \in Ab$:
   - $M \models A$ iff $\mathbb{L}(A) = \texttt{in}$, and
   - $M \models \overline{A}$ iff $\mathbb{L}(A) = \texttt{out}$.

4.2 The Translation from [30] and its limitations

In [30] a translation from ABA into adaptive logics was presented guided by two ideas: (1) a three-valued logic is used to give an appropriate semantical interpretation of the rules in the rule base $\mathcal{R}$ and the contrariness operator $\overline{}$ and (2) the abnormalities $\{\neg A \mid A \in \Lambda\}$ assure that as many assumptions will be made true as possible in view of Γ and $\mathcal{R}$.

In more detail, the translation makes use of some connectives from Kleene's well-known 3-valued logic $\mathbf{K}_3$ (see Table 1) and superimposes them on a logic that is characterised by the rules in $\mathcal{R}$. In more detail, the 3-valued logic $\mathbf{L}^3_{\text{ABF}}$ is defined semantically in the following way: the operators $\sim$ and $\vee$ (which are supposed not to occur in the alphabet of $\mathcal{L}$) are superimposed on the language $\mathcal{L}$ resulting in the set of well-formed formulas $\mathcal{L}^3_{\text{ABF}}$. The operators are characterised by the truth tables in Table 1.[13]

A	$\overline{A}$
1	0
0	1
u	u

A	$\sim A$
1	0
0	1
u	1

$\vee$	1	0	u
1	1	1	1
0	1	0	u
u	1	u	u

Table 1: Truth Tables for $\overline{}$, $\neg$ and $\vee$ in $\mathbf{L}^3_{\text{ABF}}$.

Definition 14 (Valuations in $\mathcal{L}^3_{\text{ABF}}$). *Where $v : \mathcal{L} \to \{0, 1, u\}$ is a assignment which respects the truth-table for $\overline{}$ (i.e., $v(\overline{A}) = 1$ iff $v(A) = 0$, $v(\overline{A}) = 0$ iff $v(A) = 1$,*

[13] In the terminology of [49], the negation $\sim$ corresponds to Bochvar's 'external negation' whereas $\overline{}$ corresponds to Kleene's negation in his $\mathbf{K}_3$. The disjunction $\vee$ is Kleene's strong disjunction.

and $v(\overline{A}) = u$ *iff* $v(A) = u$*). The valuation function* $v_M : \mathcal{L}^3_{\mathsf{ABF}} \to \{0, u, 1\}$ *is defined inductively as follows:*

1. *where* $A \in \mathcal{L}$, $v_M(A) = v(A)$*;*
2. $v_M({\sim}A) = 0$ *iff* $v_M(A) = 1$, *and* $v_M({\sim}A) = 1$ *else;*
3. $v_M(A \vee B) = \max(v_M(A), v_M(B))$ *where* $0 < u < 1$.

As usual, $M \models A$ *iff* $v_M(A) = 1$ *(so* 1 *is the only designated value). I write* $\Vdash_{\mathbf{L}^3_{\mathsf{ABF}}}$ *for the resulting consequence relation.*

$\mathbf{L}^3_{\mathsf{ABF}}$ is used as a lower limit logic for an adaptive logic with the set of abnormalities:

$$\Omega^{\sim}_{\Lambda} = \{{\sim}A \mid A \in \Lambda\}$$

The rules of $\mathcal{R}$ are translated as follows: $A_1, \ldots, A_n \to B$ is translated to ${\sim}A_1 \vee \ldots \vee {\sim}A_n \vee B$. Where $\mathcal{R}$ is a set of rules, I write $\tau_3(\mathcal{R})$ for the set of translated rules. It is important to note that the connectives $\vee$ and $\sim$ are not assumed to be part of $\mathcal{L}$ but are *superimposed* on $\mathcal{L}$.

Remark 1. *It is important to note that [30] does not assume that* $\mathcal{L}$ *contains a disjunction* $\vee$ *or a Kleene-negation* $\sim$*: these connectives are* superimposed *on the language* $\mathcal{L}$ *resulting in the language* $\mathcal{L}^3_{\mathsf{ABF}}$*. What this means is that the members of* $\mathcal{L}$ *are regarded as atoms in the construction of* $\mathcal{L}^3_{\mathsf{ABF}}$*: when* $A, B \in \mathcal{L}$, $A \vee B \in \mathcal{L}^3_{\mathsf{ABF}}$ *and* ${\sim} A \in \mathcal{L}^3_{\mathsf{ABF}}$*. This means that, when for example* $\mathcal{L}$ *already contains a disjunction, it is important to ensure that the two disjunctions are distinguishable. Furthermore, the connectives of* $\mathcal{L}$ *should* not *function as connectives in* $\mathcal{L}^3_{\mathsf{ABF}}$*. For example, where* $\dot{\vee}$ *is a disjunction used in* $\mathcal{L}$*, and* $A, B, C \in \mathcal{L}$, $(A \dot{\vee} B) \vee C \in \mathcal{L}^3_{\mathsf{ABF}}$ *yet* $(A \vee B) \dot{\vee} C \notin \mathcal{L}^3_{\mathsf{ABF}}$*.*[14]

The main representational result for this translation is the following:[15]

Theorem 3 ([30]). *Where* $\Gamma \cup \{A\} \subseteq \mathcal{L}$, *and* $\mathsf{sem} = \mathcal{N}$,

1. $\mathsf{ABF} \vdash^{\cup}_{\mathsf{sem}} A$ *iff* $\Gamma \cup \tau_3(\mathcal{R}) \Vdash^{\Omega^{\sim}_{\Lambda}, \mathbf{L}^3_{\mathsf{ABF}}}_{\cup} A$
2. $\mathsf{ABF} \vdash^{\cap}_{\mathsf{sem}} A$ *iff* $\Gamma \cup \tau_3(\mathcal{R}) \Vdash^{\Omega^{\sim}_{\Lambda}, \mathbf{L}^3_{\mathsf{ABF}}}_{\cap} A$

[14] For more on the importance and subtleties of superimposing connectives in adaptive logics see [42, p.50–53].

[15] [30] also shows extensional adequacy but I restrict myself to recalling the faithfulness results on the level of the consequence relations.

3. $\texttt{ABF} \vdash^{\Cap}_{\mathsf{sem}} A$ *iff* $\Gamma \cup \tau_3(\mathcal{R}) \Vdash^{\Omega^{\sim}_{\Lambda}, \mathbf{L}^3_{\texttt{ABF}}}_{\Cap} A$

Proof. Follows immediately from Theorem 8 in [30] and Theorem 14 (Appendix A).[16] □

As noted in [30], the result can be strengthened if the rule system based on $\mathcal{R}$ satisfies the following requirement (adapted to labelling-based semantics): where $\texttt{ABF} = ((\mathcal{L}, \mathcal{R}), \Gamma, \Lambda, \overline{})$ and $\Gamma \cup \{A\} \subseteq \mathcal{L}$,

EX Where $\mathbb{L}$ is a naive labelling in $\texttt{ABF}$ and $\mathbb{L}(A) \neq \texttt{in}$, $\Gamma \cup \texttt{in}(\mathbb{L}) \vdash \overline{A}$.

This criterion ensures that every naive set is stable. It is interesting to note that contraposition (in short, CPOS) ensures (EX). An $\texttt{ABF} = ((\mathcal{L}, \mathcal{R}), \Gamma, \Lambda, \overline{})$ satisfies CPOS if for every $\Delta \cup \{A\} \subseteq \Lambda$:

CPOS $\Gamma \cup \Delta \vdash_{\mathcal{R}} \overline{A}$ (where Δ is minimal) implies $\Gamma \cup (\Delta \cup \{A\}) \setminus \{B\} \vdash_{\mathcal{R}} \overline{B}$ for every $B \in \Delta$.

It is not hard to show that (CPOS) implies (EX) (proven in Appendix B).[17]

Lemma 1. *If* ABF *satisfies (CPOS) then* ABF *satisfies (EX).*

The faithfulness result for ABFs which satisfy (EX) is the following:

Theorem 4 ([30]). *Where* $\Gamma \cup \{A\} \subseteq \mathcal{L}$*: if* ABF *satisfies (EX), items 1–3 in Theorem 3 hold for* $\mathsf{sem} \in \{\mathcal{N}, \mathcal{P}, \mathcal{S}\}$.

However, this translation is not adequate for admissibility-based semantics, as witnessed by the following example.[18]

Example 5. *I use the* ABF *from Example 1 to show that in general the translation from [30] is not extensionally faithful for admissibility-based semantics. In particular, I show that there exist some models that are minimally abnormal but do not correspond to any preferred, grounded or stable labelling.*

The following table characterizes all $\mathbf{L}^3_{\texttt{ABF}}$*-models of* $\tau_3(\mathcal{R}) \cup \Gamma = \{\sim p \vee \overline{q}\}$:

[16] Indeed, Theorem 3 is exactly the same as Theorem 8 in [30] but phrased for labelling-based consequence operations instead of extension-based semantics). This should also take away any suspicion of circularity, since the proof of Theorem 14 does not depend on any results from [30].

[17] This observation was not made in [30].

[18] It should be noted that when an ABF satisfies (EX) preferred, stable and naive labellings coincide and stable labellings are guaranteed to exist (see [27, Corollary 1] and [29, Theorem 6]). Consequently for such ABFs, the translation from [30] is adequate.

i	$v_{M_i}(p)$	$v_{M_i}(q)$	i	$v_{M_i}(p)$	$v_{M_i}(q)$
1	0	0	*5*	u	u
2	0	u	*6*	u	1
3	0	1	*7*	1	0
4	u	0			

Observe that $\min_{\Omega^{\sim}_{\Lambda}}(\tau(\mathcal{R})\cup\Gamma)) = \{M_3, M_6, M_7\}$ *is the set of minimally abnormal models of* $\tau(\mathcal{R})\cup\Gamma$. *This means that there are some models (namely* M_3 *and* M_6*) for which* $M \models q$*, even though for the only preferred labelling, which is also grounded and semi-stable,* $\mathbb{L}_9(q) = \texttt{out}$. *Thus, this translation is not faithful for preferred, grounded and stable labellings. Exactly what is going on here?* $\{q\}$ *is conflict-free but not admissible since there is one attacker,* $\{p\}$ *that* $\{q\}$ *cannot defend itself from. This is, however, information that cannot be expressed "within"* M_3 *and* M_6*. All that is expressible within these models, is that once* q *is accepted,* p *cannot be accepted (since then* $\overline{q}$ *would have to be accepted as well in view of the rule* $p \to \overline{q}$*). What is needed, however, in order to notice* within M_3 *(or* M_6*) that it is not admissible is some kind of hypothetical reasoning:* p *allows to derive* $\overline{q}$*. Consequently if* q *is to be acceptable, a counterargument against* p *is needed.*

A Logic for Reasoning about Acceptability: To obtain this additional expressibility, the language is supplemented with a modal operator $\diamond$. Informally, $\diamond A$ will mean that A is unattacked. Recall that A is unattacked in ABA (with respect to a set of assumptions Δ) iff $\overline{A}$ is not derivable from Δ. Accordingly, $\diamond A$ will be true iff $\overline{A}$ is false. In the modal logic, this comes down to *defining* $\overline{A}$ as $\neg \diamond A$.[19]

In more detail, given an $\mathsf{ABF} = ((\mathcal{L}, \mathcal{R}), \Gamma, \Lambda, \overline{})$, $\mathcal{L}_{\mathsf{ABF}}$ is obtained by superimposing $\diamond, \wedge, \vee, \neg, \supset, \overline{}$ over $\mathcal{L}$ (see Note 1). Using this language, I define the modal logic $\mathbf{L}_{\mathsf{ABF}}$ as follows:

Definition 15 ($\mathbf{L}_{\mathsf{ABF}}$-model). *Given an* $\mathsf{ABF} = ((\mathcal{L}, \mathcal{R}), \Gamma, \Lambda, \overline{})$*, and an assignment* $v : \mathcal{L} \to \{0, 1\}$*, a* $\mathbf{L}_{\mathsf{ABF}}$*-model is a structure*

$$(v_M, W, R, a)$$

where:

- $W = \{a, w, w'\}$ *is a set of possible worlds where* a *is the actual world;*
- $R = \{(a, w), (w, w')\}$ *is an accessibility relation;*
- $v_M : \mathcal{L}_{\mathsf{ABF}} \times \{a, w, w'\} \to \{0, 1\}$ *is a valuation function respecting the following conditions (where* $x \in \{a, w, w'\}$*):*

[19]Notice the similarity with intuitionistic negation [23].

- $v_M(\neg A, x) = 1$ *iff* $v_M(A, x) = 0$;
- $v_M(A \wedge B, x) = 1$ *iff* $v_M(A, x) = 1$ *and* $v_M(B, x) = 1$;
- $v_M(\diamond A, x) = 1$ *iff there is a* $x' \in W$ *s.t.* xRx' *and* $v_M(A, x') = 1$.

As usual, I will say that if M is an $\mathbf{L}_{\mathsf{ABF}}$-model and $x \in \{a, w, w'\}$: $M, x \models A$ iff $v_M(x, A) = 1$. Furthermore $M \in \mathcal{M}_{\mathbf{L}_{\mathsf{ABF}}}(\tau(\mathtt{ABF}))$ iff $M, a \models A$ for every $A \in \Gamma$, in which case I'll also write $M \models A$. Finally $\Gamma \Vdash_{\mathbf{L}_{\mathsf{ABF}}} A$ iff $M \models A$ for every $M \in \mathcal{M}_{\mathbf{L}_{\mathsf{ABF}}}(\Gamma)$. $\supset$ and $\vee$ are defined as usual: $A \vee B := \neg(\neg A \wedge \neg B)$ and $A \supset B := \neg(A \wedge \neg B)$. The connective $\overline{}$ for representing contrariness in $\mathbf{L}_{\mathsf{ABF}}$ is defined as follows: $\overline{A} := \diamond\neg A$.

I thus consider a modal logic[20] based on a rather simple modal framework consisting of three worlds and an accessibility relation consisting of just two pairs of worlds. It has several properties which are worth noticing. First, notice that $\Gamma \Vdash_{\mathbf{L}_{\mathsf{ABF}}} \diamond A$ iff $\Gamma \Vdash_{\mathbf{L}_{\mathsf{ABF}}} \neg\diamond\neg A$ for any $\Gamma \cup \{A\} \subseteq \mathcal{L}_{\mathsf{ABF}}$. Furthermore, in modal logic it is common to define a necessity operator as follows (relative to a modal structure $(v_M, , W, R, a)$ and given some $x \in W$): $v_M(\Box A, x) = 1$ iff for every is a $x' \in W$ s.t. xRx' and $v_M(A, x') = 1$. It is easy to see that in that case, for any $\Gamma \cup \{A\} \subseteq \mathcal{L}_{\mathsf{ABF}}$ where A contains no modal operators:

$$\Gamma \Vdash_{\mathbf{L}_{\mathsf{ABF}}} \diamond A \quad \text{iff} \quad \Gamma \Vdash_{\mathbf{L}_{\mathsf{ABF}}} \Box A, \quad \text{and}$$
$$\Gamma \Vdash_{\mathbf{L}_{\mathsf{ABF}}} \diamond\diamond A \quad \text{iff} \quad \Gamma \Vdash_{\mathbf{L}_{\mathsf{ABF}}} \Box\Box A$$

However, this correspondence breaks down once one looks at formulas with three or more possibility respectively necessity operators (i.e. it can be the case that $\Gamma \not\Vdash_{\mathbf{L}_{\mathsf{ABF}}} \diamond\diamond\diamond A$ even though $\Gamma \Vdash_{\mathbf{L}_{\mathsf{ABF}}} \Box\Box\Box A$). The reason for this is that for any $\mathbf{L}_{\mathsf{ABF}}$-model M and any $A \in \mathcal{L}_{\mathsf{ABF}}$, $M, w' \not\models \diamond A$ yet $M, w' \models \Box A$.

The idea behind the semantics is that every model corresponds to a single labelling. In more detail, given a labelling $\mathbb{L}$, the actual world a is meant to contain all of the propositions that are contained in or follow from $\mathtt{in}(\mathbb{L})$. The possible world w, accessible from a, contains all the propositions that are contained in or follow from the unattacked assumptions $\Lambda \setminus \mathtt{out}(\mathbb{L})$. To see this, note that if $A \in \Lambda \setminus \mathtt{out}(\mathbb{L})$, $\Gamma \cup \mathtt{in}(\mathbb{L}) \nvdash_{\mathcal{R}} \overline{A}$ and thus in the translated model it should hold that $M, a \not\models \overline{A}$. By definition, this means that $M, a \not\models \diamond\neg A$, i.e. $M, w \models A$. The possible world w' has the same function with respect to w as w has with respect to a: if $\Lambda \setminus \mathtt{out}(\mathbb{L}) \nvdash \overline{A}$ for some $A \in \Lambda$ then A will be contained in w'. This way, it is possible that $M, w \models \overline{A}$, which would not be possible in a modal frame containing only a and w. Thus, the function of w' is in a sense merely technical: it avoids that $\diamond\neg\overline{A}$ is a theorem, which would mean that for no model, $\diamond\overline{A}$ is the case and thus one can never derive $\overline{A}$

[20] It is perhaps interesting to note that $\mathbf{L}_{\mathsf{ABF}}$ is not a normal modal logic. This is seen by observing that e.g. $\emptyset \vdash_{\mathbf{L}_{\mathsf{ABF}}} \diamond\diamond\top$ yet $\emptyset \nvdash_{\mathbf{L}_{\mathsf{ABF}}} \Box\diamond\diamond\top$.

Informal concept	Meaning in ABA	Translation in $\mathbf{L}_{\mathtt{ABF}}$
Accepted Assumption	$\mathbb{L}(A) = \mathtt{in}$	$M \models A$
Rejected Assumption	$\mathbb{L}(A) = \mathtt{out}$	$M \models \overline{A}$
Undecided Assumption	$\mathbb{L}(A) = \mathtt{undec}$	$M \models \neg\overline{A} \wedge \neg A$
Unattacked Assumption	$\mathbb{L}(A) \neq \mathtt{out}$	$M \models \diamond A$
Existence of an Unattacked Attacker of A	$\Lambda \setminus \mathtt{out}(\mathbb{L}) \vdash_{\mathcal{R}} \overline{A}$	$M \models \diamond\overline{A}$

Table 2: Overview of the most important concepts of assumption-based argumentation and their translation in $\mathbf{L}_{\mathtt{ABF}}$.

from an unattacked set, meaning that every model would be automatically made admissible by the semantics. In Table 2, the reader finds an overview of the most important concepts from assumption-based argumentation and their translation in $\mathbf{L}_{\mathtt{ABF}}$.

With the modal logic $\mathbf{L}_{\mathtt{ABF}}$ in place, the rules $\mathcal{R}$ and strict premises Γ can now be translated. This can be done in a very straightforward way: rules $A_1, \ldots, A_n \to A \in \mathcal{R}$ are translated as material implications $A_1 \wedge \ldots \wedge A_n \supset A$. Since these rules should be applicable both in the actual world a and the possible world w, it is necessary to also include $\diamond(A_1 \wedge \ldots \wedge A_n \supset A)$. The strict premises Γ can be taken over without any translation. Finally, care has to be taken that in the actual world $\overline{A}$ acts as a consistent negation in the sense that $M, a \not\models A \wedge \overline{A}$ (since for no conflict-free or admissible labelling can it be the case that $\mathbb{L}(A) = \mathtt{in}$ and $\Gamma \cup \mathtt{in}(\mathbb{L}) \vdash_{\mathcal{R}} \overline{A}$). This is done by requiring that every model satisfies $\neg(A \wedge \overline{A})$ for every $A \in \Lambda$. Note that $\overline{A}$ is not a *complete* negation: it can be the case that $M, a \models \neg A \wedge \neg\overline{A}$. Furthermore, it should be noted that $\overline{}$ is both paraconsistent and paracomplete in w and w'. This corresponds to the fact that it can be the case that there is a labelling $\mathbb{L}$ for which $\Gamma \cup (\Lambda \setminus \mathtt{out}(\mathbb{L})) \vdash_{\mathcal{R}} \overline{A}$ and $\mathbb{L}(A) \neq \mathtt{out}$. A case in point is $\mathbb{L}_1$ in Example 1, where $\mathbb{L}_1(q) = \mathtt{undec}$ yet $\Lambda \setminus \mathtt{out}(\mathbb{L}) \vdash_{\mathcal{R}} \overline{q}$ (since $p \in \Lambda \setminus \mathtt{out}(\mathbb{L}) = \Lambda$ and $\{p\} \vdash_{\mathcal{R}} \overline{q}$). However, $\overline{}$ is the only non-classical connective in the logic: $\neg$ behaves completely classical. Its role, just like the other classical connectives is to be able to reason *about* what is or is not derivable from a given set of assumptions (see Table 2 for some examples).

Definition 16 (Translation $\tau(\mathtt{ABF})$). *Given* $\mathtt{ABF} = ((\mathcal{L}, \mathcal{R}), \Gamma, \Lambda, \overline{})$*:*

- $\tau(\mathcal{R}) = \{\bigwedge_{i=1}^{n} A_i \supset A \mid A_1, \ldots, A_n \to A \in \mathcal{R}\} \cup$
  $\{\diamond(\bigwedge_{i=1}^{n} A_i \supset A) \mid A_1, \ldots, A_n \to A \in \mathcal{R}\}$*;*

- $\bot(\texttt{ABF}) = \{\neg(A \wedge \overline{A}) \mid A \in \Lambda\}$;
- $\tau(\texttt{ABF}) = \tau(\mathcal{R}) \cup \Gamma \cup \bot(\texttt{ABF})$.

We now illustrate this definition by translating the ABF from Example 1.

Example 6 (Example 1 continued). *For the* ABF *of Example 1, we have that:*

- $\tau(\mathcal{R}) = \{p \supset \overline{q}\}$;
- $\bot(\texttt{ABF}) = \{\neg(p \wedge \overline{p}), \neg(q \wedge \overline{q})\}$.

Consequently, since $\Gamma = \emptyset$*, the translation of* ABF *is:* $\tau(\texttt{ABF}) = \tau(\mathcal{R}) \cup \bot(\texttt{ABF})$.

Defining the Abnormalities for Admissible Labellings: The extended language allows for a straightforward definition of abnormalities according to which the selection of models that correspond to admissible labellings becomes possible. Recall that an admissible labelling assigns `in` only to assumptions that are defended: for every attacking set of assumptions Δ there is at least one member of Δ that is labelled `out`. Phrased negatively, this means that there is no attacking set of assumptions Δ which is acceptable. Consequently, in the modal logic $\mathbf{L}_{\texttt{ABF}}$, for a model M to correspond to an admissible labelling, for no $A \in \Lambda$ should it be the case that $M \models A$ and $M \models \diamond\overline{A}$ (see also Table 2). In other words, $M \models A \wedge \diamond\overline{A}$ should be avoided. I accordingly define:

$$\Omega^{\mathcal{A}}_{\Lambda} = \{A \wedge \diamond\overline{A} \mid A \in \Lambda\}.$$

Defining the Abnormalities for Complete Labellings: For an assumption labelling to be admissible, it has to defend every assumption that it labels as `in`. Complete labellings are admissible extensions that additionally label every argument that they defend `in`. Consequently, in addition to including only defended assumptions (thus avoiding $A \wedge \diamond\overline{A}$), whenever it is the case that $M \models \neg \diamond \overline{A}$ (i.e. an assumption is defended), an assumption should be accepted (i.e. $M \models A$). Consequently, models for which $M \models \neg \diamond \overline{A} \supset A$ holds as much as possible should be selected and thus $\neg \diamond \overline{A} \wedge \neg A$ should be avoided. The set of abnormalities for complete semantics $\Omega^{\mathcal{C}}_{\Lambda}$ is accordingly defined as follows:

$$\Omega^{\mathcal{C}}_{\Lambda} = \{\neg A \wedge \neg \diamond \overline{A} \mid A \in \Lambda\} \cup \Omega^{\mathcal{A}}_{\Lambda}.$$

Defining the Abnormalities for Preferred, Grounded and Semi-Stable Labellings: Once there is a way to select the models that correspond to admissible respectively complete labellings, it becomes rather straightforward to select from one of these sets the models that correspond to preferred, semi-stable or grounded labelling.

Semantics	Condition	Abnormalities
Admissible	Def. 7	$\Omega_\Lambda^{\mathcal{A}} = \{A \wedge \diamond\overline{A} \mid A \in \Lambda\}$
Complete	Def. 8	$\Omega_\Lambda^{\mathcal{C}} = \{\neg A \wedge \neg\diamond\overline{A} \mid A \in \Lambda\} \cup \Omega_\Lambda^{\mathcal{A}}$
Preferred	Maximize $\texttt{in}(\mathbb{L})$	$\Omega_\Lambda^{\mathcal{P}} = \{\neg A \mid A \in \Lambda\}$
Grounded	Minimize $\texttt{in}(\mathbb{L})$	$\Omega_\Lambda^{\mathcal{G}} = \Lambda$
Semi-Stable	Minimize $\texttt{out}(\mathbb{L})$	$\Omega_\Lambda^{\mathcal{S}} = \{\neg A \wedge \neg\overline{A} \mid A \in \Lambda\}$

Table 3: Abnormalities for different assumption labellings.

For the preferred labellings, for example it suffices to select from the models that correspond to admissible labellings (in view of Theorem 1) those models that validate a maximal (w.r.t. set inclusion) set of assumptions. Phrased negatively, the models that correspond to preferred extensions will be the ones that validate $\neg A$ for some $A \in \Lambda$ as little as possible. The following set of abnormalities can consequently be defined for preferred labellings:

$$\Omega_\Lambda^{\mathcal{P}} = \{\neg A \mid A \in \Lambda\}.$$

For semi-stable labellings, i.e. admissible labellings (again in view of Theorem 1) that are $\texttt{undec}(\mathbb{L})$-minimal, the models selected are those that minimize $\neg A \wedge \neg\overline{A}$:

$$\Omega_\Lambda^{\mathcal{S}} = \{\neg A \wedge \neg\overline{A} \mid A \in \Lambda\}.$$

In a similar way, since the grounded labelling is the one that minimizes $\texttt{in}(\mathbb{L})$ among all complete labellings, the models that correspond to the grounded labelling can be selected by selecting models that minimize the number of assumptions validated:

$$\Omega_\Lambda^{\mathcal{G}} = \Lambda.$$

The abnormalities defined until here are summarized in Table 3.

4.3 A Technical Problem and its Solution

At this point one could wonder if e.g. models in $\min_{\Omega_\Lambda^{\mathcal{P}}}(\min_{\Omega_\Lambda^{\mathcal{A}}}(\tau(\texttt{ABF})))$ are extensionally faithful for the desired correspondence to the set of preferred labellings. Unfortunately, an additonal, rather technical problem needs to be taken care of.

Example 7. *Let* ABF *be defined as in Example 1. Then* $\tau(\mathcal{R}) = \{p \supset \overline{q}; \diamond(p \supset \overline{q})\}$*. The following table lists all the* $\mathbf{L}_{\texttt{ABF}}$ *models of* $\tau(\texttt{ABF})$ *(where an omission of* $M, x \models A$ *means that* $M, x \not\models A$ *for any* $A \in \mathcal{L}$ *and any* $x \in \{a, w\}$*):*

i	$M_i, a \models$	$M_i, w \models$	i	$M_i, a \models$	$M_i, w \models$
1	$\emptyset$	$p, q, \overline{q}$	7	$q, \overline{p}$	q
2	$\emptyset$	$p, q, \overline{p}, \overline{q}$	8	$q, \overline{p}$	$q, \overline{p}$
3	$p, \overline{q}$	$p, \overline{q}$	9	$q, \overline{p}$	$q, \overline{q}$
4	$p, \overline{q}$	$p, \overline{q}, \overline{p}$	10	$q, \overline{p}$	$q, p, \overline{q}$
5	q	$q, p, \overline{q}$	11	$q, \overline{p}$	$q, p, \overline{q}, \overline{p}$
6	q	$q, p, \overline{p}, \overline{q}$			

*Note that $v_{M_7}(p, a) = v_{M_7}(\overline{q}, a) = 0$ and $v_{M_7}(q, a) = v_{M_7}(\overline{p}, a) = 1$. This means that M_7 validates no single abnormality in Ω^{A}_{Λ} and accordingly $M_7 \in \min_{\Omega^{A}_{\Lambda}}(\tau(\mathtt{ABF}))$. However, M_7 does not correspond to any admissible labelling: indeed, there is no admissible labelling that labels q **in**. The problem is that $M_7, a \models \overline{p}$ even though $\Gamma \cup \{q\} \nvdash_{\mathcal{R}} \overline{p}$ which means that, q is defended from $\{p\}$ in M_7 (even though there is no admissible labelling that labels q **in**). This problem occurs because the model M_7 is not a* minimal *model of $\mathcal{R}$ in a sense familiar from logic programming: it validates some consequences that cannot be derived from $\Gamma \cup \{A \in \Lambda \mid M, a \models A\}$ using $\mathcal{R}$.*

To filter out problematic models like M_7, two extra selections have to be carried out to ensure that only models M are taken into account for which it holds that: $M, x \models A$ only if $\Gamma \cup \{B \in \Lambda \mid M, x \models B\} \vdash_{\mathcal{R}} A$ (for $x \in \{a, w\}$ and $A \in \mathcal{L}$). In particular, $\{\overline{A} \mid A \in \Lambda\}$ has to be minimized *locally* or within the set of models that validate exactly the same assumptions. In order to do this, I formulate the following set of abnormalities:[21]

$$\Omega^1_{\Lambda} = \Lambda \cup \{\neg A \mid A \in \Lambda\} \cup \{\overline{A} \mid A \in \Lambda\}$$

To see how exactly the above set of abnormalities minimizes $\{\overline{A} \mid A \in \Lambda\}$ *within* sets of models that validate exactly the same assumptions, it suffices to observe that as soon as two models differ on the valuation of an assumption $A \in \Lambda$, their respective sets of abnormalities will be incomparable. Indeed, consider two models M and M' such that $M \models A$ whereas $M' \models \neg A$. It is easy to see that in that case $\Omega^1_{\Lambda}(M) \not\subseteq \Omega^1_{\Lambda}(M')$, since $A \in \Omega^1_{\Lambda}(M) \setminus \Omega^1_{\Lambda}(M')$. Similarly, $\neg A \in \Omega^1_{\Lambda}(M') \setminus \Omega^1_{\Lambda}(M)$ ensures that $\Omega^1_{\Lambda}(M') \not\subseteq \Omega^1_{\Lambda}(M)$. However, selecting the models that validate only $\mathcal{R}$-consequences of $\{A \in \Lambda \mid M, a \models A\}$ does not suffice: it is also necessary to make a similar selection with respect to the the w-world modelling the unattacked assumptions and their consequences. The set Ω^1_{Λ} can, however, be straightforwardly adapted to a second set Ω^2_{Λ} for this purpose:

$$\Omega^2_{\Lambda} = \{\diamond A \mid A \in \Lambda\} \cup \{\diamond\neg A \mid A \in \Lambda\} \cup \{\diamond\overline{A} \mid A \in \Lambda\}$$

[21] I thank Christian Straßer for suggesting this idea.

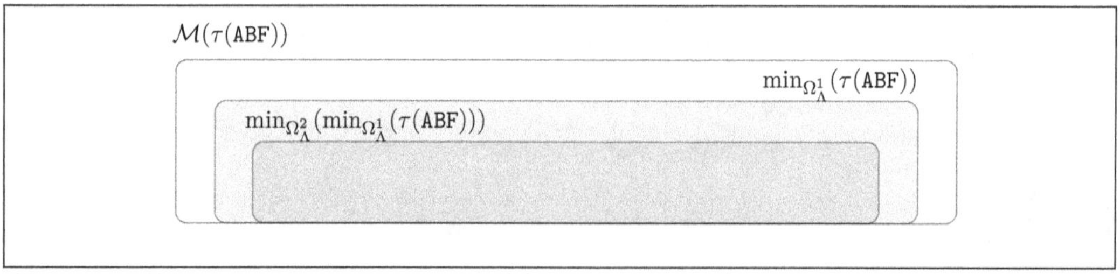

Figure 2: Schematic representation of the semantics of the sequential adaptive logic $\mathbf{L}_{\mathtt{ABF}\cap}^{\Omega^2_\Lambda}(\mathbf{L}_{\mathtt{ABF}\cap}^{\Omega^1_\Lambda}(\tau(\mathtt{ABF})))$.

Example 8 (Example 7 continued)**.** *To see that the above solution solves the problem described in Example 7, it suffices to note that there is a model, M_5, in $\mathcal{M}(\tau(\mathtt{ABF}))$ with $\Omega^1_\Lambda(M_5) = \{q, \neg p\}$. Since $\Omega^1_\Lambda(M_7) = \{q, \neg p, \overline{p}\}$, it is clear that $M_7 \notin \min_{\Omega^1_\Lambda}$ $(\tau(\mathtt{ABF}))$.*

That the solution above solves the problem in Example 7 is no coincidence: for any $\mathtt{ABF}$, the sequential adaptive logic $\mathbf{L}_{\mathtt{ABF}}{}^{\cap}_{\Omega^2_\Lambda}(\mathbf{L}_{\mathtt{ABF}}{}^{\cap}_{\Omega^1_\Lambda}(\tau(\mathtt{ABF})))$ returns as a consequence set of $\tau(\mathtt{ABF})$ exactly those members of the original language $\mathcal{L}$ that are a consequence of $\mathtt{in}(\mathbb{L})$ for every labelling $\mathbb{L}$:

$$Cn_{\cap}^{\Omega^2_\Lambda,\mathbf{L}_{\mathtt{ABF}}}\left(Cn_{\cap}^{\Omega^1_\Lambda,\mathbf{L}_{\mathtt{ABF}}}(\tau(\mathtt{ABF})\right) \cap \mathcal{L} = \{A \mid \bigcap_{\mathbb{L}\text{ is a labelling}} \{A \mid \mathtt{in}(\mathbb{L}) \vdash A\}$$

This follows from the following more general result (proven in Appendix C):

Theorem 5. *Given an $\mathtt{ABF}$, $\mathbb{L}$ is a labelling of $\mathtt{ABF}$ iff there is a $M \in \min_{\Omega^2_\Lambda}\left(\min_{\Omega^1_\Lambda}(\tau(\mathtt{ABF}))\right)$ where:*

- $v_M(A, a) = 1$ *iff* $\mathtt{in}(\mathbb{L}) \vdash_{\mathcal{R}} A$
- $v_M(A, w) = 1$ *iff* $\Lambda \setminus \mathtt{out}(\mathbb{L}) \vdash_{\mathcal{R}} A$

Remark 2. *The reader might wonder if a simpler adaptive logic might not be obtained by replacing the sequential adaptive logic $\mathbf{L}_{\mathtt{ABF}\cap}^{\Omega^2_\Lambda}(\mathbf{L}_{\mathtt{ABF}\cap}^{\Omega^1_\Lambda}(.))$ by the adaptive logic $\mathbf{L}_{\mathtt{ABF}\cap}^{\Omega^2_\Lambda\cup\Omega^1_\Lambda}(.)$. The latter would not solve the problem described in Example 7. To see this, consider the models M_5 and M_7 (see Example 7). Notice that $\Omega^1_\Lambda(M_5) \subset \Omega^1_\Lambda(M_7)$ and thus $M_7 \notin \min_{\Omega^1_\Lambda}(\tau(\mathtt{ABF}))$. However, $\Omega^1_\Lambda \cup \Omega^2_\Lambda(M_5) = \{q, \neg p, \diamond q, \diamond p, \diamond\overline{q}\}$ is incomparable with $\Omega^1_\Lambda \cup \Omega^2_\Lambda(M_7) = \{q, \neg p, \overline{p}, \diamond q, \diamond\neg p\}$ which means that $M_7 \in \min_{\Omega^1_\Lambda\cup\Omega^2_\Lambda}(\tau(\mathtt{ABF}))$. This would result again in a model validating q ending up in $\min_{\Omega^A_\Lambda}(\min_{\Omega^1_\Lambda\cup\Omega^2_\Lambda}(\tau(\mathtt{ABF})))$ and thus give in an inadequate translation.*

4.4 Representation Results

With the solution to this technical problem in place, I can now state faithfulness results both on the level of labellings (i.e. extensional faithfulness, see Section 4.1) and consequence relations. All of the theorems in this section are proven in Appendix D.

The results central to proving the faithfulness results for grounded, complete and semi-stable labellings are similar faithfulness results for admissible and complete labellings:

Theorem 6. *For any* ABF, $\min_{\Omega^{\mathcal{A}}_{\Lambda}}\left(\min_{\Omega^{2}_{\Lambda}}\left(\min_{\Omega^{1}_{\Lambda}}(.)\right)\right)$ *is extensionally faithful for* $\tau(\mathtt{ABF})$ *and the admissible labellings* $\mathcal{A}$.

Theorem 7. *For any* ABF, $\min_{\Omega^{\mathcal{C}}_{\Lambda}}\left(\min_{\Omega^{2}_{\Lambda}}\left(\min_{\Omega^{1}_{\Lambda}}(.)\right)\right)$ *is extensionally faithful for* $\tau(\mathtt{ABF})$ *and the complete labellings* $\mathcal{C}$.

I now state the extensional faithfulness of the translation for preferred, grounded and semi-stable labellings on the level of labellings:

Theorem 8. *For any* ABF, $\min_{\Omega^{\mathcal{P}}_{\Lambda}}\left(\min_{\Omega^{\mathcal{A}}_{\Lambda}}\left(\min_{\Omega^{2}_{\Lambda}}\left(\min_{\Omega^{1}_{\Lambda}}(.)\right)\right)\right)$ *is extensionally faithful for* $\tau(\mathtt{ABF})$ *and the preferred labellings* $\mathcal{P}$.

Theorem 9. *For any* ABF, $\min_{\Omega^{\mathcal{G}}_{\Lambda}}\left(\min_{\Omega^{\mathcal{C}}_{\Lambda}}\left(\min_{\Omega^{2}_{\Lambda}}\left(\min_{\Omega^{1}_{\Lambda}}(.)\right)\right)\right)$ *is extensionally faithful for* $\tau(\mathtt{ABF})$ *and the grounded labelling* $\mathcal{G}$.

Theorem 10. *For any* ABF, $\min_{\Omega^{\mathcal{S}}_{\Lambda}}\left(\min_{\Omega^{\mathcal{A}}_{\Lambda}}\left(\min_{\Omega^{2}_{\Lambda}}\left(\min_{\Omega^{1}_{\Lambda}}(.)\right)\right)\right)$ *is extensionally faithful for* $\tau(\mathtt{ABF})$ *and the semi-stable labellings* $\mathcal{S}$.

From these results the faithfulness results on the level of consequence relation follow immediately.

Theorem 11. *Where* $\dagger \in \{\cup, \cap, \Cap\}$,

$$\mathtt{ABF} \mid\!\sim^{\dagger}_{\mathcal{P}} A \textit{ iff } A \in \mathcal{L} \cap Cn_{\dagger}^{\Omega^{\mathcal{P}}_{\Lambda},\mathbf{L}_{\mathtt{ABF}}}\left(Cn_{\cap}^{\Omega^{\mathcal{A}}_{\Lambda},\mathbf{L}_{\mathtt{ABF}}}\left(Cn_{\cap}^{\Omega^{2}_{\Lambda},\mathbf{L}_{\mathtt{ABF}}}\left(Cn_{\cap}^{\Omega^{1}_{\Lambda},\mathbf{L}_{\mathtt{ABF}}}(\tau(\mathtt{ABF}))\right)\right)\right).$$

Theorem 12. *Where* $\dagger \in \{\cup, \cap, \Cap\}$,

$$\mathtt{ABF} \mid\!\sim^{\dagger}_{\mathcal{G}} A \textit{ iff } A \in \mathcal{L} \cap Cn_{\dagger}^{\Omega^{\mathcal{G}}_{\Lambda},\mathbf{L}_{\mathtt{ABF}}}\left(Cn_{\cap}^{\Omega^{\mathcal{C}}_{\Lambda},\mathbf{L}_{\mathtt{ABF}}}\left(Cn_{\cap}^{\Omega^{2}_{\Lambda},\mathbf{L}_{\mathtt{ABF}}}\left(Cn_{\cap}^{\Omega^{1}_{\Lambda},\mathbf{L}_{\mathtt{ABF}}}(\tau(\mathtt{ABF}))\right)\right)\right).$$

Theorem 13. *Where* $\dagger \in \{\cup, \cap, \Cap\}$,

$$\mathtt{ABF} \mid\!\sim^{\dagger}_{\mathcal{S}} A \textit{ iff } A \in \mathcal{L} \cap Cn_{\dagger}^{\Omega^{\mathcal{S}}_{\Lambda},\mathbf{L}_{\mathtt{ABF}}}\left(Cn_{\cap}^{\Omega^{\mathcal{A}}_{\Lambda},\mathbf{L}_{\mathtt{ABF}}}\left(Cn_{\cap}^{\Omega^{2}_{\Lambda},\mathbf{L}_{\mathtt{ABF}}}\left(Cn_{\cap}^{\Omega^{1}_{\Lambda},\mathbf{L}_{\mathtt{ABF}}}(\tau(\mathtt{ABF}))\right)\right)\right).$$

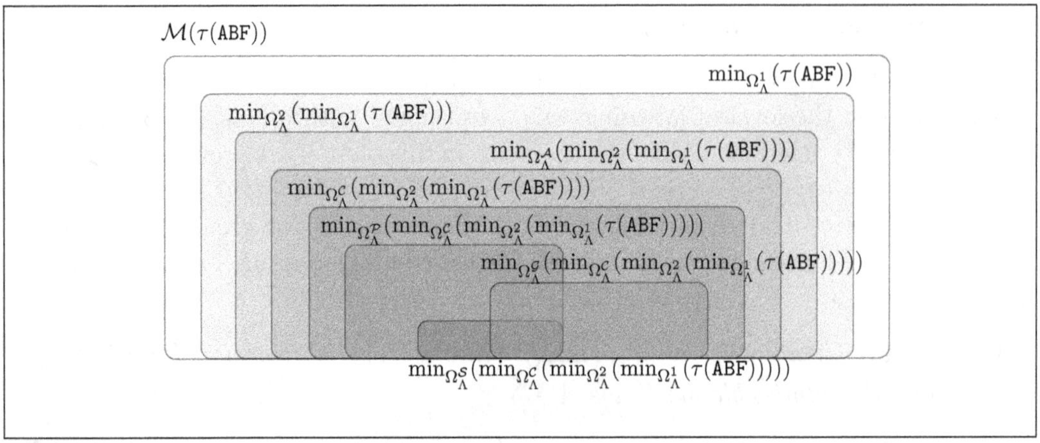

Figure 3: Schematic representation of the semantics of the translations presented in this paper.

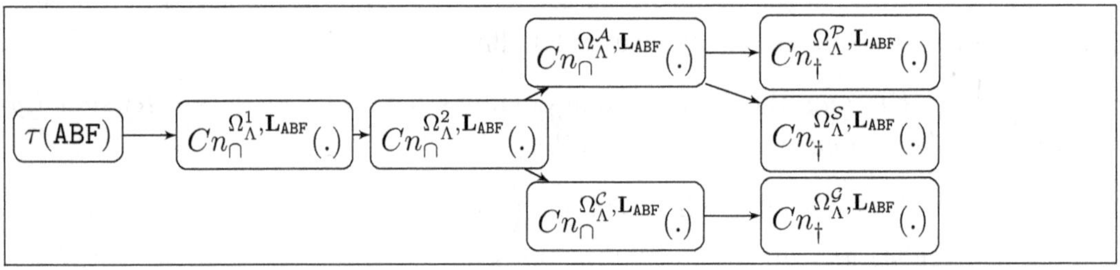

Figure 4: Schematic representation of the order of consequence relations for the adaptive logics described in this paper (where $\dagger \in \{\cup, \cap, \Cap\}$).

4.5 On Universal Definability: Stable Labellings and Non-Flat ABFs

The critical reader might wonder why another useful and well-studied argumentation labellings, the stable labellings, is not translated in this work. The reason for not considering stable labellings for a translation is that, in contrast to preferred, grounded or semi-stable labellings, stable labellings are not *universally defined* [3]: there are ABFs which do not have a stable labellings. First I recall the definition of stable labellings:

Definition 17 (Stable Assumption Labellings [40]). *Given an assumption-based framework* $\texttt{ABF} = ((\mathcal{L}, \mathcal{R}), \Gamma, \Lambda, \overline{})$, *a complete assumption labelling* $\mathbb{L}$ *is* stable *iff* $\texttt{undec}(\mathbb{L}) = \emptyset$.

The following example shows that stable labellings do not necessarily exist:

Example 9. *Consider* $\texttt{ABF} = ((\{p, \overline{p}\}, \{p \to \overline{p}\}), \emptyset, \{p\}, \overline{})$. *There is just one admissible labelling for this* ABF*:* $\mathbb{L}(p) = \texttt{undec}$. *To see this, suppose first for a contradiction that there is an admissibe labelling* $\mathbb{L}'$ *s.t.* $\mathbb{L}'(p) = \texttt{in}$. *Note that* p *attacks* p. *Thus, for* $\mathbb{L}'(p) = \texttt{in}$, $\mathbb{L}'(p) = \texttt{out}$ *would be needed, contradiction. Suppose now that there is an admissibe labelling* $\mathbb{L}'$ *s.t.* $\mathbb{L}'(p) = \texttt{out}$. *Then there has to be an attacker that is labelled in. But the only attacker of* p *is* p *itself, contradiction. However,* $\mathbb{L}$ *is clearly not stable.*

The fact that stable labellings are not universally defined makes it hard to translate them into adaptive logics in a way set out in Section 4.1 since minimally abnormal models do not suffer from similar problems. Indeed, for adaptive logics it is well-known that as soon as a premise set Γ has models in the lower limit logic, it will have minimally abnormal models (this property is known as reassurance, cf. [42, p.24]).

Likewise, the reader might ask why attention is restricted to flat frameworks. For non-flat frameworks, many of the argumentation theoretic properties that were used in the meta-theory (for example Fact 4 and Fact 10) fail (see [18]). Furthermore, there are non-flat frameworks for which there exist no preferred, semi-stable or grounded labellings. Just as with stable labellings, this makes it hard to translate non-flat ABFs into adaptive logics in view of the reassurance property of the latter.

5 Related Work

I will compare the work done in this paper with three main strands of related work. The first one is the use of modal logics for reasoning about formal argumentation frameworks [26]. The second approach I discuss is a translation of abstract argumentation into adaptive logics [43]. The third one is the use of modal logics to obtain a semantics for various non-monotonic reasoning formalisms (see [11] for an overview).

In [26] modal logic is applied to formalize fragments of formal argumentation theory. In particular, [26] establishes a correspondence between a given argumentation framework and a modal logic frame. The idea is that the argumentation framework and the modal frame will have the same number of nodes: for every argument there will be exactly one corresponding world. The meaning of the relation is, in a sense, inversed: if a attacks b then the world corresponding to b will be an accessible from the world corresponding to a. Consequently, the way modal semantics are used in this paper is quite different from how it is used in [26]. This paper provides a (non-monotonic) logic for every argumentation framework and proves correspondence between the selected models and the preferred, grounded or semi-stable labellings. On the other hand, [26] provides a way to embed argumentation theory within modal logic and then makes use of well-established results and techniques from modal logic to benefit formal argumentation (e.g. determining indistinguishability of argumentation frameworks using bisimulation). Furthermore, it should be noticed that [26] deals with *abstract* argumentation whereas this work deals with a model of structured argumentation.

Another approach which is related to the work presented here is found in [43], where various adaptive logics for representing and reasoning with abstract argumentation frameworks are provided. In this work (just as in [26]), the language allows to express the basic notions of *abstract* argumentation: arguments and attacks. In more detail, they use propositional letters to represent arguments and a binary logical operator $\twoheadrightarrow$, where $A \twoheadrightarrow B$ means that A attacks B. Furthermore, if a propositional letter that represents an argument is derivable, this means that the argument represented by the propositional letter is accepted. Based on this interpretation, they formulate two lower limit logics that allow to reason about admissibility respectively completeness. The semantics of this lower limit logic is a standard bivalent one (i.e. they don't make use of modal notions) and as the authors themselves concede, some of the valuation functions are "of a rather complex form" [43]. Even though the requirements for admissibility respectively completeness are hard-wired in the monotonic lower limit logic as inference rules, to get a translation that is adequate for admissible respectively complete semantics, they first need to do an adaptive selection in order to avoid some interpretative surplus (in more detail, to avoid that some models validate attacks which are not part of the given premises, a problem not unlike the problem described in Section 4.3). They then combine these adaptive logics for admissible respectively complete labellings with a second selection to obtain adaptive logics that adequately represent preferred, semi-stable and grounded semantics. This combination is done in a sequential way, just like in this paper and the abnormalities are very similar to the sets $\Omega^{\mathcal{P}}_{\Lambda}$, $\Omega^{\mathcal{S}}_{\Lambda}$ and $\Omega^{\mathcal{G}}_{\Lambda}$. Even though the goal of [43] and this paper is very similar, there are thus quite some differences in the way this goal is achieved. First, this paper is concerned with assumption-based argumentation, a branch of *structured* argumentation as opposed to [43] which treats *abstract* argumentation frameworks. Because of this, in the translation formulated in this paper, one can formulate attacks indirectly (based on derivable contraries) instead of having to encode the attack relation as a primitive connective. Furthermore, I use a modal semantics to give a more intuitive representation of the reasoning on the basis of argumentative notions such as attack, admissibility and acceptance. Finally, I do not encode admissibility or completeness in the monotonic lower limit logic but use the non-monotonic selection semantics to select admissible and complete sets. Note that this is closer to the way formal argumentation deals with these notions, since they are non-monotonic concepts (e.g. an admissible labelling might turn out to be inadmissible once new information is added to the knowledge base, see e.g. [17]).

There is a large amount of work on modal logic for non-monotonic reasoning and logic programming and detailed comparisons of all the work in this area and the translation given here are beyond the scope of this paper. To the best of my knowledge, however, this is the first time a translation from structured argumentation in a modal logic or a preferential logic was investigated. Given the connection between logic programming and assumption-based argumentation [15, 16, 39], one might perhaps wonder why not just use a well-studied modal logic for logic-programming like equilibrium logic [36] instead of defining a new one. To answer this objection, something has to be said first about the exact relation between assumption-based argumentation and logic-programming. Without going into formal details, the following relation exists between logic programming semantics and ABA labellings (cf. [15]):

- well-founded models correspond to grounded assumption labellings.
- regular models correspond to preferred assumption labellings.
- L-stable models correspond to semi-stable assumption labellings.
- stable models correspond to stable assumption labellings.

Stable models can be viewed as a 2-valued semantics for logic programs whereas all the other models are 3-valued semantics (just like assumption-based labellings that use the three values `in`, `out` and `undec`). As for assumption-based argumentation, there might be logic programs for which there is no stable model. Equilibrium logic is a non-monotonic formalism based on a monotonic modal logic, the logic of here and there, which uses modal frames that contain two worlds, h ("here") and t ("there") where t is accessible from h but not the other way around. Just like in this paper, h is meant to contain all the propositions that are true, whereas t contains all the propositions that are not false. Roughly speaking, a model will be an equilibrium model if all the propositions true in t will also be true in h, i.e. every non-false proposition is true. It is shown in [36] that equilibrium models of a logic program correspond to its stable models. For the 3-valued semantics, however, "a more complex notion of frame" [14] is required. In fact, to characterize 3-valued logic programming semantics [14] proposes *partial equilibrium logic* that is based on modal frames with four possible worlds and two distinct accessibility relations. Accordingly, one can say that the approach in this paper simplifies things since the logic $\mathbf{L}_{\mathsf{ABF}}$ works with modal frameworks consisting of three worlds and one accessibility relation. I plan to investigate the exact relations between the adaptive logics presented here and partial equilibrium logic in future work.

6 Conclusion and Future Work

This paper contributed to the unification of formal models of defeasible reasoning by providing a translation from assumption-based argumentation into adaptive logics. Indeed, the translation provided in this paper is based on a modal logic in order to avoid the problems suffered by a translation provided in [30] and in this way provided a translation that is faithful for preferred, grounded and semi-stable semantics, complementing the (simpler) translation from [30] faithful for naive semantics. Furthermore, [30] has shown that adaptive logics can be translated straightforwardly into ABA. Both adaptive logics and assumption-based argumentation have been shown to be able to capture several other formalisms for defeasible reasoning before. Adaptive logics can be seen as a special but rich subclass of prefential semantics [32, 41] as was shown by [6]. Furthermore, in [51], the connection between ALs and Makinson's Default Assumption Consequence Relations (in short, DACRs) [33, chapter 2] was established. DACRs give formal substance to the idea that, in many situations, non-monotonic reasoning makes use of a set Δ of defeasible background assumptions in combination with the strict and explicit premises in Γ. These background assumptions are used to the extent that they are consistent with Γ. Accordingly, DACRs make use of the notion of maximal consistent subset. In [33, chapter 2], it is also shown that many other non-monotonic consequence relations, such as Reiter's Closed World Assumption, Poole's Background Constraints, etc. can be expressed as DACRs. Finally, in [45] adaptive logic has been shown to admit input/output logic [34]. Assumption-based argumentation, on the other hand, was shown to be able to capture default logic [12], logic programming [16], autoepistemic logic [12] and ASPIC$^+$ [30]. The translation in this work now shows how all these formalisms are interrelated.

To the reader familiar with adaptive logics, it might perhaps be suprising to see there was no mention of the dynamic proof theory of adaptive logics. Adaptive logics were originally introduced in [4] using their dynamic proof theories to give a formalisation of some aspects of reasoning with inconsistencies. The driving idea behind ALs is to apply defeasible inference rules under explicit normality assumptions. More specifically, whenever the core logic gives rise to $\Gamma \vdash_{\mathbf{L}} A \vee \mathsf{ab}$ where $\mathsf{ab} \in \Omega$, A can be derived in the adaptive logic (based on $\mathbf{L}$ and Ω) on the (defeasible) assumption that ab is false. To keep track of normality assumptions, proof lines in the Hilbert-style proofs of adaptive logics are equipped with an additional column in which the abnormalities are listed which are assumed to be false. Different retraction mechanisms for lines with abnormality assumptions that turn out mistaken are implemented in terms of the *adaptive strategies*. These dynamic proof theories have received a dialogic or game-theoretical interpretation in [7]. Unfortunately, no proof theories for sequential combinations of adaptive logics using $\cup$-consequence relations (in the terminology of adaptive logics: the normal selections strategy) have been devised until now, thus hindering the study of a proof theory for some of the translations presented in this paper. Likewise, several computational tools have been defined for ABA [21, 22, 47]. These tools are all constructed as a dispute between a proponent and an opponent. The proponent tries to show a given proposition is acceptable while the opponent tries to prevent this proposition coming out as acceptable. Such disputes combine the construction of arguments from a claim (by deriving it from other claims) and the identifaction of attacks between these arguments. In future work, I plan to investigate the relation between these two kinds of proof theories.

Furthermore, the results of this paper can be generalized in several directions, such as dropping the restriction to flat frameworks and incorporating priorities ([19, 28]). Additionally, this translation, and the translation from adaptive logics into ABA ([30]), can be used to transfer meta-theoretic insights. Research on computability issues has been done both for adaptive logics [53, 46, 35] and ABA [20]. This research can be used to e.g. obtain conditions under which reasoning becomes more or less complex. Furthermore, for adaptive logics there have been investigations into the effect of logical operations on the abnormalities (e.g. closing the set of abnormalities under conjunction) on the consequence relation [51]. This research will be helpful to obtain similar results for ABA.

Acknowledgements

I thank Christian Straßer and the anonymous reviewers for helpful comments. My research was supported by a Sofja Kovalevskaja award of the Alexander von Humboldt-Foundation, funded by the German Ministry for Education and Research.

A Conflict-Free and Naive Assumption Labellings

To the best of my knowledge, conflict-free and naive assumption labellings have not been introduced before. I prove here that these labellings are sound and complete with respect to their respective extensional definitions.

I first recall the definitions of conflict-free and naive argumentation extensions:

Definition 18 (Conflict-free Assumption Extension, [12]). *Given an assumption-based framework* $\mathtt{ABF} = ((\mathcal{L}, \mathcal{R}), \Gamma, \Lambda, \overline{})$, $\Delta \subseteq \Lambda$ *is:*

- conflict-free *iff* Δ *does not attack* Δ.
- naive *iff* Δ *is conflict-free and for every* $\Lambda \supseteq \Delta' \supset \Delta$, Δ' *is not conflict-free.*

Theorem 14. *Given an assumption-based framework* $\mathtt{ABF} = ((\mathcal{L}, \mathcal{R}), \Gamma, \Lambda, \overline{})$*:*

- *if* $\Delta \subseteq \Lambda$ *is conflict-free[naive] then there is a conflict-free[naive] assumption labelling* $\mathbb{L}$ *with* $\mathtt{in}(\mathbb{L}) = \Delta$.
- *if* $\mathbb{L}$ *is a conflict-free[naive] assumption labelling then* $\mathtt{in}(\mathbb{L})$ *is a conflict-free[naive] assumption extension.*

Proof. I first prove the claim for conflict-freeness.

Suppose first that $\Delta \subseteq \Lambda$ is conflict-free. Take $\mathbb{L}$ as follows: $\mathbb{L}(A) = \mathtt{in}$ iff $A \in \Delta$, $\mathbb{L}(A) = \mathtt{undec}$ otherwise. Suppose now that there is a $\Theta \subseteq \Lambda$ that Θ attacks Δ. Since Δ is conflict-free, $\Theta \not\subseteq \Delta$. Consequently, $\mathbb{L}(\Theta) \neq \mathtt{in}$ and thus $\mathbb{L}$ is a conflict-free labelling.

Suppose now that $\mathbb{L}$ is a conflict-free assumption-labelling. Suppose for a contradiction that $\mathtt{in}(\mathbb{L})$ attacks $\mathtt{in}(\mathbb{L})$. But then there is an A with $\mathbb{L}(A) = \mathtt{in}$ such that there is an attacker $\Delta \subseteq \mathtt{in}(\mathbb{L})$ of A for which $\mathbb{L}(\Delta) = \mathtt{in}$, contradiction to $\mathbb{L}$ being conflict-free.

I now prove the claim for naive semantics and labellings.

Suppose first that $\Delta \subseteq \Lambda$ is naive. Take $\mathbb{L}$ as follows: $\mathbb{L}(A) = \mathtt{in}$ iff $A \in \Delta$, $\mathbb{L}(A) = \mathtt{undec}$ otherwise. Suppose now there is a conflict-free labelling $\mathbb{L}'$ such that $\mathtt{in}(\mathbb{L}) \subset \mathtt{in}(\mathbb{L}')$. By the previous result, $\mathtt{in}(\mathbb{L}')$ is conflict-free, contradicting Δ being naive.

The other direction is analogous. □

It is perhaps interesting to note that there is no bijection between the set of conflict-free extensions and the set of conflict-free labellings, as witnessed by Example 10.

Example 10. *Let* $\mathcal{R} = \{p \to \overline{q}\}$ *and* $\Lambda = \{p, q\}$. *Then there are the following labellings:*

i	$\mathbb{L}_i(p)$	$\mathbb{L}_i(q)$	i	$\mathbb{L}_i(p)$	$\mathbb{L}_i(q)$
1	undec	undec	2	undec	out
3	out	undec	4	out	out
5	undec	in	6	out	in
7	in	undec	8	in	out

Even though there are eight conflict-free assumption labellings, there are only 3 conflict-free assumption extensions: $\emptyset$, $\{p\}$ and $\{q\}$.

B Proof of Lemma 1

Lemma 1. If $\mathtt{ABF}$ satisfies (CPOS) then $\mathtt{ABF}$ satisfies (EX).

Proof. Suppose that $\mathtt{ABF}$ satisfies (CPOS) and suppose that some $\mathbb{L}$ is a naive labelling. Since $\mathbb{L}$ is naive, by Theorem 14, $\mathtt{in}(\mathbb{L})$ is a naive assumption extension. We prove now that $\mathbb{L}$ is stable (cf. Definition 17) by showing that (for some $A \in \Lambda$), if $\mathbb{L}(A) \neq \mathtt{in}$ then $\Gamma \cup \mathtt{in}(\mathbb{L}) \vdash_{\mathcal{R}} \overline{A}$. Suppose indeed that $\mathbb{L}(A) \neq \mathtt{in}$ for some $A \in \Lambda$. Consequently, $\Gamma \cup \{A\} \cup \mathtt{in}(\mathbb{L}) \vdash_{\mathcal{R}} \overline{B}$ for some $B \in \mathtt{in}(\mathbb{L}) \cup \{A\}$. Let $\Delta \subseteq \{A\} \cup \mathtt{in}(\mathbb{L})$ be minimal such that $\Gamma \cup \Delta \vdash_{\mathcal{R}} \overline{B}$. If $A \notin \Delta$ we consider two cases: (1) $A = B$. In that case we immediately get that, $\Gamma \cup \Delta \vdash_{\mathcal{R}} \overline{A}$ and we are done. (2) Suppose $A \neq B$. In that case, $\Delta \subseteq \mathtt{in}(\mathbb{L})$. This would contradict $\mathbb{L}(B) = \mathtt{in}$ and $\mathbb{L}$ being a conflict-free labelling. Suppose now $A \in \Delta$. If $A = B$ then by (CPOS), $\Gamma \cup (\Delta \cup \{A\}) \setminus \{B\} \vdash_{\mathcal{R}} \overline{B}$ for any $B \in \Delta$. Again by (CPOS), $\Gamma \cup (\Delta \setminus \{A\}) \vdash_{\mathcal{R}} \overline{A}$ and thus $\Gamma \cup \mathtt{in}(\mathbb{L}) \vdash_{\mathcal{R}} \overline{A}$. Suppose now that $B \in \mathtt{in}(\mathbb{L})$. Then by (CPOS), $\Gamma \cup (\Delta \cup \{B\}) \setminus \{A\} \vdash_{\mathcal{R}} \overline{A}$ and consequently $\Gamma \cup \mathtt{in}(\mathbb{L}) \vdash_{\mathcal{R}} \overline{A}$. □

C Proof of Theorem 5

In this Section I will assume that $\mathtt{ABF} = ((\mathcal{L}, \mathcal{R}), \Gamma, \Lambda, \overline{})$.

Definition 19 (Consistent Sets of Assumptions). *Where $\Delta \subseteq \Lambda$, Δ is consistent in* $\mathtt{ABF}$ *iff for no $A \in \Delta$, $\Gamma \cup \Delta \vdash_{\mathcal{R}} \overline{A}$.*

Lemma 2. *If $\Delta \subseteq \Lambda$ is consistent in* $\mathtt{ABF}$ *and $\Gamma \cup \Delta \vdash_{\mathcal{R}} B$ then for every $M \in \mathcal{M}(\tau(\mathtt{ABF}))$, $v_M(\bigwedge \Delta \supset B, a) = 1$.*

Proof. I prove this by induction on the number of rules used in deriving B. Suppose first that $B \in \Delta$. Then obviously the Lemma holds. Suppose now that for every $1 \leqslant i \leqslant n$ (where $\bigcup_{i=1}^{n} \Delta_i \subseteq \Lambda$), $\Gamma \cup \Delta_i \vdash_{\mathcal{R}} B_i$ implies that $v_M(\bigwedge \Delta_i \supset B_i, a) = 1$. Since Δ is consistent in $\mathtt{ABF}$, $\Gamma \cup \bigcup_{i=1}^{n} \Delta_i \nvdash_{\mathcal{R}} \overline{A}$ for any $A \in \bigcup_{i=1}^{n} \Delta_i$. Suppose now that B was derived using $B_1, \ldots, B_n \to B \in \mathcal{R}$. Since $M \in \mathcal{M}(\tau(\mathtt{ABF}))$ and $v_M(\bigwedge_{i=1}^{n} B_i \supset B, a) = 1$, this means that $v_M(B, a) = 1$. □

Lemma 3. *Where $\Delta \subseteq \Lambda$, if $\Delta \vdash_{\mathcal{R}} B$ then for every $M \in \mathcal{M}(\tau(\mathtt{ABF})$, $v_M(\bigwedge \Delta \supset B, w) = 1$.*

Proof. Analogous to the proof of Lemma 2. □

Fact 1. *If $M, M' \in \mathcal{M}(\Gamma)$ (for some $\Gamma \subseteq \mathcal{L}_{\texttt{ABF}}$) and $\Omega^1_\Lambda(M) \supset \Omega^1_\Lambda(M')$ then $v_M(A, a) = 1$ iff $v_{M'}(A, a) = 1$ for every $A \in \Lambda$.*

Proof. Suppose that $M, M' \in \mathcal{M}(\Gamma)$ (for some $\Gamma \subseteq \mathcal{L}_{\texttt{ABF}}$) and $\Omega^1_\Lambda(M) \supset \Omega^1_\Lambda(M')$ and suppose furthermore for a contradiction that there is an $A \in \Lambda$ such that $v_M(A, a) = 1$ yet $v_{M'}(A, a) = 0$. This means that $\neg A \notin \Omega_1(M)$ yet $\neg A \in \Omega^1_\Lambda(M')$ which contradicts $\Omega^1_\Lambda(M) \supset \Omega_1(M')$. □

Lemma 4. *For every $M \in \min_{\Omega^1_\Lambda}(\tau(\texttt{ABF}))$, $A \in \mathcal{L}$ and $\Delta \subseteq \Lambda$, if $v_M(\bigwedge \Delta \supset A, a) = 1$ then $\Gamma \cup \Delta \vdash_{\mathcal{R}} A$.*

Proof. Suppose that $\Gamma \cup \Delta \nvdash_{\mathcal{R}} B$. I construct a model $M \in \min_{\Omega^1_\Lambda}(\tau(\texttt{ABF}))$ such that $v_M(\bigwedge \Delta \supset B, a) = 0$. Let $v_M(A, a) = 1$ if $\Gamma \cup \Delta \vdash_{\mathcal{R}} A$ and $v(A, a) = 0$ else. Furthemore let $v_M(\overline{A}, a) = 1$ imply $v_M(A, w) = 0$ and $v_M(A, w) = 1$ otherwise. Finally let $v_M(\overline{A}, w) = 1$ imply $v_M(A, w') = 0$ and $v_M(A, w') = 1$ otherwise.

The following things have to be verified:

1. M is an $\mathbf{L}_{\texttt{ABF}}$-model;
2. $v_M(\bigwedge_{i=1}^n A_i \supset A, x) = 1$ for every $A_1, \ldots, A_n \to A \in \mathcal{R}$ (for $x \in \{a, w\}$);
3. $v_M(B, a) = 0$;
4. $M \in \min_{\Omega^1_\Lambda}(\tau(\texttt{ABF}))$.

Ad 1: it is easy to see that the clauses of the connectives $\neg$, $\wedge$ and $\supset$ are satisfied by v_M. Suppose now that $v_M(\overline{A}, a) = 1$. In that case (by definition of v_M) $v_M(A, w) = 0$, which means that $v_M(\diamond\neg A, a) = 1$. Likewise, $v_M(\overline{A}, a) = 0$ implies $v_M(\diamond A, a) = 1$. Likewise, it is easy to see that $v_M(\overline{A}, w) = 1$ iff $v_M(A, w') = 0$. Thus, the definition of $\overline{}$ is respected as well. Altogether, this suffices to show that M is an $\mathbf{L}_{\texttt{ABF}}$-model.

Ad 2: I prove the claim for $x = a$, for $x = w$ the proof is analogous. Suppose that $v_M(A_i, a) = 1$ for every $1 \leqslant i \leqslant n$ and that $A_1, \ldots, A_n \to A \in \mathcal{R}$. This means that $\Gamma \cup \Delta \vdash_{\mathcal{R}} A_i$ for every $1 \leqslant i \leqslant n$. Consequently, $\Gamma \cup \Delta \vdash_{\mathcal{R}} A$. But then $v_M(A, a) = 1$. So $v_M(\bigwedge_{i=1}^n A_i \supset A, a) = 1$.

Ad 3: this is immediately clear from how v_M is defined.

Ad 4: Suppose that $M \notin \min_{\Omega^1_\Lambda}(\tau(\texttt{ABF}))$, i.e. there is an $M' \in \mathcal{M}(\tau(\texttt{ABF}))$ such that $\Omega_1(M') \subset \Omega_1(M)$. This means that $v_M(A, a) = 1$ iff $v_{M'}(A, a) = 1$ for every $A \in \Lambda$ by Fact 1. Also, $v_{M'}(\overline{A}, a) = 0$ whereas $v_M(\overline{A}, a) = 1$ for some $A \in \Lambda$. By Lemma 2, this means that $\Delta \nvdash_{\mathcal{R}} \overline{A}$ which contradicts $v_M(\overline{A}, a) = 1$ (which is known in view of the construction of the model). □

Lemma 5. *For every* $M \in \min_{\Omega^2_\Lambda}(\min_{\Omega^1_\Lambda}(\tau(\mathtt{ABF})))$, $A \in \mathcal{L}$ *and* $\Delta \subseteq \Lambda$, *if* $v_M(\bigwedge \Delta \supset A, w) = 1$ *then* $\Gamma \cup \Delta \vdash_{\mathcal{R}} A$.

Proof. Similar to the proof of Lemma 4. □

Theorem 5. Given ABF, $\mathbb{L}$ is a labelling iff there is a $M \in \min_{\Omega^2_\Lambda}(\min_{\Omega^1_\Lambda}(\tau(\mathtt{ABF})))$ where (for any $A \in \mathcal{L}$):

- $v_M(A, a) = 1$ iff $\Gamma \cup \mathtt{in}(\mathbb{L}) \vdash_{\mathcal{R}} A$
- $v_M(A, w) = 1$ iff $\Gamma \cup (\Lambda \setminus \mathtt{out}(\mathbb{L})) \vdash_{\mathcal{R}} A$

Proof. [⇒]: Suppose that $\mathbb{L}$ is a labelling. I construct M as follows:

- $v_M(A, a) = 1$ iff $\Gamma \cup \mathtt{in}(\mathbb{L}) \vdash_{\mathcal{R}} A$.
- $v_M(A, w) = 1$ iff $\Gamma \cup (\Lambda \setminus \mathtt{out}(\mathbb{L})) \vdash_{\mathcal{R}} A$

Similarly as in the proof of Lemma 4, it can be easily verified that $M \in \mathcal{M}(\tau(\mathtt{ABF}))$.

I first show that $M \in \min_{\Omega^1_\Lambda}(\tau(\mathtt{ABF}))$. Indeed suppose for a contradiction that $M \notin \min_{\Omega^1_\Lambda}(\tau(\mathtt{ABF}))$. Then there is a $M' \in \mathcal{M}(\tau(\mathtt{ABF}))$ such that $\Omega^1_\Lambda(M') \subset \Omega^1_\Lambda(M)$. Without loss of generality (by the so-called stopperedness or strong reassurance property, see [2, Theorem 4.3]), we can suppose that $M' \in \min_{\Omega^1_\Lambda}(\tau(\mathtt{ABF}))$. By Fact 1, for any $A \in \Lambda$, $v_M(A, a) = 1$ iff $v_{M'}(A, a) = 1$. Consequently, $\Lambda \cap (\Omega^1_\Lambda(M) \setminus \Omega^1_\Lambda(M')) = \emptyset$. Since $M' \in \min_{\Omega^1_\Lambda}(\tau(\mathtt{ABF}))$, by Lemma 2, $\Gamma \cup \mathtt{in}(\mathbb{L}) \vdash_{\mathcal{R}} A$ implies that $v_{M'}(A, a) = 1$. But this contradicts $\Omega^1_\Lambda(M') \subset \Omega^1_\Lambda(M)$. Thus, $M \in \min_{\Omega^1_\Lambda}(\tau(\mathtt{ABF}))$.

To show that $M \in \min_{\Omega^2_\Lambda}(\min_{\Omega^1_\Lambda}(\tau(\mathtt{ABF})))$, an argument analogous to the one for $M \in \min_{\Omega^1_\Lambda}(\tau(\mathtt{ABF}))$ can be used, substituting the reference to Lemma 2 with a reference to Lemma 3.

[⇐]: Suppose that $M \in \min_{\Omega^2_\Lambda}(\min_{\Omega^1_\Lambda}(\tau(\mathtt{ABF})))$. $\mathbb{L}$ is obtained as follows:

- $\mathbb{L}(A) = \mathtt{in}$ iff $v_M(A, a) = 1$ and $A \in \Lambda$
- $\mathbb{L}(A) = \mathtt{out}$ iff $v_M(\overline{A}, a) = 1$
- $\mathbb{L}(A) = \mathtt{undec}$ otherwise

Since $v_M(A \wedge \overline{A}, a) = 0$ for every $A \in \Lambda$ (since $M \in \mathcal{M}(\tau(\mathtt{ABF}))$ and $\tau(\mathtt{ABF}) \ni \neg(A \wedge \overline{A})$ for every $A \in \Lambda$), it is clear that $\mathbb{L}$ is a labelling.

I now prove that $v_M(A, a) = 1$ iff $\Gamma \cup \mathtt{in}(\mathbb{L}) \vdash_{\mathcal{R}} A$ for any $A \in \mathcal{L}$. I first show the [⇒]-direction. The case for $A \in \Lambda$ is clear. Suppose now that $v_M(A, a) = 1$ for some $A \in \mathcal{L} \setminus \Lambda$. By Lemma 4, this means that $\Gamma \cup \mathtt{in}(\mathbb{L}) \vdash_{\mathcal{R}} A$. Consequently,

$v_M(A, a) = 1$ implies $\Gamma \cup \mathtt{in}(\mathbb{L}) \vdash_{\mathcal{R}} A$. For the [$\Leftarrow$]-direction, suppose now that $\Gamma \cup \mathtt{in}(\mathbb{L}) \vdash_{\mathcal{R}} A$. By Lemma 2 and since $M \in \min_{\Omega^1_\Lambda}(\tau(\mathtt{ABF}))$, this implies that $v_M(A, a) = 1$.

Finally I prove that $v_M(A, w) = 1$ iff $\Gamma \cup (\Lambda \setminus \mathtt{out}(\mathbb{L})) \vdash_{\mathcal{R}} A$. Observe first that $\mathtt{out}(\mathbb{L}) = \{A \in \Lambda \mid v_M(\overline{A}, a) = 1\}$. By definition of $\overline{A}$ in $\mathbf{L}_{\mathtt{ABF}}$, it follows that $A \in \mathtt{out}(\mathbb{L})$ iff $v_M(A, w) = 0$. The rest follows immediately from Lemma 3 and Lemma 5. □

D Proofs of Theorems 8, 9 and 10

I first introduce some technical notions that will prove useful in the following.

Definition 20 (Undefeated Assumptions). *Where $\mathbb{L}$ is a labelling, $[\mathbb{L}] := \Lambda \setminus \mathtt{out}(\mathbb{L})$ and $[[\mathbb{L}]] = \{A \in \Lambda \mid \Gamma \cup [\mathbb{L}] \nvdash_{\mathcal{R}} \overline{A}\}$.*[22]

Fact 2. *If $\mathbb{L}$ is an admissible labelling then $\Gamma \cup [\mathbb{L}] \nvdash_{\mathcal{R}} \overline{A}$ for any $A \in \mathtt{in}(\mathbb{L})$.*

Proof. Suppose that $\mathbb{L}$ is an admissible labelling yet $\Gamma \cup [\mathbb{L}] \vdash_{\mathcal{R}} \overline{A}$ for some $A \in \Lambda$. This means that there is a $\Delta \subsetneq \Lambda$ such that $\mathbb{L}(B) \neq \mathtt{out}$ for any $B \in \Delta$ and $\Gamma \cup \Delta \vdash_{\mathcal{R}} \overline{A}$. This contradicts $\mathbb{L}$ being admissible. □

Lemma 6. *If $M \in \min_{\Omega^2_\Lambda}(\min_{\Omega^1_\Lambda}(\tau(\mathtt{ABF})))$, where for any $A \in \Lambda$:*

- *$\mathbb{L}(A) = \mathtt{in}$ iff $v_M(A, a) = 1$,*
- *$\mathbb{L}(A) = \mathtt{out}$ iff $v_M(\overline{A}, a) = 1$ and*
- *$\mathbb{L}(A) = \mathtt{undec}$ otherwise*

then for any $B \in \Lambda$, $B \in [[\mathbb{L}]]$ iff $v_M(\overline{B}, w) = 0$.

Proof. Suppose that $M \in \min_{\Omega^2_\Lambda}(\min_{\Omega^1_\Lambda}(\tau(\mathtt{ABF})))$ and $B \in \Lambda$. Define $\mathbb{L}$ such that for any $A \in \Lambda$:

- $\mathbb{L}(A) = \mathtt{in}$ iff $v_M(A, a) = 1$,
- $\mathbb{L}(A) = \mathtt{out}$ iff $v_M(\overline{A}, a) = 1$ and
- $\mathbb{L}(A) = \mathtt{undec}$ otherwise

[22] These concepts were inspired by [10, Chapter 7].

[⇒]. Suppose that $B \in [[\mathbb{L}]]$. This means that $\Gamma \cup [\mathbb{L}] \nvdash_{\mathcal{R}} \overline{B}$. By Theorem 5 (since $M \in \min_{\Omega^2_\Lambda}(\min_{\Omega^1_\Lambda}(\tau(\texttt{ABF}))))$ and the construction of $\mathbb{L}$, it follows that $v_M(C, w) = 1$ iff $C \in [\mathbb{L}]$ (for any $C \in \Delta$). With Lemma 5 this means that $v_M(\overline{B}, w) = 0$.

[⇐]: Suppose that $v_M(\overline{B}, w) = 0$. Since $M \in \min_{\Omega^2_\Lambda}(\min_{\Omega^1_\Lambda}(\tau(\texttt{ABF})))$ and by Theorem 5 it follows that $v_M(C, w) = 1$ iff $C \in [\mathbb{L}]$ (for any $C \in \Delta$). But then by Lemma 2, $\Gamma \cup [\mathbb{L}] \nvdash_{\mathcal{R}} \overline{B}$ and consequently $B \in [[\mathbb{L}]]$. □

Fact 3. *Where $\mathbb{L}$ is admissible, $\mathbb{L}(A) = 1$ implies that $\Gamma \cup [\mathbb{L}] \vdash_{\mathcal{R}} A$.*

Lemma 7. *If $\mathbb{L}$ is admissible, there is a $M \in \min_{\Omega^{\mathcal{A}}_\Lambda}(\min_{\Omega^2_\Lambda}(\min_{\Omega^1_\Lambda}(\tau(\texttt{ABF}))))$ where:*

- $v_M(A, a) = 1$ *iff* $\Gamma \cup \texttt{in}(\mathbb{L}) \vdash_{\mathcal{R}} A$,
- $v_M(A, w) = 1$ *iff* $\Gamma \cup [\mathbb{L}] \vdash_{\mathcal{R}} A$.

Proof. Suppose $\mathbb{L}$ is admissible and let M be defined as follows: $v_M(A, a) = 1$ iff $\Gamma \cup \texttt{in}(\mathbb{L}) \vdash_{\mathcal{R}} A$ and $v_M(A, w) = 1$ iff $\Gamma \cup [\mathbb{L}] \vdash_{\mathcal{R}} A$. By Theorem 5, $M \in \min_{\Omega^2_\Lambda}(\min_{\Omega^1_\Lambda}(\tau(\Gamma)))$. Now take $A \in \texttt{in}(\mathbb{L})$. By admissibility and Fact 3, $\mathbb{L}(A) = 1$ implies that $\Gamma \cup [\mathbb{L}] \nvdash_{\mathcal{R}} \overline{A}$. By Lemma 6, this means that $v_M(\overline{A}, w) = 0$. Consequently, $v_M(\diamond\overline{A}, a) = 0$. This shows that $\Omega^{\mathcal{A}}_\Lambda(M) = \emptyset$ and consequently, $M \in \min_{\Omega^{\mathcal{A}}_\Lambda}(\min_{\Omega^2_\Lambda}(\min_{\Omega^1_\Lambda}(\tau(\texttt{ABF}))))$. □

Fact 4. *For every* ABF, *there is an admissible labelling $\mathbb{L}$ such that $\texttt{in}(\mathbb{L}) = \emptyset$.*

Proof. Observe that where $\mathbb{L}(A) = \texttt{undec}$ for every $A \in \Lambda$, $\mathbb{L}$ is admissible. □

Lemma 8. *If $M \in \min_{\Omega^{\mathcal{A}}_\Lambda}(\min_{\Omega^2_\Lambda}(\min_{\Omega^1_\Lambda}(\tau(\texttt{ABF}))))$, $\Omega^{\mathcal{A}}_\Lambda(M) = \emptyset$.*

Proof. Suppose that $M \in \min_{\Omega^2_\Lambda}(\min_{\Omega^1_\Lambda}(\tau(\texttt{ABF})))$ and suppose that $\Omega^{\mathcal{A}}_\Lambda(M) \neq \emptyset$. By the Fact 4, there is an admissible labelling $\mathbb{L}$ such that $\texttt{in}(\mathbb{L}) = \emptyset$. By Theorem 5, there is a $M' \in \min_{\Omega^2_\Lambda}(\min_{\Omega^1_\Lambda}(\tau(\texttt{ABF})))$ such that $\{A \in \Lambda \mid v_{M'}(A, a) = 1\} = \emptyset$. But then $\Omega^{\mathcal{A}}_\Lambda(M') \subset \Omega^{\mathcal{A}}_\Lambda(M)$, contradiction to $M \in \min_{\Omega^{\mathcal{A}}_\Lambda}(\min_{\Omega^2_\Lambda}(\min_{\Omega^1_\Lambda}(\tau(\texttt{ABF}))))$. □

Theorem 6. For any ABF, $\min_{\Omega^{\mathcal{A}}_\Lambda}\left(\min_{\Omega^2_\Lambda}\left(\min_{\Omega^1_\Lambda}(.)\right)\right)$ is extensionally adequate for $\tau(\texttt{ABF})$ and the set of admissible labellings $\mathcal{A}$.

Proof. [⇐]: follows immediately from Lemma 7.

[⇒]: Suppose now that $M \in \min_{\Omega^{\mathcal{A}}_\Lambda}(\min_{\Omega^2_\Lambda}(\min_{\Omega^1_\Lambda}(\tau(\texttt{ABF}))))$ and take $\mathbb{L}$ as follows:

- $\mathbb{L}(A) = 1$ iff $M, a \models A$,

- $\mathbb{L}(A) = \texttt{out}$ iff $M, a \models \overline{A}$ and
- $\mathbb{L}(A) = \texttt{undec}$ otherwise

I first show that for every $A \in \Lambda$, $\mathbb{L}(A) = \texttt{in}$ implies that for every $\Delta \subseteq \Lambda$ such that Δ attacks A, there is a $B \in \Delta$ such that $\mathbb{L}(B) = \texttt{out}$. Suppose for a contradiction that there is a $\{A\} \cup \Delta \subseteq \Lambda$ such that $\mathbb{L}(A) = \texttt{in}$ and Δ attacks A, yet there is no $B \in \Delta$ such that $\mathbb{L}(B) = \texttt{out}$. Then by Theorem 5 $v_M(\overline{A}, w) = 1$, i.e. $M, a \models \diamond\overline{A} \wedge A$. This contradicts Lemma 8. That $\mathbb{L}(A) = \texttt{out}$ implies there is a $\Delta \subseteq \Lambda$ that attacks A such that $\mathbb{L}(\Delta) = \texttt{in}$ follows immediately from Theorem 5. Suppose finally for a contradiction that $\mathbb{L}(A) = \texttt{undec}$ yet there is a $\Delta \subseteq \Lambda$ such that $\mathbb{L}(\Delta) = \texttt{in}$ and $\Gamma \cup \Delta \vdash_{\mathcal{R}} \overline{A}$. Then by Theorem 5, $M, a \models \overline{A}$, contradiction to the construction of the model. □

Theorem 8. For any ABF, $\min_{\Omega^{\mathcal{P}}_{\Lambda}}\left(\min_{\Omega^{\mathcal{A}}_{\Lambda}}\left(\min_{\Omega^{2}_{\Lambda}}\left(\min_{\Omega^{1}_{\Lambda}}(.)\right)\right)\right)$ is extensionally adequate for $\tau(\texttt{ABF})$ and the preferred labellings $\mathcal{P}$.

Proof. [$\Rightarrow$]: Suppose that $M \in \min_{\Omega^{\mathcal{P}}_{\Lambda}}(\min_{\Omega^{\mathcal{A}}_{\Lambda}}(\min_{\Omega^{2}_{\Lambda}}(\min_{\Omega^{1}_{\Lambda}}(\tau(\texttt{ABF})))))$ yet $\mathbb{L}$ is not preferred (with $\mathbb{L}$ defined as in the statement of this Theorem). This means that there is an admissible labelling $\mathbb{L}'$ such that $\texttt{in}(\mathbb{L}) \subset \texttt{in}(\mathbb{L}')$. By Lemma 6, there is a $M' \in \min_{\Omega^{\mathcal{A}}_{\Lambda}}(\min_{\Omega^{2}_{\Lambda}}(\min_{\Omega^{1}_{\Lambda}}(\tau(\texttt{ABF}))))$ such that $\texttt{in}(\mathbb{L}') = \{A \in \Lambda \mid v_M(A, a) = 1\}$. But $\texttt{in}(\mathbb{L}) \subset \texttt{in}(\mathbb{L}')$ implies that for some $A \in \texttt{in}(\mathbb{L}') \setminus \texttt{in}(\mathbb{L})$, $v_M(A, a) = 0$. This contradicts $M \in \min_{\Omega^{\mathcal{A}}_{\Lambda}}(\min_{\Omega^{2}_{\Lambda}}(\min_{\Omega^{1}_{\Lambda}}(\tau(\texttt{ABF}))))$.

[$\Leftarrow$]: similar. □

Fact 5. *Where* $\mathbb{L}$ *is complete, if* $\mathbb{L}(A) \neq \texttt{in}$ *then* $\Gamma \cup [\mathbb{L}] \vdash_{\mathcal{R}} \overline{A}$.

Proof. Suppose that for some complete labelling $\mathbb{L}$ and some $A \in \Lambda$, $\mathbb{L}(A) \neq \texttt{in}$. Suppose first that $\mathbb{L}(A) = \texttt{out}$. Then there is a Δ that attacks A for which $\mathbb{L}(\Delta) = \texttt{in}$. Thus, $\Gamma \cup \texttt{in}(\mathbb{L}) \vdash_{\mathcal{R}} \overline{A}$ and since $\texttt{in}(\mathbb{L}) \subseteq [\mathbb{L}]$, $\Gamma \cup [\mathbb{L}] \vdash_{\mathcal{R}} \overline{A}$. Suppose now that $\mathbb{L}(A) = \texttt{undec}$. Then for some $\Delta \subseteq \Lambda$ that attacks A, $\mathbb{L}(\Delta) \neq \texttt{out}$ and consequently, $\Gamma \cup [\mathbb{L}] \vdash_{\mathcal{R}} \overline{A}$. □

Lemma 9. *If* $\mathbb{L}$ *is complete then there is a* $M \in \min_{\Omega^{\mathcal{C}}_{\Lambda}}(\min_{\Omega^{2}_{\Lambda}}(\min_{\Omega^{1}_{\Lambda}}(\tau(\texttt{ABF}))))$ *where:*

- $v_M(A, a) = 1$ *iff* $\Gamma \cup \texttt{in}(\mathbb{L}) \vdash_{\mathcal{R}} A$,
- $v_M(A, w) = 1$ *iff* $\Gamma \cup [\mathbb{L}] \vdash_{\mathcal{R}} A$

Proof. Suppose $\mathbb{L}$ is complete, and let M be defined as follows: $v_M(A, a) = 1$ iff $\Gamma \cup \texttt{in}(\mathbb{L}) \vdash_{\mathcal{R}} A$ and $v_M(A, w) = 1$ iff $\Gamma \cup [\mathbb{L}] \vdash_{\mathcal{R}} A$. Take $A \in \texttt{in}(\mathbb{L})$. Since $\mathbb{L}$ is also admissible and Fact 3, $\mathbb{L}(A) = 1$ implies that $\Gamma \cup [\mathbb{L}] \nvdash_{\mathcal{R}} \overline{A}$. By Lemma 6, this means that $v_M(\overline{A}, w) = 0$. Consequently, $v_M(\diamond\overline{A} \wedge A, a) = 0$. Suppose now that $v_M(A, a) = 0$. By Fact 5 and since $\mathbb{L}$ is complete, this means that $\Gamma \cup [\mathbb{L}] \vdash_{\mathcal{R}} \overline{A}$ and by Theorem 5 this means that $v_M(\overline{A}, w) = 1$. Consequently, $v_M(\neg A \wedge \neg\diamond\overline{A}, a) = 0$. Altogether this shows that $\Omega^{\mathcal{C}}_{\Lambda}(M) = \emptyset$. □

Lemma 10. *For every* ABF*, there is at least one complete labelling* $\mathbb{L}$.

Proof. This follows from [40, Theorem 4] and [18, Theorem 2.12]. □

Lemma 11. *If* $M \in \min_{\Omega^{\mathcal{C}}_{\Lambda}}(\min_{\Omega^2_{\Lambda}}(\min_{\Omega^1_{\Lambda}}(\tau(\texttt{ABF}))))$, $\Omega^{\mathcal{C}}_{\Lambda}(M) = \emptyset$.

Proof. Suppose that $M \in \min_{\Omega^{\mathcal{C}}_{\Lambda}}(\min_{\Omega^2_{\Lambda}}(\min_{\Omega^1_{\Lambda}}(\tau(\texttt{ABF}))))$ and suppose that $\Omega^{\mathcal{C}}_{\Lambda}(M) \neq \emptyset$. By Lemma 10, there is a complete labelling $\mathbb{L}$. By Theorem 5, there is an $M' \in \min_{\Omega^2_{\Lambda}}(\min_{\Omega^1_{\Lambda}}(\tau(\texttt{ABF})))$ such that $\{A \in \Lambda \mid v_{M'}(A, a) = 1\} = \texttt{in}(\mathbb{L})$. But this means that $\Omega^{\mathcal{C}}_{\Lambda}(M') \subset \Omega^{\mathcal{C}}_{\Lambda}(M)$, which contradicts $M \in \min_{\Omega^{\mathcal{C}}_{\Lambda}}(\min_{\Omega^2_{\Lambda}}(\min_{\Omega^1_{\Lambda}}(\tau(\texttt{ABF}))))$. □

Theorem 7 For any ABF, $\min_{\Omega^{\mathcal{C}}_{\Lambda}}\left(\min_{\Omega^2_{\Lambda}}\left(\min_{\Omega^1_{\Lambda}}(.)\right)\right)$ is extensionally adequate for $\tau(\texttt{ABF})$ and the complete labellings $\mathcal{C}$.

Proof. [$\Leftarrow$]: follows from Lemma 9.

[$\Rightarrow$]: Suppose now that $M \in \min_{\Omega^{\mathcal{C}}_{\Lambda}}(\min_{\Omega^2_{\Lambda}}(\min_{\Omega^1_{\Lambda}}(\tau(\texttt{ABF}))))$ and take $\mathbb{L}$ as follows: $\mathbb{L}(A) = 1$ iff $M, a \models A$, $\mathbb{L}(A) = \texttt{out}$ iff $M, a \models \overline{A}$ and $\mathbb{L}(A) = \texttt{undec}$ otherwise. I first show that for every $A \in \Lambda$, $\mathbb{L}(A) = \texttt{in}$ implies that for every $\Delta \subseteq \Lambda$ such that Δ attacks A, there is a $B \in \Delta$ such that $\mathbb{L}(B) = \texttt{out}$. Suppose for a contradiction that there is a $\{A\} \cup \Delta \subseteq \Lambda$ such that $\mathbb{L}(A) = \texttt{in}$ and Δ attacks A, yet there is no $B \in \Delta$ such that $\mathbb{L}(B) = \texttt{out}$. Then by Theorem 5: $v_M(\overline{A}, w) = 1$, i.e. $M \models \diamond\overline{A} \wedge A$. This contradicts Lemma 11 and the assumption that $M \in \min_{\Omega^{\mathcal{C}}_{\Lambda}}(\min_{\Omega^2_{\Lambda}}(\min_{\Omega^1_{\Lambda}}(\tau(\texttt{ABF}))))$. That $\mathbb{L}(A) = \texttt{out}$ implies there is a $\Delta \subseteq \Lambda$ that attacks A such that $\mathbb{L}(\Delta) = \texttt{in}$ follows immediately from Theorem 5. Suppose now for a contradiction that $\mathbb{L}(A) = \texttt{undec}$ yet there is a $\Delta \subseteq \Lambda$ such that $\mathbb{L}(\Delta) = \texttt{in}$ and $\Gamma \cup \Delta \vdash_{\mathcal{R}} \overline{A}$. Then by Theorem 5, $M, a \models \overline{A}$, contradiction to the construction of the model. Suppose finally that $\mathbb{L}(A) = \texttt{undec}$ yet for every attacker $\Delta \subseteq \Lambda$ of A, $\mathbb{L}(\Delta) = \texttt{out}$. By Theorem 5 this means that $v_M(\overline{A}, w) = 0$ and consequently, $v_M(\neg A \wedge \neg\diamond\overline{A}, a) = 1$, contradicting Lemma 11. □

Theorem 9. For any $\mathtt{ABF}$, $\min_{\Omega_\Lambda^{\mathcal{G}}}\left(\min_{\Omega_\Lambda^{\mathcal{C}}}\left(\min_{\Omega_\Lambda^2}\left(\min_{\Omega_\Lambda^1}(.)\right)\right)\right)$ is extensionally adequate for $\tau(\mathtt{ABF})$ and the set of grounded labelling $\mathcal{G}$.

Proof. [$\Rightarrow$]: Suppose that $M \in \min_{\Omega_\Lambda^{\mathcal{G}}}(\min_{\Omega_\Lambda^{\mathcal{C}}}(\min_{\Omega_\Lambda^2}(\min_{\Omega_\Lambda^1}(\tau(\mathtt{ABF})))))$ yet $\mathbb{L}$ is not grounded (with: $\mathbb{L}(A) = \mathtt{in}$ iff $v_M(A,a) = 1$). This means that there is a complete labelling $\mathbb{L}'$ such that $\mathtt{in}(\mathbb{L}) \supset \mathtt{in}(\mathbb{L}')$. By Lemma 7, we know that there is a $M' \in \min_{\Omega_\Lambda^{\mathcal{C}}}(\min_{\Omega_\Lambda^2}(\min_{\Omega_\Lambda^1}(\tau(\mathtt{ABF}))))$ such that $\mathtt{in}(\mathbb{L}') = \{A \in \Lambda \mid v_{M'}(A,a) = 1\}$. But $\mathtt{in}(\mathbb{L}) \supset \mathtt{in}(\mathbb{L}')$ implies that $\{A \in \Lambda \mid v_{M'}(A,a) = 1\} \supset \{A \in \Lambda \mid v_{M'}(A,a) = 1\}$. In other words, $\Omega_\Lambda^{\mathcal{G}}(M) \supset \Omega_\Lambda^{\mathcal{G}}(M')$, which contradicts $M \in \min_{\Omega_\Lambda^{\mathcal{C}}}(\min_{\Omega_\Lambda^2}(\min_{\Omega_\Lambda^1}(\tau(\mathtt{ABF}))))$.

[$\Leftarrow$]: similar. □

Theorem 10. For any $\mathtt{ABF}$, $\min_{\Omega_\Lambda^{\mathcal{S}}}\left(\min_{\Omega_\Lambda^{\mathcal{A}}}\left(\min_{\Omega_\Lambda^2}\left(\min_{\Omega_\Lambda^1}(.)\right)\right)\right)$ is extensionally adequate for $\tau(\mathtt{ABF})$ and the set of semi-stable labellings $\mathcal{S}$.

Proof. Analogous to the proof of Theorem 8. □

References

[1] Leila Amgoud, Martin Caminada, Claudette Cayrol, Marie-Christine Lagasquie, and Henry Prakken. Towards a consensual formal model: inference part. *Deliverable of ASPIC project*, 2004.

[2] Arnon Avron and Iddo Lev. Formula-preferential systems for paraconsistent non-monotonic reasoning. In *International Conference on Artificial Intelligence (ICAIâĂŹ01)*, pages 816–820, 2001.

[3] Pietro Baroni, Martin Caminada, and Massimiliano Giacomin. An introduction to argumentation semantics. *The Knowledge Engineering Review*, 26(4):365–410, 2011.

[4] Diderik Batens. Dialectical dynamics within formal logics. *Logique et Analyse*, 29(114):161–173, 1986.

[5] Diderik Batens. Inconsistency-adaptive logics. *Logic at Work, Essays dedicated to the memory of Helena Rasiowa*, pages 445–472, 1999.

[6] Diderik Batens. A universal logic approach to adaptive logics. *Logica universalis*, 1(1):221–242, 2007.

[7] Diderik Batens. Towards a dialogic interpretation of dynamic proofs. In Cédric Dégremont (et al.), editor, *Dialogues, Logics and Other Strange Things. Essays in Honour of Shahid Rahman*, pages 27–51. College Publications, 2009.

[8] Philippe Besnard, Alejandro Garcia, Anthony Hunter, Sanjay Modgil, Henry Prakken, Guillermo Simari, and Francesca Toni. Introduction to structured argumentation. *Argument and Computation*, 5(1):1–4, 2014.

[9] Philippe Besnard and Anthony Hunter. A logic-based theory of deductive arguments. *Artificial Intelligence*, 128(1):203–235, 2001.

[10] Alexander Bochman. *Explanatory nonmonotonic reasoning*, volume 4. World scientific, 2005.

[11] Alexander Bochman. Logic in nonmonotonic reasoning. *Nonmonotonic Reasoning. Essays Celebrating its 30th Anniversary. College Publ*, pages 25–61, 2011.

[12] Andrei Bondarenko, Phan Minh Dung, Robert A Kowalski, and Francesca Toni. An abstract, argumentation-theoretic approach to default reasoning. *Artificial Intelligence*, 93(1):63–101, 1997.

[13] Gerhard Brewka, Thomas Eiter, and Mirosław Truszczyński. Answer set programming at a glance. *Communications of the ACM*, 54(12):92–103, 2011.

[14] Pedro Cabalar, Sergei Odintsov, David Pearce, and Agustín Valverde. Partial equilibrium logic. *Annals of Mathematics and Artificial Intelligence*, 50(3-4):305, 2007.

[15] Martin Caminada, Samy Sá, João Alcântara, and Wolfgang Dvořák. On the equivalence between logic programming semantics and argumentation semantics. *International Journal of Approximate Reasoning*, 58:87–111, 2015.

[16] Martin Caminada and Claudia Schulz. On the equivalence between assumption-based argumentation and logic programming. In *The first international Workshop on Argumentation and Logic Programming*, 2015.

[17] Claudette Cayrol, Florence Dupin de Saint-Cyr, and M Lagasquie-Schiex. Change in abstract argumentation frameworks: Adding an argument. *Journal of Artificial Intelligence Research*, 38:49–84, 2010.

[18] Kristijonas Čyras, Xiuyi Fan, Claudia Schulz, and Francesca Toni. Assumption-based argumentation: Disputes, explanations, preferences. *Handbook of Formal Argumentation*, 2018.

[19] Kristijonas Čyras and Francesca Toni. ABA+: assumption-based argumentation with preferences. In *Principles of Knowledge Representation and Reasoning: Proceedings of the Fifteenth International Conference, KR 2016, Cape Town, South Africa, April 25-29, 2016.*, pages 553–556, 2016.

[20] Yannis Dimopoulos, Bernhard Nebel, and Francesca Toni. On the computational complexity of assumption-based argumentation for default reasoning. *Artificial Intelligence*, 141(1-2):57–78, 2002.

[21] Phan Minh Dung, Robert A Kowalski, and Francesca Toni. Dialectic proof procedures for assumption-based, admissible argumentation. *Artificial Intelligence*, 170(2):114–159, 2006.

[22] Phan Minh Dung, Paolo Mancarella, and Francesca Toni. Computing ideal sceptical argumentation. *Artificial Intelligence*, 171(10-15):642–674, 2007.

[23] Dov Gabbay and Michael Gabbay. The attack as intuitionistic negation. *Logic Journal of the IGPL*, 24(5):807–837, 2016.

[24] Alejandro J García and Guillermo R Simari. Defeasible logic programming: An argumentative approach. *Theory and Practice of Logic Programming*, 4(1+ 2):95–138,

2004.

[25] Georg Gottlob. The power of beliefs or translating default logic into standard autoepistemic logic. In *Foundations of Knowledge Representation and Reasoning*, pages 133–144. Springer, 1994.

[26] Davide Grossi. On the logic of argumentation theory. In *Proceedings of the 9th International Conference on Autonomous Agents and Multiagent Systems: Volume 1*, pages 409–416. International Foundation for Autonomous Agents and Multiagent Systems, 2010.

[27] Jesse Heyninck and Ofer Arieli. On the semantics of simple contrapositive assumption-based argumentation frameworks. *Proc. COMMAâĂŹ18*, 2018.

[28] Jesse Heyninck, Pere Pardo, and Christian Straßer. Assumption-based approaches to reasoning with priorities. In *AI*3 2017.*, pages 58–72, 2017.

[29] Jesse Heyninck and Christian Straßer. A comparative study of assumption-based approaches to reasoning with priorities. In *Second Chinese Conference on Logic and Argumentation.*

[30] Jesse Heyninck and Christian Straßer. Relations between assumption-based approaches in nonmonotonic logic and formal argumentation. *Proceedings of the 16th International Workshop on Non-Monotonic Reasoning (NMR'16)*, 2016.

[31] Jesse Heyninck and Christian Straßer. Revisiting unrestricted rebut and preferences in structured argumentation. In *Proceedings of the 26th International Joint Conference on Artificial Intelligence*, pages 1088–1092. AAAI Press, 2017.

[32] Sarit Kraus, Daniel Lehmann, and Menachem Magidor. Nonmonotonic reasoning, preferential models and cumulative logics. *Artificial Intelligence*, 44(1):167–207, 1990.

[33] David Makinson. *Bridges from classical to nonmonotonic logic.* College Publications, 2005.

[34] David Makinson and Leendert Van Der Torre. Input/output logics. *Journal of philosophical logic*, 29(4):383–408, 2000.

[35] Sergei P Odintsov and Stanislav O Speranski. Computability issues for adaptive logics in multi-consequence standard format. *Studia Logica*, 101(6):1237–1262, 2013.

[36] David Pearce. Equilibrium logic. *Annals of Mathematics and Artificial Intelligence*, 47(1-2):3, 2006.

[37] Henry Prakken. An abstract framework for argumentation with structured arguments. *Argument and Computation*, 1(2):93–124, 2010.

[38] Raymond Reiter. A logic for default reasoning. *Artificial Intelligence*, 13(1):81–132, 1980.

[39] Claudia Schulz and Francesca Toni. Logic programming in assumption-based argumentation revisited-semantics and graphical representation. In *Proceedings of the Twenty-Ninth AAAI Conference on Artificial Intelligence*, pages 1569–1575, 2015.

[40] Claudia Schulz and Francesca Toni. Labellings for assumption-based and abstract argumentation. *International Journal of Approximate Reasoning*, 84:110–149, 2017.

[41] Yoav Shoham. Reasoning about change. Technical report, Yale Univ., New Haven, CT

(USA), 1987.

[42] Christian Straßer. *Adaptive Logics for Defeasible Reasoning*. Springer, 2014.

[43] Christian Straßer. Towards the proof-theoretic unification of dungâĂŹs argumentation framework: an adaptive logic approach. In *Adaptive Logics for Defeasible Reasoning*, pages 209–241. Springer, 2014.

[44] Christian Straßer and Ofer Arieli. Sequent-based argumentation for normative reasoning. In *International Conference on Deontic Logic in Computer Science*, pages 224–240. Springer, 2014.

[45] Christian Straßer, Mathieu Beirlaen, and Frederik Van De Putte. Adaptive logic characterizations of input/output logic. *Studia Logica*, 104(5):869–916, 2016.

[46] Christian Straßer and Frederik Van De Putte. Adaptive strategies and finite-conditional premise sets. *Journal of Logic and Computation*, 26(5):1517–1539, 2014.

[47] Francesca Toni. A generalised framework for dispute derivations in assumption-based argumentation. *Artificial Intelligence*, 195:1–43, 2013.

[48] Francesca Toni. A tutorial on assumption-based argumentation. *Argument and Computation*, 5(1):89–117, 2014.

[49] Alasdair Urquhart. Basic many-valued logic. In *Handbook of philosophical logic, Volume III*, pages 249–295. Springer, 2001.

[50] Frederik Van De Putte. *Generic formats for prioritized adaptive logics, with applications in deontic logic, abduction and belief revision.* PhD thesis, Ghent University, 2012.

[51] Frederik Van De Putte. Default assumptions and selection functions: a generic framework for non-monotonic logics. In *MICAI 2013*, pages 54–67. Springer, 2013.

[52] Frederik Van De Putte and Christian Straßer. Three formats of prioritized adaptive logics: a comparative study. *Logic Journal of IGPL*, 21(2):127–159, 2013.

[53] Peter Verdée. Adaptive logics using the minimal abnormality strategy are π_1^1-complex. *Synthese*, 167(1):93–104, 2009.

Received 28 February 2018

Reasoning about Covering-based Rough Sets Using Three Truth Values

Beata Konikowska
Institute of Computer Science, Polish Academy of Sciences, ul. Jana Kazimierza 5, 01-248 Warsaw, Poland
Beata.Konikowska@ipipan.waw.pl

Arnon Avron
School of Computer Science, Tel Aviv University, Tel Aviv, Israel
aa@cs.tau.ac.i

Abstract

The paper presents a natural three-valued logic for reasoning about covering-based rough sets. Atomic formulas of the logic represent membership of objects of the universe in rough sets, and complex formulas are built out of the atomic ones using three-valued Kleene connectives. To reflect the structure of rough sets, semantics of the logic employs three truth values: **t** — representing truth and corresponding to membership of an object in the positive region of a set, **f** — representing falsity and corresponding to membership in the negative region, and **u** — representing undefinedness (lack of information) and corresponding to membership in the boundary region of the set. In the paper we provide a finitely strongly sound and complete Gentzen-style sequent calculus for the described logic.

1 Introduction

Rough sets, developed by Pawlak in the early 1980s [26, 27], represent a simple and yet very powerful notion designed to model vague or imprecise information. Unlike Zadeh's fuzzy sets, they are not based on any numerical measure of the degree of membership of an object in an imprecisely defined set. Instead, they employ a much more universal and versatile idea of an indiscernibility relation, which groups together into disjoint equivalence classes objects having the same properties from the viewpoint of a certain application.

This concept has proved extremely useful in practice. Since their introduction in the early 1980s, rough sets have found numerous applications in areas like control of manufacturing processes [18], development of decisions tables [28], data mining [18], data analysis [29], knowledge discovery [21], and so on. They have also been the subject of an impressive body of research. Though the research has been mainly focused on algebraic properties of rough sets, a number of logicians have also explored this area, presenting and studying various brands of logics connected with rough sets [8, 35, 7, 24, 25, 31].

Later, the original notion of rough sets was generalized by replacing the indiscernibility relation (representing a partition of the universe of objects) underlying Pawlak's approach with other, less restrictive concepts. They included e.g. the similarity relation [32, 16, 12], and the exhaustiveness and complementarity relations [11, 7, 13]. However, the broadest generalization were covering-based rough sets [38, 30], defined based on an arbitrary covering of the universe of objects instead of a partition, like in Pawlak's original approach. By now, this notion has also been examined in various papers (see e.g. [36, 40, 14, 41, 19, 37, 20]). The main focus has again been on the algebraic properties of such generalized rough sets — with interesting links to mathematics of vagueness [19] and probability theory [20].

In turn, the logically oriented papers have often aimed to present a minimum equational axiomatization of covering-based approximation operations, see e.g. [39]. However, other approaches have also appeared. They include, like in case of Pawlak's rough sets, many-valued logics [4, 17] or modal logics [22, 23].

In this paper we pursue the approach based on many-valued logic. We explore the logical aspects of covering-based rough sets from the viewpoint of membership of objects in such rough sets. For that purpose, we employ a three-valued logic with an analytic proof system based on a Gentzen style sequent calculus. The motivation for using three truth values stems from the fact that, exactly in case of the ordinary rough sets, a covering-based approximation space defines three regions of any set X of objects:

- positive region, containing all objects of the universe which *certainly belong to* X in the light of the information provided by the covering;
- negative region, containing all objects which *certainly do not belong to* X;
- boundary, containing all objects which *cannot be said for sure to either belong or not to belong to* X.

Hence a natural idea for reasoning about membership of objects in covering-based rough sets is to use a logic with semantics based on three truth values **t** – true,

f – false, **u** – unknown, corresponding to membership of an object in the positive region, the negative region and the boundary of the set X, respectively. Note that though the use of three-valued logics for reasoning about rough sets themselves is fraught with the pitfalls analysed in depth in [10], they do not occur in reasoning about membership of the objects (elements) of universe in rough sets rather than about rough sets themselves.

Such an idea was first exploited in [3] for the original rough sets based on an equivalence relation on the universe of objects. However, the logic developed there was just a simple propositional logic generated by a three-valued non-deterministic matrix (see [5, 2]), shortly: Nmatrix, which only reflected some properties of set-theoretic operations on rough sets.

Then next attempt was made in [17] for covering-based rough sets. There the semantics of the logics was based on the natural frameworks for such sets, i.e. covering-based approximations spaces. Atomic formulas of the logic represented either membership of objects of the universe in rough sets or the subordination relation[1] between such objects. However, strong completeness of the proof system in the form of sequent calculus was only proved for a reduced subset of the language.

The line of research was continued in [4], where a finitely strongly sound and complete Gentzen calculus was presented for the logic of rough sets defined as in [17], but without the subordination relation. The present paper continues the direction of [4] by presenting a new version of the logic, corresponding to weak semantics of the same language. In such semantics we have two designated values: **t** and **u**, so elements of the boundary of a set X are treated as belonging to X. The motivation has been to tailor the logic to the applications where we do not want to miss any possible element of X in our considerations (e.g. when looking for a drug that might be effective in treating a give disease, we surely want to examine all that have shown any promise of that). It should be noted that such an approach results in an inherently paraconsistent logic — for the boundaries of X and $-X$ are identical[2].

The paper is organized as follows. Section 2 presents the fundamentals of covering-based rough sets. Section 3 defines the syntax and semantics of the logic $\mathcal{L}_{RS}$ examined in the paper, including satisfaction and consequence relations for formulas and sequents. Section 4 presents a strongly sound and complete Gentzen sequent calculus for $\mathcal{L}_{RS}$, and finally Section 5 presents the conclusions and outlines future work.

[1] Given a covering $\mathcal{C}$ of a universe U, the subordination relation generated by $\mathcal{C}$ is the binary relation $\prec_{\mathcal{C}}$ on U such that $x \prec_{\mathcal{C}} y \Leftrightarrow (\forall C \in \mathcal{C})(y \in C \rightarrow x \in C)$. The relation $\prec_{\mathcal{C}}$ is reflexive and symmetric, and it is an equivalence relation iff $\mathcal{C}$ is a partition, i.e. in case of the original Pawlak's rough sets.

[2] Other paraconsistent approaches to rough sets were dicusssed in [34, 33].

2 Covering-based rough sets

In what follows, for any set X, by $\mathcal{P}(X)$ we denote the powerset of X, i.e. the set of all subsets of X, and by $\mathcal{P}^+(X)$ — the set of all nonempty subsets of X.

Definition 1. By a *covering-based approximation space*, or shortly *approximation space*, we mean any ordered pair $\mathcal{A} = (U, \mathcal{C})$, where U is a non-empty universe of *objects*, and $\mathcal{C} \subseteq \mathcal{P}^+(U)$ is a covering of U, i.e. $\bigcup\{C \mid C \in \mathcal{C}\} = \mathcal{U}$.

Definition 2. For any approximation space $\mathcal{A} = (U, \mathcal{C})$:

- The *lower approximation* of a set $X \subseteq U$ in $\mathcal{A}$ is defined as

$$L_{\mathcal{C}}(X) = \{x \in U \mid \forall C \in \mathcal{C}(x \in C \Rightarrow C \subseteq X)\}$$

- The *upper approximation* of a set $X \subseteq U$ in $\mathcal{A}$ is defined as

$$H_{\mathcal{C}}(X) = \bigcup\{C \in \mathcal{C} \mid C \cap X \neq \emptyset\}$$

In view of the above definition, one can say that, given the approximate knowledge about objects of the universe available in the approximation space $\mathcal{A}$:

- $L_{\mathcal{C}}(X)$ is the set of all the objects in U which *certainly* belong to X,

- $H_{\mathcal{C}}(X)$ is the set of all the objects in U which *might* belong to X.

However, it should be stressed that though the above choice of the approximation operations was one of those first introduced and studied most, at present it is just one of over 30 versions considered in the literature. Their extensive survey is given e.g. in [37, 19]. The choice of the most appropriate pair of approximations is made based on the application where they are to be used — see [19, 14] for explanatory examples.

Our choice corresponds here to the approximations $\underline{apr''_{C}}(A), \overline{apr''_{C}}(A)$, mentioned in [37] as one of the pairs satisfying the principle of duality of lower and upper approximations. It should be noted that the lower approximation we have selected is the smallest element-based one [3] and the upper approximation — the largest granular one.

[3]Note that $L_{\mathcal{C}}(X) = \{x : \bigcup MD(x) \subset X\}$, where $MD(x)$ is the maximum description of x considered e.g. in [19, 37].

Such a choice of two extremes is not only interesting as delimiting a kind of border circumscribing all other selections, but also has a practical interpretation. Namely, assume that each set $C \in \mathcal{C}$ represents the set of objects which are equivalent according to some expert e_C — and that we know that one of these experts is surely right, but we do not know which one. As one neither can or should distinguish between equivalent objects, then an object x can classified to $L_{\mathcal{C}}X$ only if all objects equivalent to it in the opinion of any expert are included in X (classification accuracy). Analogously, not to miss any object which might belong to X, we must include in $H_{\mathcal{C}}X$ all objects equivalent in the opinion of some expert to some object in X (classification adequacy). And these conditions correspond exactly to Definition 2.

A rough set based on the covering $\mathcal{C}$ is defined in the traditional way — as a pair consisting of the lower and upper approximations of some subset of the universe.

Definition 3. *By a covering-based rough set over $\mathcal{A} = (U, \mathcal{C})$ we mean a pair of the form $(L_{\mathcal{C}}(X), H_{\mathcal{C}}(X))$, where X is a subset of the universe U.*

The operations of lower and upper approximation defined above have the same basic properties, as in case of "classic" Pawlak's rough sets:

Fact 1. *For any $X, Y \subseteq U$, we have:*

$$L_{\mathcal{C}}(X) \subseteq X \subseteq H_{\mathcal{C}}(X)$$

$$\begin{array}{ll} H_{\mathcal{C}}(X \cup Y) = H_{\mathcal{C}}X \cup H_{\mathcal{C}}Y & L_{\mathcal{C}}(X \cup Y) \supseteq L_{\mathcal{C}}X \cup L_{\mathcal{C}}Y \\ L_{\mathcal{C}}(X \cap Y) = L_{\mathcal{C}}X \cap L_{\mathcal{C}}Y & H_{\mathcal{C}}(X \cap Y) \subseteq H_{\mathcal{C}}X \cap H_{\mathcal{C}}Y \\ L_{\mathcal{C}}(-X) = -H_{\mathcal{C}}X & H_{\mathcal{C}}(-X) = -L_{\mathcal{C}}X \end{array} \tag{1}$$

where none of the inequalities in (1) can be replaced by the equality.

Note that, by the last equality, the operation of higher approximation is dual to the lower one. In consequence, in what follows we shall use the upper approximation only in places where it is commendable for methodological or practical reasons.

Following the example of Pawlak's rough sets, with any subset of a universe U of an approximation space we can associate three regions of that universe: positive, negative and boundary, representing the three basic statuses of membership of an object of the universe U in X:

Definition 4. Let $\mathcal{A} = (U, \mathcal{C})$ be an approximation space, and let $X \subseteq U$.

- The *positive region* of X in the approximation space $\mathcal{A}$ is

$$POS_{\mathcal{C}}(X) = L_{\mathcal{C}}(X)$$

- The *negative region* of X in the approximation space $\mathcal{A}$ is

$$NEG_{\mathcal{C}}(X) = L_{\mathcal{C}}(U - X)$$

- The *boundary region* of X in the approximation space $\mathcal{A}$ is

$$BND_{\mathcal{C}}(X) = U - (POS_{\mathcal{C}}(X) \cup NEG_{\mathcal{C}}(X))$$

Corollary 1. *For any approximation space $\mathcal{A} = (U, \mathcal{C})$ and any $X \subseteq U$:*

$$POS_{\mathcal{C}}(X) = \{x \in U \mid \forall C \in \mathcal{C}(x \in C \Rightarrow C \subseteq X)\}$$

$$NEG_{\mathcal{C}}(X) = \{x \in U \mid \forall C \in \mathcal{C}(x \in C \Rightarrow C \subseteq U - X)\}$$

$$BND_{\mathcal{C}}(X) = \{x \in U \mid \exists C \in \mathcal{C}(x \in C \wedge C \cap X \neq \emptyset \wedge C \cap (U - X) \neq \emptyset\}$$

The regions defined as above are obviously disjoint, and constitute a partition of the universe U. Accordingly, any subset X of the universe can be identified with the three-valued set $(POS_{\mathcal{C}}(X), NEG_{\mathcal{C}}(X), BND_{\mathcal{C}}(X))$, which allows us to build a natural truth-functional logic for reasoning about membership of objects in covering-based rough sets.

Moreover, we can say that, according to the approximate knowledge regarding the objects in U available in the approximation space A:

- Elements of $POS_{\mathcal{C}}(X)$ *certainly belong* to X;
- Elements of $NEG_{\mathcal{C}}(X)$ *certainly do not belong* to X;
- *We cannot tell if* elements of $BND_{\mathcal{C}}(X)$ *belong* to X or not.

As a result, a natural solution for a logic for reasoning about covering-based rough set is — exactly like in case of Pawlak's rough sets — to base its semantics on three truth values: **t** — true, **f** – false, **u** — unknown, corresponding to the positive, negative and boundary region of a set, respectively.

3 Syntax and semantics of the language $\mathfrak{L}_{RS}$

Now we shall introduce the language $\mathfrak{L}_{RS}$ of the three-valued logic for reasoning about covering-based rough sets described in the introduction. Like in [4], formulas of $\mathfrak{L}_{RS}$ will contain expressions representing sets of objects (built out of set variables and set constants by using symbols of set-theoretic operators), variables representing individual objects of the universe, the symbol $\hat{\in}$ of a three-valued binary predicate representing membership of an object in a rough set, and the logical connectives $\neg, \wedge, \vee$, which will be interpreted as negation, conjunction and disjunction in the 3-valued Kleene calculus.

3.1 Syntax of $\mathfrak{L}_{RS}$

Definition 5. Assume that:

- OV is a non-empty denumerable set of *object variables*;
- SV is a non-empty denumerable set of *set variables*;
- $\mathbf{0}$ and $\mathbf{1}$ are *set constants*
- $SV^+ = SV \cup \{\mathbf{0}, \mathbf{1}\}$.

The syntax of the language $\mathfrak{L}_{RS}$ is defined as follows:

1. The set SE of *set expressions* of $\mathfrak{L}_{RS}$ is the least set containing SV^+ and closed under the set-theoretic operators $-, \cup, \cap$;
2. The set of *atomic formulas* of $\mathfrak{L}_{RS}$ is

$$\mathcal{A}_{RS} = \{x \,\hat{\in} A \mid x \in OV, A \in SE\}$$

3. The set $\mathcal{F}_{RS}$ of *formulas* of $\mathfrak{L}_{RS}$ is the least set containing $\mathcal{A}_{RS}$ and closed under the connectives $\neg, \vee, \wedge$.

3.2 Semantic frameworks for $\mathfrak{L}_{RS}$ and interpretation of formulas

The semantics of $\mathfrak{L}_{RS}$ is based on interpreting the formulas of $\mathfrak{L}_{RS}$ in semantic frameworks for that language, built on covering-based approximation spaces and including valuations of set variables, set constants and object variables.

Definition 6. A *semantic framework*, or shortly *framework*, for $\mathfrak{L}_{RS}$ is an ordered triple $\mathcal{R} =< \mathcal{A}, v, w >$, where:

- $\mathcal{A} = (U, \mathcal{C})$ is a covering-based approximation space;
- $v : OV \to U$ is a valuation of object variables;
- $w : SV^+ \to \mathcal{P}(U)$ is a valuation of set variables and constants such that $w(\mathbf{0}) = \emptyset, w(\mathbf{1}) = U$.

For any valuation $w : SV^+ \to \mathcal{P}(U)$, by w^* we shall denote the extension of w to SE obtained by interpreting $-, \cup, \cap$ as the set-theoretic operations of complement, union and intersection. In other words, for any $X \in SV^+$ and any $A, B \in SE$:

$$\begin{array}{ll} w^*(X) = w(X) & w^*(-A) = U - w(A) \\ w^*(A \cup B) = w^*(A) \cup w^*(B) & w^*(A \cap B) = w^*(A) \cap w^*(B) \end{array}$$

Definition 7. An *interpretation* of $\mathfrak{L}_{RS}$ in a framework $\mathcal{R} =< \mathcal{A}, v, w >$, where $\mathcal{A} = (U, \mathcal{C})$, is a mapping $\iota_{\mathcal{R}} : \mathcal{F}_{RS} \to \{\mathbf{t}, \mathbf{f}, \mathbf{u}\}$ defined as follows:

1. For any $x, y \in OV$ and any $A \in SE$,

$$\iota_{\mathcal{R}}(x \mathrel{\hat{\in}} A) = \begin{cases} \mathbf{t} & \text{if } v(x) \in Pos_{\mathcal{C}}(w^*(A)) \\ \mathbf{f} & \text{if } v(x) \in Neg_{\mathcal{C}}(w^*(A)) \\ \mathbf{u} & \text{if } v(x) \in Bnd_{\mathcal{C}}(w^*(A)) \end{cases}.$$

2. For any $\varphi, \psi \in \mathcal{F}$,

- $$\iota_{\mathcal{R}}(\neg\varphi) = \begin{cases} \mathbf{t} & \text{if } \iota_{\mathcal{R}}(\varphi) = \mathbf{f} \\ \mathbf{f} & \text{if } \iota_{\mathcal{R}}(\varphi) = \mathbf{t} \\ \mathbf{u} & \text{if } \iota_{\mathcal{R}}(\varphi) = \mathbf{u} \end{cases}$$

- $$\iota_{\mathcal{R}}(\varphi \vee \psi) = \begin{cases} \mathbf{t} & \text{if either } \iota_{\mathcal{R}}(\varphi) = \mathbf{t} \text{ or } \iota_{\mathcal{R}}(\psi) = \mathbf{t} \\ \mathbf{f} & \text{if } \iota_{\mathcal{R}}(\varphi) = \mathbf{f} \text{ and } \iota_{\mathcal{R}}(\psi) = \mathbf{f} \\ \mathbf{u} & \text{otherwise} \end{cases}$$

- $$\iota_{\mathcal{R}}(\varphi \wedge \psi) = \begin{cases} \mathbf{t} & \text{if } \iota_{\mathcal{R}}(\varphi) = \mathbf{t} \text{ and } \iota_{\mathcal{R}}(\psi) = \mathbf{t} \\ \mathbf{f} & \text{if either } \iota_{\mathcal{R}}(\varphi) = \mathbf{f} \text{ or } \iota_{\mathcal{R}}(\psi) = \mathbf{f} \\ \mathbf{u} & \text{otherwise} \end{cases}$$

It can be easily seen that the interpretation $\iota_{\mathcal{R}}$ is a well-defined mapping of the set of formulas into $\{\mathbf{t}, \mathbf{f}, \mathbf{u}\}$. Indeed, as the regions of a rough set are disjoint, Point 1 provides a well-defined interpretation of atomic formulas. Note that $\hat{\in}$ is interpreted as a three-valued relation of membership of an object x in a set A, with the values $\mathbf{t}, \mathbf{f}, \mathbf{u}$ assigned to objects belonging to the positive region, the negative region and the boundary of that set, respectively, which is compliant with the character of those regions discussed in the foregoing. Further, the clauses for $\neg, \vee, \wedge$ in Point 2 extend $\iota_{\mathcal{R}}$ uniquely to complex formulas by interpreting those conjunctives as negation, disjunction and conjunction in Kleene's three-valued calculus. In the sequel we will drop the subscript $\mathcal{R}$ in $\iota_{\mathcal{R}}$ if $\mathcal{R}$ is arbitrary or understood.

3.3 Satisfaction and consequence relations for formulas and sequents

To complete the definition of the semantics of $\mathfrak{L}_{RS}$, we need to define the notions of satisfaction and the consequence relation. Since the proof system we are going to develop for $\mathfrak{L}_{RS}$ will be a sequent calculus, we will define both the notions for formulas as well as for sequents.

Definition 8.
- *By a* sequent *we mean a structure of the form* $\Gamma \Rightarrow \Delta$*, where* Γ *and* Δ *are finite* sets *of formulas. The set of all sequents over the language* $\mathfrak{L}_{RS}$ *is denoted by* Seq_{RS}*.*
- *A sequent* $\Sigma \in Seq_{RS}$ *is called* atomic *if each formula in* Σ *is atomic.*

Depending on the specific application of rough sets, one can choose either the strong version of the three-valued semantics of $\mathfrak{L}_{RS}$ — with $\mathbf{t}$ as the only designated value, or its weak version — with two designated values: $\mathbf{t}, \mathbf{u}$. The strong version was examined in [3], [4] and [17]. In this paper we turn to the weak semantics, which gives rise to a paraconsistent logic. Consequently, we adopt the following definitions of satisfaction and consequence:

Definition 9.
1. *A formula* $\varphi \in \mathcal{F}_{RS}$ *is* satisfied *by an interpretation* ι *of* $\mathfrak{L}_{RS}$*, in symbols* $\iota \models \varphi$*, if* $\iota(\varphi) \neq \mathbf{f}$.
2. *A formula* $\varphi \in \mathcal{F}_{RS}$ *is* valid*, in symbols* $\models_{RS} \varphi$*, if* $\iota \models \varphi$ *for any interpretation* ι *of* $\mathfrak{L}_{RS}$*.*
3. *A set of formulas* $T \subseteq \mathcal{F}_{RS}$ *is* satisfied *by an interpretation* ι *, in symbols* $\iota \models T$*, if* $\iota \models \varphi$ *for all* $\varphi \in T$*.*
4. *A sequent* $\Sigma = (\Gamma \Rightarrow \Delta)$ *is* satisfied *by an interpretation* ι *, in symbols* $\iota \models \Sigma$*, iff either* $\iota \models \varphi$ *for some* $\varphi \in \Delta$*, or* $\iota \not\models \psi$ *for some* $\psi \in \Gamma$*.*
5. *A sequent* $\Sigma = (\Gamma \Rightarrow \Delta)$ *is* valid*, in symbols* $\models_{RS} \Sigma$*, if* $\iota \models \Sigma$ *for any interpretation* ι *of* $\mathfrak{L}_{RS}$
6. *The* formula consequence relation *in* $\mathfrak{L}_{RS}$ *is the relation* $\vdash_{RS}$ *on* $\mathcal{P}(\mathcal{F}_{RS}) \times \mathcal{F}_{RS}$ *such that, for every* $T \subset \mathcal{F}_{RS}$ *and every* $\varphi \in \mathcal{F}_{RS}, T \vdash_{RS} \varphi$ *if each interpretation* ι *of* $\mathfrak{L}_{RS}$ *which satisfies* T *satisfies* φ *too.*
7. *The* sequent consequence relation *in* $\mathfrak{L}_{RS}$ *is the relation* $\vdash_{RS}$ *on* $\mathcal{P}(Seq_{RS}) \times Seq_{RS}$ *such that, for every* $Q \subseteq Seq_{RS}$*, and every* $\Sigma \in Seq_{RS}, Q \vdash_{RS} \Sigma$ *iff, for any interpretation* ι *of* $\mathfrak{L}_{RS}, \iota \models_{RS} Q$ *implies* $\iota \models_{RS} \Sigma$.

Note that the use of the same symbol for the formula and sequent consequence relations will not lead to misunderstanding, for the meaning of the symbol will always be clear from the context.

It should be also noted that on the atomic formula level the above definition of formula satisfaction is underpinned by the notion of upper approximation:

Corollary 2. *For any framework $\mathcal{R}$, any object variable x and any set expression A, it obtains $\iota_{\mathcal{R}} \models x \mathrel{\hat{\in}} A$ iff $v(x) \in H_{\mathcal{C}}(w^*(A))$.*

The proof follows directly from Definition 7 and from the fact that $H_{\mathcal{C}}(X) = POS_{\mathcal{C}}(X) \cup BND_{\mathcal{C}}(X)$ for any set X.

4 Proof system for the logic $\mathcal{L}_{RS}$

Now we shall present a proof system for the logic $\mathcal{L}_{RS}$ with the language $\mathfrak{L}_{RS}$, corresponding to the consequence relation $\vdash_{RS}$ defined in the preceding section. The deduction formalism we use for $\mathcal{L}_{RS}$ is a Gentzen-style sequent calculus.

Sequent calculus CRS

Axioms: $(A1)\ \ \varphi \Rightarrow \varphi \qquad (A2)\ \ x \mathrel{\hat{\in}} \mathbf{0} \Rightarrow \qquad (A3)\ \ \Rightarrow x \mathrel{\hat{\in}} \mathbf{1}$

Structural rules: weakening, cut

Boolean tautology rules: for any $A, B \in SE$ such that $A = B$ is a Boolean tautology

$$(taut-l)\ \frac{\Gamma, x \mathrel{\hat{\in}} A \Rightarrow \Delta}{\Gamma, x \mathrel{\hat{\in}} B \Rightarrow \Delta} \qquad (taut-r)\ \frac{\Gamma \Rightarrow \Delta, x \mathrel{\hat{\in}} A}{\Gamma \Rightarrow \Delta, x \mathrel{\hat{\in}} B}$$

Union rules:

$$(\cup \Rightarrow)\ \frac{\Gamma, x \mathrel{\hat{\in}} B \Rightarrow \Delta \quad \Gamma, x \mathrel{\hat{\in}} C \Rightarrow \Delta}{\Gamma, x \mathrel{\hat{\in}} B \cup C \Rightarrow \Delta} \qquad (\Rightarrow \cup)\ \frac{\Gamma \Rightarrow \Delta, x \mathrel{\hat{\in}} B, x \mathrel{\hat{\in}} C}{\Gamma \Rightarrow \Delta, x \mathrel{\hat{\in}} B \cup C}$$

Inference rules for Kleene connectives:

$$(\neg \mathrel{\hat{\in}} \Rightarrow)\ \frac{\Gamma, x \mathrel{\hat{\in}} -A \Rightarrow \Delta}{\Gamma, \neg(x \mathrel{\hat{\in}} A) \Rightarrow \Delta} \qquad (\Rightarrow \neg \mathrel{\hat{\in}})\ \frac{\Gamma \Rightarrow \Delta, x \mathrel{\hat{\in}} -A}{\Gamma \Rightarrow \Delta, \neg(x \mathrel{\hat{\in}} A)}$$

$$(\neg\neg \Rightarrow)\ \frac{\Gamma, \varphi \Rightarrow \Delta}{\Gamma, \neg\neg\varphi \Rightarrow \Delta} \qquad (\Rightarrow \neg\neg)\ \frac{\Gamma \Rightarrow \Delta, \varphi}{\Gamma \Rightarrow \Delta, \neg\neg\varphi}$$

$$(\vee \Rightarrow)\ \frac{\Gamma, \varphi \Rightarrow \Delta \quad \Gamma, \psi \Rightarrow \Delta}{\Gamma, \varphi \vee \psi \Rightarrow \Delta} \qquad (\Rightarrow \vee)\ \frac{\Gamma, \Rightarrow \Delta, \varphi, \psi}{\Gamma \Rightarrow \Delta, \varphi \vee \psi}$$

$$(\neg\vee \Rightarrow)\ \frac{\Gamma, \neg\varphi, \neg\psi \Rightarrow \Delta}{\Gamma, \neg(\varphi \vee \psi) \Rightarrow \Delta} \qquad (\Rightarrow \neg\vee)\ \frac{\Gamma \Rightarrow \Delta, \neg\varphi \quad \Gamma \Rightarrow \Delta, \neg\psi}{\Gamma \Rightarrow \Delta, \neg(\varphi \vee \psi)}$$

$$(\wedge \Rightarrow)\ \frac{\Gamma, \varphi, \psi \Rightarrow \Delta}{\Gamma, \varphi \wedge \psi \Rightarrow \Delta} \qquad (\Rightarrow \wedge)\ \frac{\Gamma \Rightarrow \Delta, \varphi \quad \Gamma \Rightarrow \Delta, \psi}{\Gamma \Rightarrow \Delta, \varphi \wedge \psi}$$

$$(\neg\wedge \Rightarrow)\ \frac{\Gamma, \neg\varphi \Rightarrow \Delta \quad \Gamma, \neg\psi \Rightarrow \Delta}{\Gamma, \neg(\varphi \wedge \psi) \Rightarrow \Delta} \qquad (\Rightarrow \neg\wedge)\ \frac{\Gamma \Rightarrow \Delta, \neg\varphi, \neg\psi}{\Gamma \Rightarrow \Delta, \neg(\varphi \wedge \psi)}$$

In all axioms and inference rules above we assume that $x, y, z \in OV$, $A, B \in SE$.

For better clarity, let us now explain the axioms and rules of CRS most relevant to our approach in the context of rough sets.

Consider an arbitrary semantic framework $\mathcal{R} =< \mathcal{A}, v, w >$ with $\mathcal{A} = (U, \mathcal{C})$, and let ι be the interpretation of $\mathfrak{L}_{RS}$ in $\mathcal{R}$.

A1 is the basic sequent axiom, which holds for our logic, too, but A2 and A3 require some comment. By Definition 6, $w(\mathbf{0}) = \emptyset$, so $w^*(\mathbf{0}) = \emptyset$, $NEG_{\mathcal{C}}(w^*(\mathbf{0})) = U$ and $v(x) \in NEG_{\mathcal{C}}(w^*(\mathbf{0}))$ for any x. In consequence, $\iota(x \mathrel{\hat{\in}} \mathbf{0}) = \mathbf{f}$, so $\iota_{\mathcal{R}} \not\models x \mathrel{\hat{\in}} 0$, whence $\iota \models A2$. The dual axiom A3 is justified in an analogous way by the fact that $w^*(\mathbf{1}) = U$, whence for any x we have $v(x) \in POS_{\mathcal{C}}(w^*(\mathbf{1}))$.

In turn, the tautology rules reflect the fact that under the interpretation ι the symbols $-, \cup, \cap$ in set expressions are interpreted as Boolean operations on sets.

Next, in view of the fact that $\iota \models (x \mathrel{\hat{\in}} A)$ iff $v(x) \in H(w^*(A))$ by Corollary 2 , it can be easily shown that the union rules simply reflect the equality $H_{\mathcal{C}}(B \cup C) = H_{\mathcal{C}}(B) \cup H_{\mathcal{C}}(C)$ known from Fact 1. Out of the rules for Kleene connectives, only the first two ones — $(\neg \in\Rightarrow$ and $(\Rightarrow \neg \in)$ — directly involve the rough set framework; all other rules are just standard sequent rules for the weak semantics of Kleene three-valued calculus. As to the mentioned rules, they reflect the fact that $\iota \models \neg(x \mathrel{\hat{\in}} A)$ iff $\iota \models x \mathrel{\hat{\in}} -A$. Indeed: $\iota \models \neg(x \mathrel{\hat{\in}} A)$ iff $\iota(\neg(x \mathrel{\hat{\in}} A)) \neq \mathbf{f}$, which holds iff $\iota(x \mathrel{\hat{\in}} A) \neq \mathbf{t}$. The latter is equivalent to $v(x) \in U - L_{\mathcal{C}}(w^*(A))$. Since by Fact 1 we have $U - L_{\mathcal{C}}(X) = H_{\mathcal{C}}(-X)$ for any set X, then $v(x) \in H_{\mathcal{C}}(U - w^*(A))$. As $U - w^*(A) = w^*(-A)$, we get $v(x) \in H_{\mathcal{C}}(w^*(-A))$, and by Corollary 2 our case is proved.

Reasoning along the lines sketched above, we can easily show the soundness of CRS:

Lemma 1.

1. *The axioms of the system CRS are valid.*

2. *For any inference rule ρ of CRS and any framework $\mathcal{R}$ for $\mathfrak{L}_{RS}$, if the interpretation ι of $\mathfrak{L}_{RS}$ in $\mathcal{R}$ satisfies all the premises of ρ, then ι satisfies the conclusion of ρ as well.*

Clearly, from the above Lemma we can immediately conclude that:

Corollary 3. *The inference rules of CRS are strongly sound, i.e. they preserve the validity of sequents.*

5 Strong soundness and completeness of the proof system

To prove strong completeness of CRS, we start with simple characterization of valid single-variable atomic sequents.

Definition 10. *For any $A, B \in SE$, we say that:*

1. *A is* Boolean-equivalent *to B, and write $A \equiv B$, iff $A = B$ is a Boolean tautology;*

2. *A is* Boolean-included *in B, and write $A \sqsubseteq B$, iff $A \cup B = B$ is a Boolean tautology, i.e., iff $A \cup B \equiv B$.*

Proposition 1. *A sequent $\Sigma = x \hat{\in} A_1, \ldots, x \hat{\in} A_k \Rightarrow x \hat{\in} B_1, \ldots, x \hat{\in} B_l$ is valid iff one of the following conditions is satisfied:*

1. $B_1 \cup B_2 \cup \cdots \cup B_l \equiv \mathbf{1}$

2. $A_i \equiv \mathbf{0}$ *for some* $i \leq k$

3. $A_i \sqsubseteq B_1 \cup B_2 \cup \cdots \cup B_l$ *for some* $i \leq k$

Proof. The backward implication follows easily from Definition 9 and from the semantics of $\mathcal{L}_{RS}$ given in Definition 7. To prove the forward implication, we argue by contradiction.

Assume now that a sequent Σ of the form given above is such that:

(i) $B_1 \cup B_2 \cup \cdots \cup B_l \not\equiv \mathbf{1}$

(ii) $A_i \not\equiv \mathbf{0}$ for each $i \leq k$

(iii) $A_i \not\sqsubseteq B_1 \cup B_2 \cup \cdots \cup B_l$ for each $i \leq k$

We will show that Σ is not valid by constructing a counter-model (precisely speaking, counter-framework) $\mathcal{R}$ for Σ. Define

$$SV_\Sigma = \{X \in SV \mid X \text{ occurs in } \Sigma\}$$
$$SE_\Sigma = \{A \in SE \mid A \text{ contains only set variables in } SV_\Sigma\}$$

As SV_Σ is finite, we have

$$SV_\Sigma = \{X_1, X_2, \ldots, X_n\}$$

for some n. The construction of a counter-model $\mathcal{R}$ for Σ is based on the use of the full disjunctive normal form (DNF) of an expression in SE_Σ with respect to SV_Σ. Such a DNF is the union of expressions of the form

$$\mathbf{X}^{\overline{\epsilon}} = X_1^{\epsilon_1} \cap X_2^{\epsilon_2} \cap \cdots \cap X_n^{\epsilon_n} \tag{2}$$

where $\overline{\epsilon} = (\epsilon_1, \epsilon_2, \ldots, \epsilon_n) \in \{-1, 1\}^n$ and $X_j^1 = X_j, X_j^{-1} = -X_j$.

Now let us define

$$B = B_1 \cup B_2 \cup \cdots \cup B_l$$

As $\mathrm{DNF}(E) \equiv E$ for any $E \in SE_\Sigma$, from (iii) we get $\mathrm{DNF}(A_i) \not\subseteq \mathrm{DNF}(B)$ for each $i \leq k$. Hence for each $i \leq k$ there is $\overline{\epsilon}^{j_i} \in \{-1, 1\}^n$ such that

$$\mathbf{X}^{\overline{\epsilon}_{j_i}} \in \mathrm{DNF}(A_i) \setminus \mathrm{DNF}(B) \tag{3}$$

Let us assign a unique symbol $a^{\overline{\epsilon}} \notin OV \cup SV$ to any $\overline{\epsilon} \in \{-1, 1\}^n$. As the universe of our counter-model $\mathcal{R}$ we take

$$U = \{x\} \cup \{a^{\overline{\epsilon}} \mid \overline{\epsilon} \in \{-1, 1\}^n\}$$

Define

$$\omega(\mathbf{X}^{\overline{\epsilon}}) = \begin{cases} \{x, a^{\overline{\epsilon}}\} & \text{if } \overline{\epsilon} = \overline{\epsilon}_{j_1} \\ \{a^{\overline{\epsilon}}\} & \text{otherwise} \end{cases} \tag{4}$$

Then it is easy to see that ω maps the set od all DNF components of the form $X^{\overline{\epsilon}}$ to a partition of U. In consequence, by the well-known properties of such components ω defines a unique valuation w of set variables in S_Σ such that $w^*(X^{\overline{\epsilon}}) = \omega(\mathbf{X}^{\overline{\epsilon}})$ for any $\overline{\epsilon} \in \{-1, 1\}^n$.

Define the covering underlying the approximation space of our intended counter-model by

$$\mathcal{C} = \{C(u) \mid u \in U\}$$

where

$$C(u) = \begin{cases} \{x, a^{\bar{\epsilon}_{j_1}}, \ldots, a^{\bar{\epsilon}_{j_k}}\} & \text{if } u = x \\ \{u\} & \text{otherwise} \end{cases} \tag{5}$$

Finally, define the valuation of object variables by $v(y) = x$ for any $y \in OV$. Then $\mathcal{R} =< (U, \mathcal{C}), v, w >$ is a well-defined semantic framework for $\mathcal{L}_{RS}$. We will now show that it represents the desired counter-model for Σ, i.e. that the interpretation $\iota_{\mathcal{R}}$ of $\mathcal{L}_{RS}$ in $\mathcal{R}$ does not satisfy Σ.

First, by (3) we have

$$a^{\bar{\epsilon}_{j_i}} \in w^*(A_i) \text{ for } i = 1, 2, \ldots, k$$

Thus $C(x) \cap w^*(A_i) \supseteq \{a^{\bar{\epsilon}_{j_i}}\} \neq \emptyset$ for each $i \leq k$ by (5). As $C(x) \in \mathcal{C}$ and $x \in C(x)$, from Corollary 1 we obtain $x \notin NEG(w^*(A_i))$. In consequence, by Definition 9 $\iota_{\mathcal{R}} \models x \,\hat{\in} A_i$ for $i = 1, 2, \ldots, k$.

On the other hand, as by (3) $\mathbf{X}^{\bar{\epsilon}_{j_i}}$ does not occur in DNF(B) for any $i \leq k$ and DNF$(B) \equiv B$, it obtains $a^{\bar{\epsilon}_{j_i}} \notin w^*(B)$ for $i = 1, 2, \ldots, k$ by (4). What is more, as in particular $\mathbf{X}^{\bar{\epsilon}_{j_1}}$ does not occur in DNF(B), we have $x \notin w^*(B)$ by (4) too. By (5), this yields $C(x) \cap w^*(B) = \emptyset$. Given that $C(x)$ is the only set $C \in \mathcal{C}$ with $x \in C$, by Corollary 1 we get $x \in$ NEG$(w^*(B))$, and so $\iota_{\mathcal{R}} \not\models x \,\hat{\in} B$. Thus $\iota_R \not\models \Sigma$, which ends the proof. □

Considering that $\mathfrak{L}_{RS}$ has no means for expressing relationships between object variables, Proposition 1 implies a similar result for multi-variable atomic sequents:

Corollary 4. *An atomic sequent $\Sigma \in Seq_{RS}$ is valid if and only if, for some object variable x occurring in Σ, the sequent Σ_x obtained from Σ by deleting all formulas with variables different from x satisfies one of the conditions of Proposition 1.*

The proof is analogous to that of Proposition 1, with the counter-model for a sequent Σ which does not satisfy any of conditions 1.,2.,3. of that Proposition constructed by combining the individual counter-models for all single-variable subsequents of Σ, constructed exactly like in the proof of Proposition 1.

As a crucial step towards proving strong completeness of CRS, we will now show that result for atomic sequents:

Proposition 2. *If an atomic sequent $\Sigma \in Seq_{RS}$ is valid, then it is derivable in CRS, i.e. $\vdash_{CRS} \Sigma$.*

Proof. For any variable x occurring in Σ, denote by Σ_x the atomic sequent obtained out of Σ by deleting all formulas with variables different from x. Since Σ is valid, then, by Corollary 4, there exists an x such that Σ_x satisfies one of the conditions 1., 2., 3. of Proposition 1. Hence, assuming that

$$\Sigma_x = x \mathrel{\hat{\in}} A_1, \ldots, x \mathrel{\hat{\in}} A_k \Rightarrow x \mathrel{\hat{\in}} B_1, \ldots, x \mathrel{\hat{\in}} B_l$$

we have

(1) either $B_1 \cup B_2 \cup \cdots \cup B_l \equiv \mathbf{1}$, or

(2) $A_i \equiv \mathbf{0}$ for some $i \leq k$, or

(3) $A_i \sqsubseteq B_1 \cup B_2 \cup \cdots \cup B_l$ for some $i \leq k$

If (1) holds, then from Axiom A1 and rule $(\Rightarrow \cup)$ applied $l-1$ times we obtain $\vdash_{CRS} x \mathrel{\hat{\in}} A_1, \ldots, x \mathrel{\hat{\in}} A_k \Rightarrow x \mathrel{\hat{\in}} (B_1 \cup \cdots \cup B_l)$. Considering that $B_1 \cup B_2 \cup \cdots \cup B_l \equiv \mathbf{1}$, from Axiom A3 and rule $(taut - r)$ we get $\vdash_{CRS} \Rightarrow x \mathrel{\hat{\in}} (B_1 \cup B_2 \cup \cdots \cup B_l)$. By weakening, this yields $\vdash_{CRS} x \mathrel{\hat{\in}} A_1, \ldots, x \mathrel{\hat{\in}} A_k \Rightarrow x \mathrel{\hat{\in}} (B_1 \cup B_2 \cup \cdots \cup B_l)$, whence $\vdash_{CRS} \Sigma_x$.

If (2) holds, then from Axiom A2 and rule (taut-l) we get $\vdash_{CRS} x \in A_i \Rightarrow$, whence $\vdash_{CRS} \Sigma_x$ by weakening.

Finally, let us assume that (3) holds. For simplicity, denote $B = B_1 \cup B_2 \cup \cdots \cup B_l$. Then $A_i \sqsubseteq B$, which implies (*) $A_i \cup B \equiv B$ by Definition 10.

From Axiom A1 by weakening we get $\vdash_{CRS} x \mathrel{\hat{\in}} A_i \Rightarrow x \mathrel{\hat{\in}} A_i, x \in B$. By rule $(\Rightarrow \cup)$, this yields $\vdash_{CRS} x \mathrel{\hat{\in}} A_i \Rightarrow x \mathrel{\hat{\in}} (A_i \cup B)$. In view of (*) and rule $(taut - r)$, we obtain $\vdash_{CRS} x \mathrel{\hat{\in}} A_i \Rightarrow x \mathrel{\hat{\in}} B$. By weakening, the latter implies again $\vdash_{CRS} \Sigma_x$.

Thus $\vdash_{CRS} \Sigma_x$ in all three cases. As $\Sigma_x \subset \Sigma$ in the standard sense of sequent inclusion, this implies $\vdash_{CRS} \Sigma$ by weakening. □

Proposition 2 is the cornerstone for proving the strong completeness theorem for the logic $\mathcal{L}_{RS}$:

Theorem 1. *The calculus CRS is finitely strongly sound and complete for* $\vdash_{RS}$, *i.e., for any finite set of sequents* $S \subseteq Seq_{RS}$ *and any sequent* $\Sigma \in Seq_{RS}$, $S \vdash_{RS} \Sigma$ *iff* $S \vdash_{CRS} \Sigma$.

Proof. (Sketch) As the backward implication (soundness) follows from Lemma 1 and Corollary 3, it suffices to prove the forward implication (completeness). The proof is by counter-model construction based on Proposition 2 and the maximum saturated sequent construction used e.g. in [1].

We argue by contradiction. Suppose that for a finite set of sequents S and a sequent $\Sigma = \Gamma \Rightarrow \Delta$ we have $S \vdash_{RS} \Sigma$, but Σ is not derivable from S in CRS. We shall construct a counter-model ι such that $\iota \models S$ but $\iota \not\models \Sigma$.

Denote by $F(S)$ the set of all formulae belonging to at least one of the sides in some sequent in S, and let SV^* be the set of all set variables which occur either in some $\varphi \in F(S)$ or in Σ. Since S is finite, so are $F(S)$ and SV^*. Using the method shown in in [1], we can construct a sequent $\Gamma' \subseteq \Delta'$ such that

(i) $\Gamma \subseteq \Gamma', \Delta \subseteq \Delta'$

(ii) $F(S) \subseteq \Gamma' \cup \Delta'$.

(iii) $\Gamma' \Rightarrow \Delta'$ is not derivable from S in CRS.

The construction is carried out by starting with Σ, and then adding consecutively linearly ordered formulas in $F(S)$ to either the left- or the right-hand side of the sequent constructed up to that time, depending on which option results in a sequent still not derivable from S in CRS. Such a construction is possible because if $S \not\vdash_{CRS}$ $(\Gamma_i \Rightarrow \Delta_i)$, then, for any $\varphi \in F(S)$, we cannot have both $S \vdash_{CRS} (\Gamma_i \Rightarrow \Delta_i, \varphi)$ and $S \vdash_{CRS} (\Gamma_i, \varphi \Rightarrow \Delta_i)$, since this would imply $S \vdash_{CRS} (\Gamma_i \Rightarrow \Delta_i)$ by cut.

Call a sequent *saturated* if it is closed under the inference rules in CRS applied backwards, whereby we assume that closure under the Boolean tautology rules $(taut - l), (taut - r)$ is limited only to premises with the set expression A in a full disjunctive normal form with respect to the set SV^*. By way of example, a sequent $\Gamma'' \Rightarrow \Delta''$ is closed under rule $(\vee \Rightarrow)$ applied backwards iff $\varphi \vee \psi \subseteq \Gamma''$ implies either $\varphi \in \Gamma''$ or $\psi \in \Gamma''$.

Let $\Gamma^* \Rightarrow \Delta^*$ be the extension of $\Gamma' \Rightarrow \Delta'$ to a saturated sequent which is not derivable from $F(S)$ in CRS (is is easy to see that such a sequent exists; note that the restriction on the closure under tautology rules ensures that the closure adds only a finite number of formulas to $\Gamma' \Rightarrow \Delta'$.

Then we can easily see that:

(1) $\Gamma \subseteq \Gamma^*, \Delta \subseteq \Delta^*$

(2) $F(S) \subseteq \Gamma^* \cup \Delta^*$

(3) $\Gamma^* \Rightarrow \Delta^*$ is saturated and it is not derivable from S in CRS

Now let $\Sigma_a = \Gamma_a \Rightarrow \Delta_a$ be a subsequent of $\Gamma^* \Rightarrow \Delta^*$ consisting of all atomic formulas in $\Gamma^* \Rightarrow \Delta^*$. Then by (3) $S \not\vdash_{CRS} \Sigma_a$, and hence also $\not\vdash_{CRS} \Sigma_a$. As Σ_a is atomic, by Proposition 2, this implies that Σ_a is not valid. Accordingly, there exists a framework $\mathcal{R}$ for $\mathfrak{L}_{RS}$ and an interpretation ι of $\mathfrak{L}_{RS}$ in $\mathcal{R}$ such that $\iota \not\models \Sigma_a$. We shall prove that ι is the desired counter-model for the original sequent Σ too, i.e. that:

$$\textbf{(A)}\ \iota \not\models (\Gamma \Rightarrow \Delta) \qquad \textbf{(B)}\ \iota \models \Sigma' \text{ for each } \Sigma' \in S$$

Let us start with (A). As $\Gamma \subseteq \Gamma^*, \Delta \subseteq \Delta^*$, then in order to prove (A) it suffices to show that $\iota \not\models (\Gamma^* \Rightarrow \Delta^*)$. Since the set of designated values for the weak semantics of $\mathfrak{L}_{RS}$ is $\{\mathbf{t}, \mathbf{u}\}$ and $\iota(\varphi) \in \{\mathbf{t}, \mathbf{f}, \mathbf{u}\}$ for any formula $\varphi \in F_{RS}$, this means we have to prove that:

$$\iota(\gamma) \in \{\mathbf{t}, \mathbf{u}\} \text{ for any } \gamma \in \Gamma^* \qquad \iota(\delta) = \mathbf{f} \text{ for any } \delta \in \Delta^* \tag{6}$$

As $\iota \not\models \Sigma_a$, then (6) holds for all atomic formulas $\gamma \in \Gamma^*, \delta \in \Delta^*$. To show that it holds for complex formulas too, we prove that, for any complex formula φ, the following is true:

$$\textbf{(A1)} \quad \iota(\varphi) \in \begin{cases} \{\mathbf{t}, \mathbf{u}\} & \text{if } \varphi \in \Gamma^* \\ \{\mathbf{f}, \mathbf{u}\} & \text{if } \neg\varphi \in \Gamma^* \end{cases} \qquad \textbf{(A2)} \quad \iota(\varphi) = \begin{cases} \mathbf{f} & \text{if } \varphi \in \Delta^* \\ \mathbf{t} & \text{if } \neg\varphi \in \Delta^* \end{cases}$$

The proof is by induction on the complexity of φ, and (A1) and (A2) are proved simultaneously, making use of the fact that Σ^* as a saturated sequent is closed under all rules in CRS applied backwards.

To illustrate the method used, consider first the case of $\xi = \neg(x \mathbin{\hat{\in}} A)$.

If $\xi \in \Gamma^*$, then $x \mathbin{\hat{\in}} -A$ is also in Γ^*, since Σ^* is closed under rule $(\neg \mathbin{\hat{\in}} \Rightarrow)$ applied backwards. As (6) holds for all atomic formulas and $x \mathbin{\hat{\in}} -A$ is atomic, this yields $\iota(x \mathbin{\hat{\in}} -A) \in \{\mathbf{t}, \mathbf{u}\}$. However, from Definition 7 and Corollary 1 we can easily conclude that

$$\iota(x \mathbin{\hat{\in}} A) = \begin{cases} \mathbf{t} & \text{iff } \iota(x \mathbin{\hat{\in}} -A) = \mathbf{f} \\ \mathbf{f} & \text{iff } \iota(x \mathbin{\hat{\in}} -A) = \mathbf{t} \\ \mathbf{u} & \text{iff } \iota(x \mathbin{\hat{\in}} -A) = \mathbf{u} \end{cases} \tag{7}$$

which implies $\iota(x \mathbin{\hat{\in}} A) \in \{\mathbf{f}, \mathbf{u}\}$ and $\iota(\xi) = \iota(\neg(x \mathbin{\hat{\in}} A)) \in \{\mathbf{t}, \mathbf{u}\}$ by Definition 7.

Next, if $\neg\xi \in \Gamma^*$, then $\neg\neg(x \mathbin{\hat{\in}} A)$ is in Γ^*, whence also $x \mathbin{\hat{\in}} A$ is in Γ^* by rule $(\neg\neg \Rightarrow)$. As $x \mathbin{\hat{\in}} A$ is atomic, this yields $\iota(x \mathbin{\hat{\in}} A) \in \{\mathbf{t}, \mathbf{u}\}$ by (6), implying $\iota(\xi) \in \{\mathbf{f}, \mathbf{u}\}$ by Def. 7.

In turn, if $\xi \in \Delta^*$, then $x \mathbin{\hat{\in}} -A$ is also in Δ^*, since Σ^* is closed under rule $(\Rightarrow \neg \mathbin{\hat{\in}})$ applied backward. As (6) holds for $x \mathbin{\hat{\in}} -A$, then $\iota(x \mathbin{\hat{\in}} -A) = \mathbf{f}$, whence in view of (7) we get $\iota(x \mathbin{\hat{\in}} A) = \mathbf{t}$. In consequence, $\iota(\xi) = \iota(\neg(x \mathbin{\hat{\in}} A)) = \mathbf{f}$ by Def. 7.

Finally, if $\neg\xi \in \Delta^*$, then $x \mathbin{\hat{\in}} A$ is again in Δ^* by rule $(\Rightarrow \neg\neg)$. Hence $\iota(x \mathbin{\hat{\in}} A) = \mathbf{f}$ by (6), which implies $\iota(\xi) = \mathbf{t}$ by Def. 7. Thus (A1) and (A2) hold for ξ.

As another example, assume that (A1), (A2) hold for φ, ψ, and that $\xi = \varphi \vee \psi$. If $\xi \in \Gamma^*$, then either $\varphi \in \Gamma^*$ or $\psi \in \Gamma^*$, since Σ^* is closed under rule $(\vee \Rightarrow)$ applied backwards. As a result, by the inductive assumption on φ, ψ we have either $\iota(\varphi) = \mathbf{t}$ or $\iota(\psi) = \mathbf{t}$, and consequently $\iota(\xi) = \mathbf{t}$ by Definition 7. In turn, if $\xi \in \Delta^*$, then $\varphi, \psi \in \Delta^*$, and $\iota(\varphi), \iota(\psi) \in \{\mathbf{f}, \mathbf{u}\}$ by the inductive assumption, whence $\iota(\xi) \in \{\mathbf{f}, \mathbf{u}\}$ by Definition 7, too. As a result, (A1) and (A2) hold for ξ too.

The proof of other cases is similar, and is left to the reader.

It remains to prove (B), i.e., to show that $\iota \models \Sigma_0$ for each $\Sigma_0 \in S$. So let $\Sigma_0 \in S$. Then $\Sigma_0 = \varphi_1, \ldots, \varphi_k \Rightarrow \psi_1, \ldots, \psi_l$ for some integers k, l and formulas $\varphi_i, \psi_j, i = 1, \ldots, k, j = 1, \ldots, l$. Clearly, we cannot have both $\{\varphi_1, \ldots, \varphi_k\} \subseteq \Gamma^*$ and $\{\psi_1, \ldots, \psi_l\} \subseteq \Delta^*$, for then $\Gamma^* \Rightarrow \Delta^*$ would be derivable from Σ_0, and hence from S, by weakening. Since $F(S) \subseteq \Gamma^* \cup \Delta^*$, this implies that either $\varphi_i \in \Delta^*$ for some i, or $\psi_j \in \Gamma^*$ for some j. Hence by (A1) and (A2), which we have already proved, we have either $\iota \not\models \varphi_i$ for some i, or $\iota \models \psi_j$ for some j, which implies that $\iota \models \Sigma$.

□

6 Conclusions and future work

The crucial feature of the three-valued logic presented in the paper, which distinguishes our approach from others, is the use of variables representing individual objects (elements) of the universe, and of the membership predicate representing three-valued rough membership of objects in subsets of the universe. This enables reasoning about membership of objects in rough sets rather than about rough sets themselves, i.e. on the object (universe element) level rather than on the rough set level. In this way, we avoid the problems involved in the latter approach, described in detail in [10]. Such a solution also allows us to ensure some compositionality of the semantics despite the problems indicated in [9] and to obtain a strongly sound and complete sequent calculus for the logic.

The three values $\mathbf{t}, \mathbf{f}, \mathbf{u}$ taken by the rough membership relation correspond to "crisp" membership of objects in the three basic regions of a rough set: the positive, negative and boundary one. The weak version of semantics with the two designated values $\mathbf{t}, \mathbf{u}$ adopted in the paper amounts to identifying membership of an object x in a rough set A with its belonging to either the positive region or the boundary region of A. In other words, the only elements excluded from membership in A by the logic are those located in the negative region of A.

The calculus of ordinary sequents used here allows for two-way decision making. Three-way decision rules are not truly necessary here, since — like in case of classical rough sets — the boundary region is the exact complement of the union of the negative region and the positive region. Nevertheless, such rules can be obtained by developing a calculus of three-place sequents for the considered semantics, which can be easily effected using the general method described in [6].

The use of connectives to form complex formulas enhances the expressive power of the language, but the 3-valued Kleene connectives used here are just one possible choice. Other interesting option, to be explored in the future, are Łukasiewicz 3-valued connectives (including implication), and the non-deterministic connectives observing the rough set Nmatrix considered in [3]. Exploring these choices is another direction for future work. Still another is to consider a richer language which allows for expressing relationships between objects — and here the most immediate future task will be extending the results of this paper to a language featuring the subordination relation of [17].

Finally, two interesting research directions will be to consider approximate inference rules, and to apply the approach used here to other versions of the lower and upper approximations.

Acknowledgment

The authors would like to thank the anonymous referees for their very helpful comments on the paper, including suggestions for the directions of future work — which we have included above and shall follow in future.

References

[1] Avron, A., Konikowska., B. and Ben-Naim, J.: Processing Information from a Set of Sources. In: Towards Mathematical Philosophy, Series: Trends in Logic , Vol. 28, Makinson, David; Malinowski, Jacek; Wansing, Heinrich (Eds.), pp.165–186, Springer Verlag (2008)

[2] Avron, A.: Logical Non-determinism as a Tool for Logical Modularity: An Introduction. In: We Will Show Them: Essays in Honor of Dov Gabbay, Vol 1 (S. Artemov, H. Barringer, A. S. d'Avila Garcez, L. C. Lamb, and J. Woods, eds.), pp. 105–124. College Publications (2005).

[3] Avron, A. and Konikowska., B.: Rough Sets and 3-valued Logics. Studia Logica, vol. **90** (1), pp. 69–92 (2008).

[4] Avron, A. and Konikowska, B.: Reasoning about Rough Sets Using Three Logical Values, in: Trivalent Logics and their applications, Proceedings of the ESSLLI 2012 Workshop, Paul Égré and David Ripley eds., http://paulegre.free.fr/TrivalentESSLLI/esslli_trivalent_proceedings.pdf, pp. 39–52 (2012).

[5] Avron, A. and Lev, I.: Non-deterministic Multiple-valued Structures. Journal of Logic and Computation 15, pp. 241–261 (2005).

[6] Avron, A., Konikowska, B., Zamansky, A.: Efficient reasoning with inconsistent information using C-systems, Information Sciences 296, pp. 219–236 (2015)

[7] Balbiani, P. and Vakarelov, D.: A modal Logic for Indiscernibility and Complementarity in Information Systems. Fundamenta Informaticae 45, pp. 173–194 (2001).

[8] Banerjee, M.: Rough sets and 3-valued Lukasiewicz logic. Fundamenta Informaticae 32, pp. 213–220 (1997).

[9] Dubois D., Prade H.: Can We Enforce Full Compositionality in Uncertainty Calculi? AAAI 1994: pp. 149-154.

[10] Ciucci D., Dubois D.: Three-Valued Logics, Uncertainty Management and Rough Sets. Trans. Rough Sets 17, pp.1-32 (2014)

[11] Demri, S., Orłowska, E., Vakarelov, D.: Indiscernibility and complementarity relations in information systems. In: Gerbrandy, J., Marx, M., de Rijke, M., Venema, Y. (eds.) JFAK: Esays dedicated to Johan van Benthem on the ocasion of his 50-th Birthday. Amsterdam University Press (1999).

[12] Deneva, A. and Vakarelov, D.: Modal Logics for Local and Global Similarity Relations. Fundamenta Informaticae, vol 31, No 3,4, pp. 295–304 (1997).

[13] Duentsch, I. Konikowska, B.: A multimodal logic for reasoning about complementarity. Journal for Applied Non-Classical Logics, Vol. 10, No 3–4, pp. 273-302 (2000).

[14] Inuiguchi, M.: Generalization of Rough Sets and Rule Extraction. In: J.F. Peters, A. Skowron (eds.) Transactions on Rough Sets-1, vol. LNCS-3100, Springer Verlag, pp. 96–116. (2004)

[15] Kleene, S.C.: Introduction to metamathematics, D. van Nostrad Co. (1952).

[16] Konikowska, B.: A logic for reasoning about relative similarity. Special Issue of Studia Logica, E. Orłowska, H. Rasiowa eds., Reasoning with incomplete information. Studia Logica 58, pp. 185–226 (1997).

[17] Konikowska, B.: Three-Valued Logic for Reasoning about Covering-based Rough Sets. In: Professor Zdzisław Pawlak in Memoriam, Volume 1, Andrzej Skowron and Zbigniew Suraj (eds.), Intelligent Systems Reference Library Series, Volume 42, Springer Verlag, pp. 439-461 (2012).

[18] Lin, T.Y. and Cercone, N. (eds.): Rough sets and Data Mining. Analysis of Imprecise Data, Kluwer, Dordrecht (1997).

[19] Mani, A.: Dialectics of Counting and the Mathematics of Vagueness. In: J.F. Peters, A. Skowron (eds.) Transactions on Rough Sets , LNCS 7255, vol. XV, Springer Verlag, pp. 122–180 (2012)

[20] Mani, A.: Probabilities, Dependence and Rough Membership Functions. In International Journal of Computers and Applications 39(1), pp. 17–35 (2016)

[21] Øhrn, A., Komorowski, J., Skowron, A. and Synak, P.: The design and implementation of a knowledge discovery toolkit based on rough sets — The ROSETTA system. In: Polkowski, L. and Skowron, A. (eds.), Rough Sets in Knowledge Discovery 1. Methodology and Applications. Physica Verlag, Heidelberg, pp.

[22] Minghui M., Mihir K. Chakraborty: Covering-based rough sets and modal logics. Part I. Int. J. Approx. Reasoning 77, pp. 55-65 (2016)

[23] Minghui M., Mihir K. Chakraborty: Covering-based rough sets and modal logics. Part II. Int. J. Approx. Reasoning 95, pp. 113- 123 (2018).

[24] Orłowska, E.: Reasoning with Incomplete Information: Rough Set Based Information

Logics. In: Proceedings of SOFTEKS Workshop on Incompleteness and Uncertainty in Information Systems, pp.16-33, (1993).

[25] Pagliani, P. Rough set theory and logic-algebraic structures. In: Orłowska, E. (editor), Incomplete Information: Rough Set Analysis. Studies in Fuzziness and Soft Computing, vol. 13, pp. 109–190, Physica-Verlag (1998).

[26] Pawlak, Z.: Rough Sets, Intern. J. Comp. Inform. Sci., 11, 341–356 (1982).

[27] Pawlak, Z.: Rough Sets. Theoretical Aspects of Reasoning about Data. Kluwer, Dordrecht (1991).

[28] Pawlak, Z.: Rough set approach to knowledge-based decision support, European Journal of Operational Research 29(3), pp. 1–10 (1997).

[29] Pawlak, Z.: Rough sets theory and its applications to data analysis. Cybernetics and Systems 29, pp. 661–688 (1998).

[30] Pomykała, J.A.: Approximation operations in approximation space. Bull. Pol. Acad. Sci. 35(9-10), pp. 653–662 (1987).

[31] Sen J., Chakraborty, M.K.: A study of intenconnections between rough and 3-valued Łukasiewicz logics. Fundamenta Informaticae 51, 311–324 (2002).

[32] Vakarelov, D.: Information Systems, Similarity Relations and Modal Logics. In: E. Orlowska (ed.) Incomplete Information: Rough Set Analysis, pp. 492–550. Studies in Fuzziness and Soft Computing, Physica-Verlag Heidelberg New York (1998).

[33] Viana H., Alcântara J., Martins A.T.: Paraconsistent rough description logic. In: Proccedings of CEUR-WS.org, Barcelona, Spain (2011)

[34] Vitória, A., Małuszyński, J., Szałas, A.: Modeling and Reasoning with Paraconsistent Rough Sets Fundamenta Informaticae 97(4), pp. 405-438 (2009).

[35] Yao, Y.Y.: Relational interpretations of neighborhood operators and rough set approximation operators. Information Sciences 111 (1–4), pp. 239–259 (1998).

[36] Yao, Y.Y.: On generalizing rough set theory. In: The 9th International Conference on Rough Sets, Fuzzy Sets, Data Mining and Granular Computing (RSFDGrc) 2003. LNCS vol. 2639, pp. 44–51 (2003).

[37] Yao, Y., Yao, B.: Covering based rough set approximations. Information Sciences 200, pp. 91–107 (2012)

[38] Żakowski, W.: On a concept of rough sets. Demonstratio Mathematica XV, 1129–1133 (1982).

[39] Zhang, Y.-L. Li, J.J. and Wu, W.-Z.: On axiomatic characterizations of three types of covering-based approximation operators. Information Sciences 180, pp. 174–187 (2010).

[40] Zhu, W.: Relationship among basic concepts in covering-based rough sets. Information Sciences 179, pp. 2478–2486 (2009)

[41] Zhu, W., Wang, F.-Y.: On three types of covering-based rough sets. IEEE Transactions on Knowledge and Data Engineering 19(8), pp. 1131–1144 (2007).

 Received 28 February 2018

Preservation properties of de Finetti Coherence

Daniele Mundici
Department of Mathematics and Computer Science, 50134 Florence, Italy
mundici@math.unifi.it

Abstract

Suppose elements $a_1, \ldots, a_{k-1}$ of a boolean algebra A are assigned fixed truth values $\rho_1, \ldots, \rho_{k-1} \in \{0,1\}$, and an element a_k is tentatively assigned a probability value $\rho \in [0,1]$. Let $a_{k+1} \in A$. De Finetti showed that there is a closed interval $\mathcal{I}(\rho) \subseteq [0,1]$ such that the set of probabilities of a_{k+1} which are *coherent* with the probability assignment ρ coincides with $\mathcal{I}(\rho)$. Now suppose ρ undergoes a small perturbation $\rho \to \rho + \mathrm{d}\rho$. Using the preservation properties of coherent sets of betting odds, we study the resulting modification $\mathcal{I}(\rho) \to \mathcal{I}(\rho + \mathrm{d}\rho)$.

Keywords: Dutch Book, de Finetti coherent bet, coherent probability assessment, measure on a boolean algebra, de Finetti Dutch Book theorem, de Finetti fundamental theorem on prevision

1 The dependence of a conclusion on an uncertain premise

For some polynomial $p\colon \mathbb{N} \to \mathbb{N}$ let us suppose that the nondeterministic Turing machine $\mathcal{M}$ recognizes instances x of an NP-problem M in $p(|x|)$ computation steps, with $|x|$ the number of symbols of x. Pick a string y over the alphabet of M and let $\{\phi_1, \ldots, \phi_{k+1}\}$ be the set of boolean formulas given by Cook's reduction [3] of the instance y of M to an instance of the boolean satisfiability problem SAT. Let us assume that ϕ_{k+1} states "$\mathcal{M}$ is in an accepting state, at time $p(|y|)$". If each formula $\phi_1, \ldots, \phi_k$ has been assigned the truth value 1 then the compatibility of this truth value assignment with the assignment of truth value 1 to ϕ_{k+1} is equivalent to saying "$\mathcal{M}$ accepts y within $p(|y|)$ steps".

Now suppose M is replaced by a problem N involving a variant $\mathcal{N}$ of $\mathcal{M}$ having the following properties: a certain set $\Psi = \{\psi_1, \ldots, \psi_h, \psi_{h+1}\}$ of boolean formulas gives a complete description of the computation tree of $\mathcal{N}$ over input y, the formula ψ_{h+1} states "$\mathcal{N}$ is in an accepting state", but the formula ψ_h can only be assigned a probability $\rho_h \in [0,1]$, rather than the truth value 1. Then one may naturally ask which assignments of a probability value ρ_{h+1} to b_{h+1} are "compatible" with the assignment $\psi_1 \mapsto 1, \ldots, \psi_{h-1} \mapsto 1$, $\psi_h \mapsto \rho_h$. De Finetti gave a satisfactory definition of "compatible, or coherent, probability assignment" and showed that the set of coherent probabilities of ψ_{h+1} is a closed interval $\mathcal{I} \subseteq [0,1]$, which depends on the value of ρ_h, (see Theorem 3.3). Using the preservation properties of de Finetti coherent sets of betting odds, we study how $\mathcal{I}$ is modified by a small perturbation $\rho_h \to \rho_h + \mathrm{d}\rho_h$ of the probability of ψ_h.

The prerequisites for this paper are some acquaintance with boolean algebras, [21, introductory sections], de Finetti's notion of a coherent set of betting odds [5, pp.311-312], [6, Chapter 1], and his Dutch book theorem, [5, §§8-9], [6, pp. 7–8]. To help the reader, all necessary background material will be provided in the text.

Throughout this paper, the adjective "linear" is understood in the affine sense.

2 The convex set of states of a boolean algebra

Definition 2.1. Let A be a boolean algebra. A map $\mathbf{s}\colon A \to [0,1]$ is said to be a *state* of A if it is additive on incompatible elements (in the sense that $x \wedge y = 0$ implies $\mathbf{s}(x \vee y) = \mathbf{s}(x) + \mathbf{s}(y)$), and $\mathbf{s}(1) = 1$.

Remark 2.2. In [21, §3, (C)], $\mathbf{s}$ is said to be a "measure satisfying $\mathbf{s}(1) = 1$". In [13], $\mathbf{s}$ is said to be a "measure". Also see [14]. Alternative terminologies include variants of "normalized finitely additive probability measure". Our present terminology in this paper is more customary when one regards any boolean algebra as an idempotent MV-algebra, following Chang [2, Theorem 1.17]. As a matter of fact, in [17, Theorem 2.4] (also see [19, Proposition 10.3]) is it shown that *the states of any MV-algebra A are in one-one correspondence with the states of the unital abelian ℓ-group (G,u) associated to A by the categorical equivalence* Γ established in [16, Theorem 3.9]. See [12] for a detailed account on states of unital partially ordered abelian groups. When G is countable, Elliott's classification, [4, 9, 10] shows that (G,u) *corresponds, via Grothendieck's K_0-functor, to precisely one AF C*-algebra* A, *and the states of (G,u) correspond to the (tracial) states of* A. States of AF C*-algebras provide a mathematical representation of the states of certain systems arising in quantum statistical mechanics and quantum field theory, [1]. As noted in [18, Remarks, pp. 240-241] and [19, p. 129], for events sitting in an MV-algebra A, coherence becomes equivalent to the extendability of β to a state s of A. Thus *de Finetti's Dutch Book theorem holds unchanged for MV-algebras.* Now, the Kroupa-Panti theorem,

[15, 20], [19, Theorem 10.5], shows that *the states of any MV-algebra A are in one-one correspondence with the regular Borel probability measures on the maximal spectral space of A.* Since MV-algebraic maximal spectral spaces range over all possible compact Hausdorff spaces, ([19, §4]), one may conclude that de Finetti's finitely additive probability theory is equivalent to Kolmogorov's countably additive probability theory.

Let $\mathsf{S}_A \subseteq [0,1]^A \subseteq \mathbb{R}^A$ denote the convex set of states of A with the restriction topology of the Tychonoff cube $[0,1]^A$. For B a subalgebra of a boolean algebra A, we use the notation

$$\mathsf{S}_A \restriction B = \{\mathbf{s} \restriction B \mid \mathbf{s} \in \mathsf{S}_A\},$$

where the symbol $\restriction$ denotes restriction. For every boolean algebra A we let $\mathrm{hom}(A)$ denote the set of homomorphisms of A into the boolean algebra $\{0,1\}$ and $\mathrm{at}(A) = \{o \in A \mid o \text{ is an } atom \text{ (i.e., a minimal nonzero element) of } A\}$.

Proposition 2.3. *Let A be a boolean algebra. If B is a subalgebra of A then every state of B has an extension to a state of A, in symbols, $\mathsf{S}_A \restriction B = \mathsf{S}_B$.*

Proof. For the particular case of states of boolean algebras, this classical extension theorem was proved by Horn and Tarski in [13, Theorem 1.22]. □

Proposition 2.4. *Let A be a finite boolean algebra. Let $\mathrm{at}(A) = \{o_1, \ldots, o_u\}$ and $\mathrm{hom}(A) = \{\eta_1, \ldots, \eta_u\}$, where η_i is the only homomorphism in $\mathrm{hom}(A)$ with $\eta_i(o_i) = 1$.*

(a) S_A coincides with the set of convex combinations of homomorphisms of A into $\{0,1\}$, in symbols, $\mathsf{S}_A = \mathrm{conv}(\mathrm{hom}(A)) \subseteq [0,1]^A \subseteq \mathbb{R}^A$. Thus, for every $a \in A$ and $\mathbf{s} \in \mathsf{S}_A$, upon writing $\mathbf{s} = \sum_{i=1}^{u} \lambda_i \eta_i$, for suitable $\lambda_i \geq 0$ with $\sum_{i=1}^{u} \lambda_i = 1$, we have $\mathbf{s}(a) = \sum \lambda_i \eta_i(a)$.

(b) Every $\eta \in \mathrm{hom}(A)$ is an extremal element of the convex set S_A. Conversely, every extremal state of A is an element of $\mathrm{hom}(A)$.

Proof. (a) The finiteness hypothesis ensures that every element $a \in A$ is the join of the atoms it *dominates*, $a = \bigvee\{o \in \mathrm{at}(A) \mid a \geq o\}$. Then the additivity property ensures that every state $\mathbf{s}$ of A is uniquely determined by the value it gives to the atoms of A, in symbols, $\mathbf{s}(a) = \sum\{\mathbf{s}(o) \mid o \in \mathrm{at}(A),\ o \leq a\}$. For any $o \in \mathrm{at}(A)$, let $\eta_o \in \mathrm{hom}(A)$ be specified by $\eta_o(o) = 1$. Thus $\eta_o(z) = 0$ for all $z \in \mathrm{at}(A)$ different from o. It is easy to see that η_o is a state of A. Further, η_o cannot be written as a convex combination of two or more distinct states of A. In other words,

$\boldsymbol{\eta}_o$ is extremal in the convex set $\mathsf{S}_A \subseteq \mathbb{R}^A$. Then every state $\mathbf{s}$ of A has the form $\mathbf{s} = \mathbf{s}(o_1) \cdot \boldsymbol{\eta}_{o_1} + \cdots + \mathbf{s}(o_u) \cdot \boldsymbol{\eta}_{o_u}$. In particular, every extremal state of A has the form $\boldsymbol{\eta}_o$ for some $o \in \mathrm{at}(A)$.

(b) immediately follows from (a). (Actually, the finiteness of A is not necessary for (b) to hold.) □

Proposition 2.4 now yields:

Proposition 2.5. *Let $e_1, \ldots, e_u$ be the unit basis vectors in $\mathbb{R}^u$, with the proviso that $e_i = (0, \ldots, 0, 1, 0, \ldots, 0)$ has 1 at the ith place. For A a finite boolean algebra with u atoms, let the $(u-1)$-dimensional simplex $\boldsymbol{\Delta} \subseteq \mathbb{R}^u$ be defined by $\boldsymbol{\Delta} = \mathrm{conv}(e_1, \ldots, e_u)$. Let $\mathrm{hom}(A) = \{\boldsymbol{\eta}_1, \ldots, \boldsymbol{\eta}_u\}$.*

(a) The map[1] $\xi\colon \mathbf{s} = \sum_{i=1}^{u} \lambda_i \boldsymbol{\eta}_i \in \mathsf{S}_A \mapsto s = \sum_{i=1}^{u} \lambda_i e_i \in \boldsymbol{\Delta}$, is an affine homeomorphism of S_A onto $\boldsymbol{\Delta}$.

(b) For every $a \in A$ let the function $\bar{a}\colon \boldsymbol{\Delta} \to [0,1]$ be defined by

$$\bar{a}(s) = \mathbf{s}(a), \text{ for all } s \in \mathrm{conv}(e_1, \ldots, e_u). \tag{1}$$

Then $\bar{a}$ is the only linear function over $\boldsymbol{\Delta}$ satisfying the condition

$$\bar{a}(e_j) = \boldsymbol{\eta}_j(a) \text{ for all } j = 1, \ldots, u. \tag{2}$$

(c) Upon restricting to $\{e_1, \ldots, e_u\}$ every linear function $\bar{a}$ on $\boldsymbol{\Delta}$ satisfying condition (2), we obtain the (Stone) isomorphism between A and the boolean algebra of all $\{0,1\}$-valued functions on $\{e_1, \ldots, e_u\}$, with the pointwise operations of the two-element boolean algebra $\{0,1\}$.

3 De Finetti's coherent probability assessments and applications

De Finetti gave the following definition of a coherent system of betting odds:

Definition 3.1. ([5, pp.311-312], [6, Chapter 1]) Let $E = \{a_1, \ldots a_k\}$ be a finite subset of a boolean algebra A, and $\boldsymbol{\beta}\colon E \to [0,1]$ a map. We then say that $\boldsymbol{\beta}$ is a (de Finetti) *coherent book in A* if for any $\sigma\colon E \to \mathbb{R}$ there is $\boldsymbol{\eta} \in \mathrm{hom}(A)$ with $\sum_{i=1}^{k} \sigma(a_i)(\boldsymbol{\beta}(a_i) - \boldsymbol{\eta}(a_i)) \geq 0$.

[1] s has a unique representation because the η_i are affinely independent in the vector space R^A.

Intuitively, E is a set of "events" and $\beta(a_1), \ldots, \beta(a_k)$ are their respective betting odds offered by a bookmaker. These odds form an *incoherent* set if a bettor can choose (positive or negative) "stakes" $\sigma(a_1), \ldots, \sigma(a_k)$ that guarantee her a minimum profit of one zillion euros—regardless of the outcome of the events a_i. In particular, by choosing a negative stake $\sigma(a_i)$ the bettor forces the bookmaker to pay her $|\sigma(a_i) \cdot \beta(a_i)|$ euros, in the hope of winning $|\sigma(a_i)|$ if a_i occurs. Thus a negative stake results in swapping the bookie/bettor roles.

Theorem 3.2. (De Finetti's Dutch book theorem, [5, §§8-9], [6, pp. 7–8]) *Let A be a boolean algebra, E a finite subset of A, and $\beta\colon E \to [0,1]$ a map. Then β is a coherent book (in A) iff it can be extended to a state of A.*

For $\beta\colon E \to [0,1]$ and $E = \{a_1, \ldots, a_k\} \subseteq A$, we set

$$\mathsf{S}_{A|\beta} = \{\mathbf{s} \in \mathsf{S}_A \mid \mathbf{s} \supseteq \beta\} = \{\mathbf{s} \in \mathsf{S}_A \mid s \restriction E = \beta\}$$

and for any $b \in A$,

$$\mathsf{S}_{A|\beta}(b) = \{\mathbf{s}(b) \mid \mathbf{s} \in \mathsf{S}_{A|\beta}\}.$$

In view of Remark 2.2 we may write

$$\boxed{\mathsf{S}_{A|\beta}(b) = \text{probabilities of } b \text{ which are coherent with the book } \beta \text{ in } A.}$$

Theorem 3.3. (De Finetti's "Fundamental theorem of probability", [8, 3.10 and references therein]) *Let $\beta\colon E \to [0,1]$ be a map on a subset $E = \{a_1, \ldots, a_k\}$ of a boolean algebra A. Then for any $b \in A$ the set $\mathsf{S}_{A|\beta}(b)$ is a closed interval contained in $[0,1]$. $\mathsf{S}_{A|\beta}(b)$ is nonempty iff β is a coherent book.*

Proof. Let us write $\beta\colon a_1 \mapsto \rho_1, \ldots, a_k \mapsto \rho_k$, with $k \geq 2$ to avoid trivialities. Fix $j = 1, \ldots, k$. By Proposition 2.5, the set of states $\mathbf{s}$ of A such that $\mathbf{s}(a_j) = \rho_j$ corresponds, via the map ξ, to the inverse image $\bar{a}_j^{-1}(\rho_j)$. Since $\bar{a}_j$ is linear, $\bar{a}_j^{-1}(\rho_j)$ coincides with $\mathbf{\Delta} \cap \mathcal{H}_j$ for some hyperplane $\mathcal{H}_j$ in $\mathbb{R}^u$. As a consequence, the set $\xi(\mathsf{S}_{A|\beta})$ coincides with the intersection $\mathcal{K} = \mathbf{\Delta} \cap \mathcal{H}_1 \cap \cdots \cap \mathcal{H}_k$. Again by Proposition 2.5, $\mathsf{S}_{A|\beta}(b) = \{\mathbf{s}(b) \mid \mathbf{s} \supseteq \beta\} = \{\bar{b}(s) \mid s \in \mathcal{K}\} = \text{range}(\bar{b} \restriction \mathcal{K})$. Since $\bar{b}$ is continuous over the connected set $\mathcal{K}$, then $\mathsf{S}_{A|\beta}(b)$ is a closed interval in $[0,1]$. By Theorem 3.2, $\mathsf{S}_{A|\beta}(b) \neq \emptyset$ iff $\mathsf{S}_{A|\beta} \neq \emptyset$ iff β is a coherent book. □

The ambient algebra A is virtually immaterial in Definition 3.1, as well as in Theorems 3.2 and 3.3. As a matter of fact, from Proposition 2.3 we immediately have:

Corollary 3.4. *Let $\beta\colon E \to [0,1]$ be a map defined on a finite subset E of a boolean algebra A.*

(a) Then the following conditions are equivalent:

(i) β is a coherent book in A;

(ii) β is a coherent book in the subalgebra $\mathrm{gen}(E)$ *generated by E in A;*

(iii) β is a coherent book in any subalgebra C of A containing E.

(b) Let $b \in A$, and B be a subalgebra of A containing $E \cup \{b\}$. Then $\mathsf{S}_{A|\beta}(b) = \mathsf{S}_{B|\beta}(b)$.

4 Preservation of coherence under quotients

Having just taken care of the preservation properties of (de Finetti) coherence under extensions and restrictions, we next consider preservation properties under quotients.

For A a boolean algebra and $0 \neq \theta \in A$, let $\langle\theta\rangle = \{z \in A \mid z \geq \theta\}$ denote the (principal) filter of A generated by θ. The map

$$'\colon\ p \mapsto p \wedge \theta$$

is a homomorphism of A onto the boolean algebra

$$A' = \{p \wedge \theta \mid p \in A\}$$

equipped with the operations $\neg(p') = (\neg p \wedge \theta) = (\neg p)'$ and $p' \wedge q' = p \wedge q \wedge \theta = (p \wedge q)'$. As is well known, A' can be identified with the quotient algebra $A/\langle\theta\rangle$ via the isomorphism $p \wedge \theta \mapsto p/\langle\theta\rangle$.

Corollary 4.1. *Let $E = \{a_1, \ldots, a_k, \theta\} \subseteq A$.*

(a) Suppose $\beta\colon E \to [0,1]$ is a coherent book in A with $\beta(\theta) = 1$. Let the map $\beta'\colon \{a'_1, \ldots, a'_k, \theta'\} \to [0,1]$ be defined by

$$\beta'(\theta') = 1,\ \ \beta'(a'_i) = \beta(a_i \wedge \theta),\ \ i = 1, \ldots, k.$$

Then for all $b \in A$, $\mathsf{S}_{A|\beta}(b) = \mathsf{S}_{A'|\beta'}(b')$. Thus β' is a coherent book on $\{a'_1, \ldots, a'_k, \theta'\}$, called the quotient book *of β.*

(b) *Conversely, suppose* $\gamma\colon \{a'_1, \ldots, a'_k, \theta'\} \to [0,1]$ *is a coherent book. Since* θ' *is the unit element of* A'*, then necessarily* $\gamma(\theta') = 1$*. Define the map* $\gamma^{\uparrow}\colon E \to [0,1]$ *by* $\gamma^{\uparrow}(\theta) = 1$ *and* $\gamma^{\uparrow}(a_i) = \gamma(a'_i)$*, for all* $i = 1, \ldots, k$*. Then* $\gamma^{\uparrow}$ *is a coherent book on* E*, and* $(\gamma^{\uparrow})' = \gamma$*. Thus, for all* $b \in A$*,* $\mathsf{S}_{A|\gamma^{\uparrow}}(b) = \mathsf{S}_{A'|\gamma}(b')$*.*

Proof. In view of Proposition 2.3 and Corollary 3.4 it is sufficient to argue under the assumption that A is finite. By Propositions 2.5(c) and 2.4 we may identify A with the algebra of boolean functions over $\mathrm{at}(A)$. Up to canonical isomorphisms,

the $\wedge$-map $\quad d \in A \mapsto d \wedge \theta \in \{a \wedge \theta \mid a \in A\}$,

the quotient map $\quad d \in A \mapsto d/\langle\theta\rangle$, and

the restriction map $d \in A \mapsto d\restriction\{o \in \mathrm{at}(A) \mid \theta \geq o\}$

yield the same algebra. Let

$$H = \{\eta \in \mathrm{hom}(A) \mid \eta(\theta) = 1\} = \{\eta_1, \ldots, \eta_l\}.$$

Let $\zeta\colon H \to \mathrm{hom}(A')$ be defined by

$$\zeta\colon \eta \in H \mapsto \eta' \in \mathrm{hom}(A'),$$

where $\eta'\colon A' \to [0,1]$ is uniquely determined by $\eta'(a') = \eta(a \wedge \theta)$, for all $a \in A$. Then

$$\zeta \text{ is one-one correspondence between } H \text{ and } \mathrm{hom}(A'). \tag{3}$$

(a) In view of Theorem 3.2 and of the assumed coherence of β, let us suppose the state $\mathsf{s} \in \mathsf{S}_A$ extends β. So, in particular, $\mathsf{s}(\theta) = 1$. By Proposition 2.4(b), s is a convex combination

$$\mathsf{s} = \lambda_1\eta_1 + \cdots + \lambda_l\eta_l$$

of the homomorphism in H, *with uniquely determined coefficients* λ_i. For every atom $o \in \mathrm{at}(A)$ which is not dominated by θ (in the sense that $o \not\leq \theta$, or equivalently, $o \wedge \theta = 0$), it must be $\mathsf{s}(o) = 0$. Thus for every $a \in A$,

$$\mathsf{s}(a) = \mathsf{s}(a \wedge \theta) + \mathsf{s}(a \wedge \neg\theta) = \mathsf{s}(a \wedge \theta).$$

Let s' be the convex combination

$$\mathsf{s}' = \lambda_1\eta'_1 + \cdots + \lambda_l\eta'_l.$$

Then $\mathbf{s}'$ is a state of A' extending β'. By Theorem 3.2, β' is coherent. Further, for all $a \in A$,

$$\mathbf{s}'(a') = \mathbf{s}(a \wedge \theta) = \mathbf{s}(a).$$

Thus any two $p, q \in A$ with $p' = q'$ satisfy $\mathbf{s}(p) = \mathbf{s}(q)$, whence $\mathsf{S}_{A|\beta}(b) = \mathsf{S}_{A'|\beta'}(b')$.

(b) In view of (3), the image of any $\boldsymbol{\epsilon} \in \hom(A')$ under ζ^{-1} is the only $\boldsymbol{\epsilon}^{\uparrow} \in H$ satisfying $\boldsymbol{\epsilon}^{\uparrow}(a) = \boldsymbol{\epsilon}(a') = \boldsymbol{\epsilon}(a \wedge \theta)$ for all $a \in A$. Thus every convex combination

$$\mathbf{t} = \mu_1 \epsilon_1 + \cdots + \mu_v \epsilon_l \in \mathsf{S}_{A'}$$

determines the state $\mathbf{t}^{\uparrow} = \mu_1 \boldsymbol{\epsilon}_1^{\uparrow} + \cdots + \mu_v \boldsymbol{\epsilon}_l^{\uparrow} \in \mathsf{S}_A$. The coherence of γ yields a state $\mathbf{g}$ of A' extending γ. By Proposition 2.4(b), $\mathbf{g}$ is a convex combination of the homomorphisms $\boldsymbol{\epsilon}_i$. Further, $\mathbf{g}^{\uparrow}$ is the combination of the $\boldsymbol{\epsilon}_i^{\uparrow}$ with the same coefficients. Since $\mathbf{g}^{\uparrow}$ extends the map $\gamma^{\uparrow}$, then $\gamma^{\uparrow}$ is consistent. It follows that $(\gamma^{\uparrow})' = \gamma$. Using (a), we obtain $\mathsf{S}_{A|\gamma^{\uparrow}}(b) = \mathsf{S}_{A'|\gamma}(b')$, as desired to conclude the proof. □

5 Preservation of coherence under definitions by fresh variables

Definition 5.1. [21, §13] Let A be a boolean algebra. Two subalgebras B and C of A are said to be *independent* if any pair $(b, c) \in B \times C$ with $b \neq 0$ and $c \neq 0$ satisfies $b \wedge c \neq 0$. With reference to the notation in Corollary 3.4, we say that $a \in A$ is *independent* of B if so are the subalgebras $\mathrm{gen}(a)$ and B. Finally, we say that $a, b \in A$ are *independent* if so are the subalgebras $\mathrm{gen}(a)$ and $\mathrm{gen}(b)$.

By definition we immediately have:

Lemma 5.2. *Let a, b be elements of a boolean algebra A. If either a or b belongs to $\{0, 1\}$ then a and b are independent. If both a and b do not belong to $\{0, 1\}$ then the following conditions are equivalent:*

(i) a and b are independent.

(ii) Each of $a \wedge b, a \wedge \neg b, \neg a \wedge b, \neg a \wedge \neg b$ is $\neq 0$.

Theorem 5.3. *Suppose A is a boolean algebra, β is a coherent book on a set $E = a_1, \ldots, a_k \subseteq A$, and $a, b \in A \setminus \{0, 1\}$ with b independent of $\mathrm{gen}(E \cup \{a\})$. Then:*

(i) The book $\beta^+ = \beta \cup \{(a \leftrightarrow b, 1)\}$ is coherent.

(ii) Every state $\mathbf{r}$ *extending* β^+ *satisfies* $\mathbf{r}(a) = \mathbf{r}(b)$.

Proof. As a consequence of the independence hypothesis, there is a canonical embedding of the free product

$$\mathrm{gen}(b) \amalg \mathrm{gen}(E \cup \{a\}),$$

(called "boolean product" in [21, p.40]), onto a subalgebra F of A. F and the free product will be identified. In view of Corollary 3.4, it suffices to argue assuming A finite. Let $o_1, \dots, o_l$ be the atoms of $\mathrm{gen}(E \cup \{a\})$. Since $b \notin \{0, 1\}$, $\mathrm{gen}(b)$ has precisely two atoms, namely b and $\neg b$, and is *freely* generated by b. The set Θ of atoms of $\mathrm{gen}(b) \amalg \mathrm{gen}(E \cup \{a\})$ has the form

$$\Theta = \{b, \neg b\} \times \{o_1, \dots, o_l\}.$$

Since β is coherent, Theorem 3.2 yields a state $\mathbf{s}$ of $\mathrm{gen}(E \cup \{a\})$ which extends the book β. Let $\tau \colon \Theta \to [0, 1]$ be the map specified by the following conditions:

τ agrees with $\mathbf{s}$ on every atom $(b, o_i) \in \Theta$ with $o_i \leq a$,

τ agrees with $\mathbf{s}$ on every atom $(\neg b, o_j) \in \Theta$ with $o_j \leq \neg a$, and

τ has value 0 on all remaining atoms of $\mathrm{gen}(b) \amalg \mathrm{gen}(E \cup \{a\})$.

Since $\mathbf{s}$ is a state of $\mathrm{gen}(E \cup \{a\})$,

$$\sum_{e \in \mathrm{at}(\mathrm{gen}(E \cup \{a\}))} \mathbf{s}(e) = 1 = \sum_{o \in \Theta} \tau(o).$$

As a consequence, τ extends to a unique state $\mathbf{t}$ of $\mathrm{gen}(b) \amalg \mathrm{gen}(E \cup \{a\})$, which agrees with $\mathbf{s}$ over $\mathrm{gen}(E \cup \{a\}) \subseteq \mathrm{gen}(b) \amalg \mathrm{gen}(E \cup \{a\})$.

We *claim* that $\mathbf{t}$ evaluates to 1 the element $a \leftrightarrow b$. To see this, one firstly notes that an atom $o \in \Theta$ satisfies $o \leq a \leftrightarrow b$ iff it has the form (b, o_i) for $o_i \leq a$, or else $(\neg b, o_j)$ for $o_j \leq \neg a$. Secondly, by construction, the disjunction d of all atoms $o \leq a \leftrightarrow b$ satisfies $\mathbf{s}(d) = 1$, and our claim is settled.

We have just shown that the book β^+ extends to the state $\mathbf{t}$ of $\mathrm{gen}(b) \amalg \mathrm{gen}(E \cup \{a\})$. By Proposition 2.3, β^+ extends to a state of A. By Theorem 3.2, β^+ is coherent.

(ii) Let as above, $\Theta = \mathrm{at}(\mathrm{gen}(b) \amalg \mathrm{gen}(E \cup \{a\}))$. Then

$$\{o \in \Theta \mid o \leq a \leftrightarrow b\} = \{o \in \Theta \mid o \leq a \wedge b\} \cup \{o \in \Theta \mid o \leq \neg a \wedge \neg b\}.$$

Since $\mathbf{r}$ is a state extending the book β^+, then $\mathbf{r}(a \leftrightarrow b) = 1$, and hence

$$0 = \mathbf{r}(o) \text{ for all } o \in \Theta \setminus \{o \in \Theta \mid o \leq a \leftrightarrow b\}.$$

As a consequence,

$$\mathbf{r}(a) \;=\; \sum_{o \in \Theta,\ o \leq a} \mathbf{r}(o) \;=\; \sum_{o \in \Theta,\ o \leq a \wedge b} \mathbf{r}(o) \;=\; \sum_{o \in \Theta,\ o \leq b} \mathbf{r}(o) \;=\; \mathbf{r}(b).$$

□

6 Modifying a betting odd

Lemma 6.1. *Let A be a boolean algebra and $a, b \in A$. Let $\boldsymbol{\rho}^* \colon \{a\} \to [0,1]$ be the singleton map assigning value ρ to a. Let $\mathsf{S}_{A|\boldsymbol{\rho}^*}(b)$ denote the closed interval $\{\mathbf{s}(b) \mid \mathbf{s} \in \mathsf{S}_A,\ \mathbf{s} \supseteq \boldsymbol{\rho}^*\} \subseteq [0,1]$ given by Theorem 3.3.*

(a) If $a = 1$ then $\boldsymbol{\rho}^*$ is coherent iff $\rho = 1$.

(b) If $a = 0$ then $\boldsymbol{\rho}^*$ is coherent iff $\rho = 0$.

(c) If $a \notin \{0,1\}$ then for every $\rho \in [0,1]$, $\boldsymbol{\rho}^*$ is a coherent book.

Proof. (a) and (b) are trivial. For the proof of (c), by hypothesis the subalgebra $\mathrm{gen}(a) \subseteq A$ has at least two atoms, say o_1 and o_2. Let $\mathbf{s}$ be the state of $\mathrm{gen}(a)$ uniquely determined by the stipulation $\mathbf{s}(o_1) = \rho$ and $\mathbf{s}(o_2) = 1 - \rho$. Since $\mathbf{s}$ extends $\boldsymbol{\rho}^*$ in the boolean algebra $\mathrm{gen}(a)$, then by Theorem 3.2 $\boldsymbol{\rho}^*$ is a coherent book in $\mathrm{gen}(a)$. By Corollary 3.4, $\boldsymbol{\rho}^*$ is a coherent book in A. □

Theorem 6.2. *Let A be a boolean algebra and $a, b \in A$. Arbitrarily fix $\rho \in [0,1]$ and let $\boldsymbol{\rho}^* \colon \{a\} \to [0,1]$ be the singleton map defined by $\boldsymbol{\rho}^*(a) = \rho$. Suppose $\boldsymbol{\rho}^*$ is a coherent book. Let, as above,*

$$\mathsf{S}_{A|\boldsymbol{\rho}^*}(b) = \{\mathbf{s}(b) \mid \mathbf{s} \in A,\ \mathbf{s} \supseteq \boldsymbol{\rho}^*\} = \{\mathbf{s}(b) \mid \mathbf{s} \in A,\ \mathbf{s}(a) = \rho\}.$$

We then have:

(a) If $b = 0$, $\mathsf{S}_{A|\boldsymbol{\rho}^*}(b) = \{0\}$. If $b = 1$, $\mathsf{S}_{A|\boldsymbol{\rho}^*}(b) = \{1\}$.

(b) If $b \notin \{0,1\}$ we have the following mutually incompatible exhaustive cases:

(i) If $b = a$ then $\mathsf{S}_{A|\boldsymbol{\rho}^*}(b) = \{\rho\}$.

(ii) If $b = \neg a$ *then* $\mathsf{S}_{A|\rho^*}(b) = \{1-\rho\}$.

(iii) If $b \neq a$ *and* $b \geq a$ *then* $\mathsf{S}_{A|\rho^*}(b) = [\rho, 1]$.

(iv) If $b \neq a$ *and* $b \leq a$ *then* $\mathsf{S}_{A|\rho^*}(b) = [0, \rho]$.

(v) If $b \neq \neg a$ *and* $b \leq \neg a$ *then* $\mathsf{S}_{A|\rho^*}(b) = [0, 1-\rho]$.

(vi) If $b \neq \neg a$ *and* $b \geq \neg a$ *then* $\mathsf{S}_{A|\rho^*}(b) = [1-\rho, 1]$.

(vii) If a *and* b *are independent then* $\mathsf{S}_{A|\rho^*}(b) = [0, 1]$.

Proof. In view of Corollary 3.4, we may assume A finite, say with u atoms, without loss of generality. By Theorem 3.2, $\mathsf{S}_{A|\rho^*}(b)$ is nonempty.

(a) is trivial.

(b) Parts (i) and (ii) are trivial. For (iii)-(vii), let us argue by cases:

Case 1: $a \notin \{0, 1\}$. Then let us partition the set $\{e_1, \ldots, e_u\}$ of unit basis vectors in $\mathbb{R}^u$ into the following two classes:

$$E_1 = \{e_j \mid \bar{a}(e_j) = 1\} \quad \text{and} \quad E_0 = \{e_i \mid \bar{a}(e_i) = 0\}.$$

By hypothesis, E_1 and E_0 are nonempty. Throughout the proof of the present case, the symbols $\mathcal{E}$ and $\mathcal{F}$ will denote arbitrary edges (1-faces) of $\mathbf{\Delta}$, joining a vertex $e_j \in E_1$ with a vertex $e_i \in E_0$.

To prove (iii), we will make repeated use of the identity

$$\mathsf{S}_{A|\rho^*}(b) = \{\bar{b}(s) \mid s \in \mathbf{\Delta},\ \bar{a}(s) = \rho\},$$

which follows from Proposition 2.5. By Lemma 6.1, ρ ranges over $[0, 1]$. The linearity of $\bar{a}$ yields a hyperplane $\mathcal{H}_{\rho^*}$ in $\mathbb{R}^u$ such that $\bar{a}^{-1}(\rho) = \mathbf{\Delta} \cap \mathcal{H}_{\rho^*}$. It is impossible that whenever $\bar{a}$ moves from 1 to 0 along an edge $\mathcal{E}$ also $\bar{b}$ does. For, this would mean $b \leq a$, against our standing assumption. So there is $\mathcal{E}$ such that $\bar{a}$ descends from 1 to 0 along $\mathcal{E}$, but $\bar{b} \geq \bar{a}$ keeps constant value 1 over $\mathcal{E}$. Let s be the only point lying in $\mathcal{E}$ such that $\bar{a}(s) = \rho$. Let $\mathbf{s}$ be the state corresponding to s via the affine homeomorphism ξ of Proposition 2.5. Since $\bar{b}(s) = 1$ and $\mathbf{s}$ extends $\boldsymbol{\rho}^*$ then $1 \in \mathsf{S}_{A|\rho^*}(b)$. Since $\bar{a} \leq \bar{b}$ and all states are order preserving, then $\mathbf{s}(a) \leq \mathbf{s}(b)$, whence no $\sigma < \rho$ lies in $\mathsf{S}_{A|\rho^*}(b)$. So there remains to be proved that $\rho \in \mathsf{S}_{A|\rho^*}(b)$. There is an edge $\mathcal{F}$ with $\bar{a}\restriction\mathcal{F} = \bar{b}\restriction\mathcal{F}$. For otherwise (absurdum hypothesis), the fact that $\bar{a}$ equals 1 at one of the vertices of $\mathcal{F}$ implies that $\bar{b} \geq \bar{a}$ takes the constant value 1 over $\mathcal{F}$, with $\mathcal{F}$ any possible edge in $E_1 \cup E_2$. Since the set of vertices of the edges in $E_1 \cup E_2$ coincides with $\{e_1, \ldots, e_u\}$, then $b = 1$, which is impossible. So pick $\mathcal{F}$ with $\bar{a} \restriction \mathcal{F} = \bar{b} \restriction \mathcal{F}$, together with the only point $s \in \mathcal{F}$ such that $\bar{a}(s) = \rho = \bar{b}(s)$. This shows that $\rho \in \mathsf{S}_{A|\rho^*}(b)$.

(iv)-(vi) now follow as routine variants of (iii).

(vii) By Lemma 5.2, the independence of a and b yields an edge $\mathcal{E}$ joining a vertex $e_j \in E_1$ with a vertex $e_i \in E_0$, such that $\bar{b}(e_j) = 0 = \bar{b}(e_i)$, whence $\bar{b}$ identically vanishes over $\mathcal{E}$. Pick the point $s \in \mathcal{E}$ where $\bar{a}(s) = \rho$. From $\bar{b}(s) = 0$ it follows that $0 \in \mathsf{S}_{A|\rho^*}(b)$. To prove that 1 is a member of $\mathsf{S}_{A|\rho^*}(b)$, the independence of a and b yields an edge $\mathcal{F}$ such that $\bar{b}$ is constantly equal to 1 over $\mathcal{F}$. Now let $t \in \mathcal{F}$ be the point where $\bar{a}(t) = \rho$. From $\bar{b}(s) = 1$ it follows that $1 \in \mathsf{S}_{A|\rho^*}(b)$. We conclude that $\mathsf{S}_{A|\rho^*}(b) = [0, 1]$.

The proof of Case 1 is thus complete.

Case 2: $a = 0$. By Lemma 5.2, a and b are independent. The assumed coherence of $\boldsymbol{\rho}^*$ entails $\boldsymbol{\rho}^*(a) = 0$, (Lemma 6.1). Since $b \notin \{0, 1\}$, arguing as in the proof of Lemma 6.1(c), (with b in place of a), we obtain $\mathsf{S}_{A|\rho^*}(b) = [0, 1]$, in agreement with the conclusion (vii).

Case 3: $a = 1$. Since a and b are independent, arguing as in Case 2 we again obtain $\mathsf{S}_{A|\rho^*}(b) = [0, 1]$, in agreement with the conclusion (vii). □

The preservation properties of de Finetti's notion of a coherent set of betting odds now yield:

Corollary 6.3. *Let A be a boolean algebra and $a_1, \ldots, a_{k+1} \in A$. For each $i = 1, \ldots, k-1$ let $\rho_i \in \{0, 1\}$ be fixed truth values. For every $\rho \in [0, 1]$, let the map $\beta_\rho \colon \{a_1, \ldots, a_k\} \to [0, 1]$ be defined by*

$$\beta_\rho(a_1) = \rho_1, \ldots, \beta_\rho(a_{k-1}) = \rho_{k-1}, \ \beta_\rho(a_k) = \rho.$$

Now let us perturb ρ to $\rho + \epsilon$ for some $\epsilon > 0$. Suppose $\beta_{\rho \pm y}$ is a coherent book for all y satisfying $-\epsilon \leq y \leq \epsilon$. Let the maps $\phi, \psi, \lambda \colon [\rho - \epsilon, \rho + \epsilon] \to [0, 1]$ be defined by stipulating that for all $x \in [\rho - \epsilon, \rho + \epsilon]$,

$$\phi(x) = \min(\mathsf{S}_{A|\beta_x}(a_{k+1})), \ \psi(x) = \max(\mathsf{S}_{A|\beta_x}(a_{k+1})), \ \lambda(x) = \psi(x) - \phi(x).$$

Then the values of the derivatives

$$\frac{\mathrm{d}\phi}{\mathrm{d}x}(\rho), \ \frac{\mathrm{d}\psi}{\mathrm{d}x}(\rho), \ \frac{\mathrm{d}\lambda}{\mathrm{d}x}(\rho),$$

are in the set $\{-1, 0, 1\}$.

Proof. In view of Corollary 3.4, without loss of generality we may assume A finite. By Theorem 3.2, $\mathsf{S}_{A|\beta_x}(a_{k+1}) \neq \emptyset$. Let the map $\widehat{\beta_x}$ be obtained from β_x replacing a_i by $\neg a_i$ whenever $\beta_x(a_i) = 0$ and contextually writing $\widehat{\beta_x}(\neg a_i) = 1$, $(i = 1, \dots, k-1)$. Then $\widehat{\beta_x}$ is coherent and in fact, $\mathsf{S}_{A|\beta_x}(a_{k+1}) = \mathsf{S}_{A|\widehat{\beta_x}}(a_{k+1})$. Replacing, if necessary, β_x by $\widehat{\beta_x}$, we may argue under the assumption $\beta_x(a_1) = \cdots = \beta_x(a_{k-1}) = 1$, without loss of generality. For $\theta = a_1 \wedge \cdots \wedge a_{k-1}$, let $\langle\theta\rangle$ be the filter of A generated by θ. Let $A' = A/\langle\theta\rangle$, $a'_{k+1} = a_{k+1}/\langle\theta\rangle$ and β'_x be the quotient book. By Corollary 4.1(a) we have $\mathsf{S}_{A|\beta_x}(a_{k+1}) = \mathsf{S}_{A'|\beta'_x}(a'_{k+1})$. An application of Theorem 6.2 to the quotient algebra A' now yields the desired conclusion. □

Remark 6.4. Closing a circle of ideas, whenever events are coded by boolean formulas $\phi_1, \dots, \phi_k$, and books $\boldsymbol{\beta}\colon \{\phi_1, \dots, \phi_k\} \to [0,1]$ are rational-valued, one is left with the problem of deciding whether $\boldsymbol{\beta}$ is a de Finetti coherent book. This is an important generalization, known as PSAT, of the boolean satisfiability problem SAT. PSAT is NP-complete, [11]. The proofs of the main results of the present paper are constructive, and for rational-valued books yield effective methods to compute the derivatives in Corollary 6.3.

References

[1] O. Bratteli, D.W.Robinson, Operator Algebras and Quantum Statistical Mechanics I, II, Springer, Berlin, 1979.

[2] C.C.Chang, Algebraic analysis of many valued logics, Trans. Amer. Math. Soc., 88 (1958) 467-490.

[3] S.A. Cook, The Complexity of Theorem Proving Procedures, Proceedings Third Annual ACM Symposium on Theory of Computing, STOC'71, May 1971, pp. 151-158.

[4] K.R. Davidson, C*-Algebras by Example, Fields Institute Monographs, Vol. 6, 1996.

[5] B. de Finetti, Sul significato soggettivo della probabilitá, Fundamenta Mathematicae, 17 (1931) 298–329. Reprinted in [7, pp. 191–222]. Translated into English as "On the Subjective Meaning of Probability". In: P. Monari, D. Cocchi (Eds.), Probabilitá e Induzione, Clueb, Bologna, pp. 291-321, 1993.

[6] B. de Finetti, La prévision: ses lois logiques, ses sources subjectives, Annales de l'Institut H. Poincaré, 7 (1937) 1–68. English translation by Henry E. Kyburg Jr., as "Foresight: Its Logical Laws, its Subjective Sources." In: H. E. Kyburg Jr., H. E. Smokler, "Studies in Subjective Probability", J. Wiley, New York, pp. 93–158, 1964. Second edition published by Krieger, New York, pp. 53–118, 1980. Reprinted in [7, pp.335–400].

[7] B. de Finetti, Opere Scelte. (Selected Works), Vol. 1. Unione Matematica Italiana, Edizioni Cremonese, Firenze, 2006.

[8] B. de Finetti, Theory of Probability. A Critical Introductory Treatment. Translated by A. Machí and A. Smith, John Wiley and Sons Ltd., Chichester, UK, 2017.

[9] E.G.Effros, Dimensions and C*-algebras, CBMS Regional Conf. Series in Math., Vol. 46, Amer. Math. Soc., Providence, R.I., 1981.

[10] G.A.Elliott, On the classification of inductive limits of sequences of semisimple finite-dimensional algebras, J. Algebra, 38 (1976) 29-44.

[11] G. Georgakopoulos, D. Kavvadias, C.H. Papadimitriou, Probabilistic satisfiability, Journal of Complexity, 4 (1988) 1-11.

[12] K. R. Goodearl, Partially Ordered Abelian Groups with Interpolation, American Mathematical Society, Providence, RI, 1986.

[13] A. Horn, A. Tarski, Measures in Boolean algebras, Transactions of the American Mathematical Society, 64 (1948) 467-497.

[14] J.L.Kelley, Measures on boolean algebras, Pacific Journal of Mathematics, 9 (1959) 1165–1177.

[15] T. Kroupa, Every state on a semisimple MV-algebra is integral, Fuzzy Sets and Systems, 157 (2006) 2771–2782.

[16] D. Mundici, Interpretation of AF C^*-algebras in Łukasiewicz sentential calculus, J. Functional Analysis, 65 (1986) 15–63.

[17] D. Mundici, Averaging the truth value in Łukasiewicz sentential logic, Studia Logica, Special issue in honor of Helena Rasiowa, 55 (1995) 113–127.

[18] D. Mundici, Interpretation of de Finetti coherence criterion in Łukasiewicz logic, Annals of Pure and Applied Logic, 161 (2009) 235–245.

[19] D. Mundici, Advanced Łukasiewicz calculus and MV-algebras, Trends in Logic, Vol. 35, Springer-Verlag, Berlin, 2011.

[20] G. Panti, Invariant measures in free MV-algebras, Communications in Algebra, 36 (2009) 2849–2861.

[21] R. Sikorski, Boolean algebras, Second edition, Ergeb. der Math. und ihrer Grenzgeb., Vol 25, Springer, Berlin, 1960.

Received 1 February 2018

Relevant Justification Logic

Nenad Savić and Thomas Studer
Institute of Computer Science, University of Bern, Switzerland
{savic,tstuder}@inf.unibe.ch

Abstract

We introduce a relevant justification logic, RJ4, which is a combination of the relevant logic R and the justification logic J4. We describe the corresponding class of models, provide the axiomatization and prove that our logic is sound and complete.

Keywords: Relevant Logic, Justification Logic, Soundness and completeness

1 Introduction

Relevant logics are non-classical logics that avoid the paradoxes of material and strict implication and provide a more intuitive deductive inference. The central systems of relevant logic, according to Anderson and Belnap [1], are the system of relevant implication R, as well as the logic of entailment E.

Justification logic replaces the □-operator of modal logic by explicit justifications [2, 5]. That is justification logic features formulas of the form $t : A$ meaning A *is believed for reason* t; hence we can reason with and about explicit justifications for an agent's belief. The framework of justification logic has been used to formalize and study a variety of epistemic situations [3, 6, 10, 11, 13, 14, 17].

However, traditional justification logic is based on classical logic and can lead to some paradoxical situations. One of those situations will be our running example in this paper.

Example 1. *Consider a person A visiting a foreign town, which she does not know well. In order to get to a certain restaurant, she asks two persons B and C for the way. Person B says that A can take path P to the restaurant whereas person C replies that P does not lead to the restaurant and A should take another way. Person A now has a reason s to believe P and a reason t to believe $\neg P$. We can formalize this in justification logic by saying that both*

$$s : P \qquad \textit{and} \qquad t : \neg P \tag{1}$$

hold. However, under certain natural assumptions, there exists a justification $r(s,t)$ *such that*

$$r(s,t) : (P \land \neg P)$$

holds. Now this implies that for any formula F*, there is a justification* u *such that*

$$u : F \tag{2}$$

holds. That means for any formula F*, person* A *has a reason to believe* F*, which, of course, is an undesirable consequence.*

It is the aim of this paper to introduce a justification logic, RJ4, in which situations of this kind cannot occur, in particular, that means a logic in which (2) does not follow from (1). We achieve this by combining the relevant logic R with the justification logic J4.

Meyer [18] proposed the logic NR, which is the relevant logic R equipped with an $S4$-style theory of necessity, in order to investigate whether the resulting theory coincides with the theory of entailment provided by Anderson and Belnap [1]. Adapting the semantics for the logic R [19], Routley and Meyer provided a complete semantics for the logic NR [20].

Our logic RJ4 is similar to NR but instead of the $\Box$-operator, we use explicit justifications and since we deal with beliefs, we do not include the truth principle $t : A \to A$ in the list of axioms. The choice of axioms for the relevant logic R can be varied in different ways, e.g., see [12]. We decided to use the first 12 axioms from [20].

Our relevant justification logic RJ4 is not just a simple combination of R and J4. The reason is that justification logic includes an application operation on terms, which is related to implication, i.e., we have the following axiom

$$t : (A \to B) \to (s : A \to (t \cdot s) : B).$$

Hence, if the meaning of implication changes, then also the meaning of the application operation has to change. This is hidden in the axiomatization, but it becomes evident in the semantics. There, property ($p7$) models the relation between justifications and relevant implication. It shows that there is a true interaction between those two parts and we cannot simply juxtapose the semantics for R and the one for J4 to obtain a semantics for RJ4.

Another motivation for this work, i.e., for combining relevant logic with justification logic, comes from the philosophical point of view. Namely, if an implication of the form

$$s : A \to t : B$$

holds, then, we argue, the antecedent $s : A$ should be relevant for the consequent $t : B$. Note that we do not claim that the justification s itself must be relevant for the justification t (it is a different topic), but rather that the fact that s is a justification for A should be relevant for the whole consequent, i.e., that t is a justification for B.

The contents of this paper are as follows. In Section 2 we present the syntax of our logic, in Section 3 we provide the axiomatization, while in Section 4 the semantics is explained. In Section 5 soundness and completeness theorems are proved and we conclude in Section 6.

2 Syntax

In this section we propose the syntax of the logic RJ4.
Let

$\mathsf{Con} = \{c_0, c_1, \ldots, c_n, \ldots\}$ be a countable set of constants,

$\mathsf{Var} = \{x_0, x_1, \ldots, x_n, \ldots\}$ be a countable set of variables, and

$\mathsf{Prop} = \{p_0, p_1, \ldots, p_n, \ldots\}$ a countable set of atomic propositions.

Definition 1 (Terms). *Terms are built from the sets* Con *and* Var *as follows:*

$$t ::= c \mid x \mid t \cdot t \mid t \tilde{\wedge} t \mid t + t \mid !t,$$

where $c \in \mathsf{Con}$ *and* $x \in \mathsf{Var}$. *The set of terms will be denoted by* Tm.

Note that, in comparison to the definition of terms in the justification logic J4, we have an additional operation, $\tilde{\wedge}$, on terms.

Definition 2 (Formulas). *Formulas are build from the sets* Prop *and* Tm *as follows:*

$$A ::= p \mid \neg A \mid A \to A \mid A \wedge A \mid A \vee A \mid A \circ A \mid t : A,$$

where $p \in \mathsf{Prop}$ *and* $t \in \mathsf{Tm}$. *The set of formulas is denoted by* For.

We define $A \leftrightarrow B$ as

$$A \leftrightarrow B =_{def} (A \to B) \wedge (B \to A).$$

For sets of formulas X and Y, we will use the following notation:

$$\begin{aligned} X \cdot Y &:= \{F \mid G \to F \in X \text{ and } G \in Y, \text{ for some formula } G\}, \\ X \wedge Y &:= \{F \mid F = G \wedge H, \text{ for some } G \in X \text{ and } H \in Y\}, \\ t : X &:= \{t : F \mid F \in X\}. \end{aligned}$$

3 Axiomatization

There are two groups of axioms for RJ4. The first group are the axioms of the logic R[1]:

(A1) $A \to A$

(A2) $A \to ((A \to B) \to B)$

(A3) $(A \to B) \to ((B \to C) \to (A \to C))$

(A4) $(A \to (A \to B)) \to (A \to B)$

(A5) $A \wedge B \to A$

(A6) $A \wedge B \to B$

(A7) $(A \to B) \wedge (A \to C) \to (A \to B \wedge C)$

(A8) $A \wedge (B \vee C) \to (A \wedge B) \vee (A \wedge C)$

(A9) $\neg\neg A \to A$

(A10) $(A \to \neg B) \to (B \to \neg A)$

(A11) $A \vee B \leftrightarrow \neg(\neg A \wedge \neg B)$

(A12) $A \circ B \leftrightarrow \neg(A \to \neg B)$

The second group consists of the axioms of J4 plus an additional axiom (A15):

(A13) $t : (A \to B) \to (s : A \to (t \cdot s) : B)$

(A14) $t : A \to\ !t : t : A$

(A15) $t : A \wedge s : B \to (t \tilde{\wedge} s) : (A \wedge B)$

(A16) $t : A \to (t + s) : A$ and $t : A \to (s + t) : A$

In the axiom $(A12)$ we introduced the binary connective, $\circ$, which is called *fusion*, or *intensional conjunction*. It is defined via $\neg$ and $\to$, i.e., introducing it is the conservative extension of our language. Fusion plays an important role in the

[1]There are many equivalent ways to axiomatize the logic R. We decided to take the axiomatizaiton from [20].

relevant logic R and its connection with implication is even stronger, namely the formula

$$((A \circ B) \to C) \leftrightarrow (A \to (B \to C)) \quad (3)$$

is valid formula of R. For more details about fusion see, e.g., [16].

To introduce the rules of our logic, we need the following definition:

Definition 3. Constant specification *is a set*

$$\mathsf{CS} \subseteq \{(c, A) \mid c \textit{ is a constant and } A \textit{ is an axiom of } \mathsf{RJ4}\}.$$

Constant specification CS *is called* axiomatically appropriate *if for each axiom A there exists a constant* $c \in \mathsf{Con}$*, such that* $(c, A) \in \mathsf{CS}$.

Given a constant specification CS, the deductive system $\mathsf{RJ4}_{\mathsf{CS}}$ is given by the axioms of RJ4 and the following rules:

$$(\mathsf{MP})\,\frac{F \quad F \to G}{G} \qquad (\mathsf{ADJ})\,\frac{F \quad G}{F \wedge G} \qquad (\mathsf{AN})\,\frac{(c, A) \in \mathsf{CS}}{c : A}$$

where the first rule is called *modus ponens*, the second *adjunction* and the last rule is called *axiom necessitation.*

As usual in justification logics, we can show the following analogue of the necessitation rule.

Lemma 1 (Constructive necessitation). *Let* CS *be an axiomatically appropriate constant specification. For each formula A,*

$$\mathsf{RJ4}_{\mathsf{CS}} \vdash A \quad \textit{implies} \quad \mathsf{RJ4}_{\mathsf{CS}} \vdash t : A \textit{ for some term } t.$$

4 Semantics

The semantics for RJ4 is based on a combination of possible world models for R and basic modular models for J4.

In order to motivate our semantics, let us look closer at the Example 1, i.e., let us formally prove that in the justification logic J4 with an axiomatically appropriate constant specification[2], for any formula F, there exists a justification term $u(s,t)$, such that

$$u(s,t) : F$$

does follow from

$$s : P \qquad \text{and} \qquad t : \neg P.$$

[2]for more details about J4 and axiomatically appropriate constant specification see, e.g., [5]

Example 2. *Consider a justification logic* J4 *with an axiomatically appropriate constant specification and suppose that* $s : P$ *and* $t : \neg P$ *hold. Then, we have the following derivation*[3]*:*

1) $\vdash_{\mathsf{J4}_{\mathsf{CS}}} P \to (\neg P \to (P \wedge \neg P))$ *PR*

2) $\vdash_{\mathsf{J4}_{\mathsf{CS}}} r : (P \to (\neg P \to (P \wedge \neg P)))$ *AN*

3) $\vdash_{\mathsf{J4}_{\mathsf{CS}}} r : (P \to (\neg P \to (P \wedge \neg P))) \to (s : P \to (r \cdot s : (\neg P \to (P \wedge \neg P))))$ *J*

4) $\vdash_{\mathsf{J4}_{\mathsf{CS}}} s : P \to (r \cdot s : (\neg P \to (P \wedge \neg P)))$ *2),3) MP*

5) $\vdash_{\mathsf{J4}_{\mathsf{CS}}} r \cdot s : (\neg P \to (P \wedge \neg P)) \to (t : \neg P \to ((r \cdot s) \cdot t : (P \wedge \neg P)))$ *J*

6) $\vdash_{\mathsf{J4}_{\mathsf{CS}}} s : P \to (t : \neg P \to ((r \cdot s) \cdot t) : (P \wedge \neg P))$ *4),5) PR*

7) $\vdash_{\mathsf{J4}_{\mathsf{CS}}} (s : P \wedge t : \neg P) \to ((r \cdot s) \cdot t : (P \wedge \neg P))$ *6) PR*

8) $\vdash_{\mathsf{J4}_{\mathsf{CS}}} (r \cdot s) \cdot t : (P \wedge \neg P)$ *follows from our assumption and 7) using MP*

9) $\vdash_{\mathsf{J4}_{\mathsf{CS}}} (P \wedge \neg P) \to F$*, for any formula* F*,* *PR*

10) $\vdash_{\mathsf{J4}_{\mathsf{CS}}} t' : ((P \wedge \neg P) \to F)$*,* *AN*

11) $\vdash_{\mathsf{J4}_{\mathsf{CS}}} t' : ((P \wedge \neg P) \to F) \to ((r \cdot s) \cdot t : (P \wedge \neg P) \to t' \cdot ((r \cdot s) \cdot t) : F)$*,* *J*

12) $\vdash_{\mathsf{J4}_{\mathsf{CS}}} (r \cdot s) \cdot t : (P \wedge \neg P) \to t' \cdot ((r \cdot s) \cdot t) : F$*,* *10),11) MP*

13) $\vdash_{\mathsf{J4}_{\mathsf{CS}}} t' \cdot ((r \cdot s) \cdot t) : F$*,* *8),12) MP.*

Note that, if "$\to$" represents relevant implication instead of implication of the classical propositional logic, the derivation above is not possible. Namely, the step 9) does not hold.

Remark 1. *If the constant specification is not axiomatically appropriate, then step* 10) *in the above example does not hold. Therefore, restricting the constant specification could be another approach to prevent the derivation of the formula* $t' \cdot ((r \cdot s) \cdot t) : F$*. However, it is a natural assumption to have an axiomatically appropriate constant specification, i.e., for each axiom there is a reason to believe it. Hence we do not use restricted constant specifications but employ a logic based on relevant implication.*

[3] PR stands for propositional reasoning, AN for axiom necessitation of the logic J4, J for $JAxiom$ of the logic J4 and MP for modus ponens.

Our models will be models for relevant logic R, namely we use Routley-Meyer semantics, i.e., Kripke-style structrure equipped with ternary relation[4], where each world is essentially basic modular model of justification logic (with an additional constraint for $\tilde{\wedge}$), see [4, 15].

Definition 4 (Model). *Let* CS *be an arbitrary constant specification. An* $\mathsf{RJ4}_{\mathsf{CS}}$*-model is a tuple of the form* $M = (K, 0, R, *, ♠, \nu)$ *where:*

1. K *is a set,*
2. $0 \in K$,
3. R *is a ternary relation on* K,
4. $*$ *is a function* $* : K \to K$,
5. ♠ *is a function* ♠ $: \mathsf{Tm} \times K \to \mathcal{P}(\mathsf{For})$,
6. ν *is a function* $\nu : K \to \mathcal{P}(\mathsf{Prop})$,

that satisfies the following properties:

(p1) $a \leq a$,

(p2) $Raaa$,

(p3) $R^2abcd \Rightarrow R^2acbd$,

(p4) $a \leq b \wedge Rbcd \Rightarrow Racd$,

(p5) $Rabc \Leftrightarrow Rac^*b^*$,

(p6) $a^{**} = a$,

(p7) $Rabc \Rightarrow t^♠_a \cdot s^♠_b \subseteq (t \cdot s)^♠_c$,

(p8) $a \leq b \Rightarrow t^♠_a \subseteq t^♠_b$,

(p9) $s^♠_a \cup t^♠_a \subseteq (s + t)^♠_a$,

(p10) $A \in t^♠_0$ *if* $(t, A) \in \mathsf{CS}$,

[4]There is no universally accepted intuition behind the ternary relation. For example, $Rxyz$ can be viewed as that the combination of the pieces of information x and y is a piece of information in z as well as that set-ups x and y are compatible according to z. For more details about the ternary relation and various models of R, see [8, 9, 7].

(p11) $t : (t_a^{\spadesuit}) \subseteq (!t)_a^{\spadesuit}$,

(p12) $s_a^{\spadesuit} \wedge t_a^{\spadesuit} \subseteq (s\tilde{\wedge}t)_a^{\spadesuit}$,

(p13) $a \leq b \Rightarrow \nu(a) \subseteq \nu(b)$,

where

$$a \leq b := R0ab,$$

and

$$R^2abcd := \exists x(Rabx \wedge Rxcd).$$

We write $t_a^{\spadesuit}$ for $\spadesuit(t, a)$ and we call an ordered pair $(\nu, \spadesuit)$ *valuation.*

The property ($p7$) deserves more attention since it is the only property that includes both the ternary relation R and justifications, i.e., gives us connection between Routley-Meyer semantics and justification semantics. Note that the axiom that corresponds to this property is the axiom (A13). It is the only axiom in the second group of axioms that has an implication in consequent. Therefore, we need to guarantee the validity of that implication and the property ($p7$) gives us a connection between the ternary relation R and justifications on the worlds that are related by R.

The property ($p7$) can also be regarded as a generalization of the following principle from basic modular models:

$$s^{\spadesuit} \cdot t^{\spadesuit} \subseteq (s \cdot t)^{\spadesuit}.$$

Our worlds are basic modular models, so we want that, for any $a \in K$,

$$s_a^{\spadesuit} \cdot t_a^{\spadesuit} \subseteq (s \cdot t)_a^{\spadesuit} \tag{4}$$

holds. Indeed, (4) follows from ($p7$) together with ($p2$). Hence our semantics is a true generalization of the traditional semantics for justification logic.

Note that $\mathsf{RJ4_{CS}}$-models do not feature the *justification yields belief* principle of modular models [4, 15]. As in models for NR, we could add a binary relation S on K to $\mathsf{RJ4_{CS}}$-model and require that justification yields belief in the sense of S. This construction would yield modular models for $\mathsf{RJ4}$.

Definition 5 (Satisfiability relation). *Given a model* $\mathcal{M} = (K, 0, R, *, \spadesuit, \nu)$ *and*

$a \in K$ *we define a relation* $\models$ *as follows:*

$$\begin{array}{rll} \mathcal{M}, a \models p & \textit{iff} & p \in \nu(a), \textit{for } p \in Prop \\ \mathcal{M}, a \models A \wedge B & \textit{iff} & \mathcal{M}, a \models A \textit{ and } \mathcal{M}, a \models B \\ \mathcal{M}, a \models A \vee B & \textit{iff} & \mathcal{M}, a \models A \textit{ or } \mathcal{M}, a \models B \\ \mathcal{M}, a \models A \circ B & \textit{iff} & Rxya \textit{ and } \mathcal{M}, x \models A \textit{ and } \mathcal{M}, y \models B, \textit{for some } x, y \in K \\ \mathcal{M}, a \models A \rightarrow B & \textit{iff} & Raxy \textit{ and } \mathcal{M}, x \models A \textit{ imply } \mathcal{M}, y \models B, \textit{for all } x, y \in K \\ \mathcal{M}, a \models \neg A & \textit{iff} & \mathcal{M}, a* \not\models A \\ \mathcal{M}, a \models t : A & \textit{iff} & A \in t_a^{\spadesuit}. \end{array}$$

We say that a formula A is *true* at a in $\mathcal{M}$ if $\mathcal{M}, a \models A$. Formula A is *verified* in M, iff $\mathcal{M}, 0 \models A$. Finally, formula A is CS-*valid* iff A is verified in every $\mathsf{RJ4}_{\mathsf{CS}}$-model. We will often write $a \models A$ instead of $\mathcal{M}, a \models A$ when $\mathcal{M}$ is clear from a context. Also, we say that A *entails* B if for all $a \in K$, if $a \models A$ then $a \models B$.

We need a couple of auxiliary lemmas. Let $\mathcal{M} = (K, 0, R, *, \spadesuit, \nu)$ be an arbitrary model and $a, b \in K$ and $A, B \in For$.

Lemma 2 (Hereditary Lemma). *If* $a \leq b$ *and* $a \models A$*, then* $b \models A$.

Proof. In order to prove this Lemma we need a few auxiliary claims, namely:

(i) $Rabc \Rightarrow Rbac$;

(ii) $a \leq b \Rightarrow b^* \leq a^*$;

(iii) $Rabc$ and $c \leq d \Rightarrow Rabd$.

Proof of (i). $Rabc \Rightarrow R^2 0abc \Rightarrow R^2 0bac \Rightarrow Rbac$.
First "$\Rightarrow$" holds since both $R0aa$ ($p1$) and $Rabc$ (assumption) hold. The second "$\Rightarrow$" is ($p3$) and the third follows from ($p4$), since ($p4$) can be written as $R0ab \wedge Rbcd \Rightarrow Racd$, i.e., $R^2 0acd \Rightarrow Racd$.
Proof of (ii). Directly from definition of "$\leq$" and ($p5$), the following derivation holds:

$$a \leq b \Rightarrow R0ab \Rightarrow R0b^*a^* \Rightarrow b^* \leq a^*.$$

Proof of (iii). Using ($p5$), (i), ($p4$) together with $c \leq d$ and (ii), (i), ($p5$) respectively, the following holds:

$$Rabc \Rightarrow Rac^*b^* \Rightarrow Rc^*ab^* \Rightarrow Rd^*ab^* \Rightarrow Rad^*b^* \Rightarrow Rabd.$$

Proof of Lemma 2. By induction on a length of a formula A.

1) If $A = p \in Prop$, then the condition $(p13)$, $a \leq b \Rightarrow \nu(a) \subseteq \nu(b)$, ensures the claim.

2) The cases when $A = B \wedge C$ or $A = B \vee C$ are trivial.

3) Let $A = \neg B$ and $a \models A$. That means that $a^* \not\models B$. Since $a \leq b$, from (ii) we have $b^* \leq a^*$. Therefore $b^* \not\models B$, so $b \models A$.

4) Now, let $A = B \to C$ and $a \models A$. For all x, y with $Raxy$ we have that if $x \models B$ then $y \models C$. Suppose that $Rbcd$ and $c \models B$. The question is whether $d \models C$. Since $a \leq b$ and $Rbcd$, from $(p4)$ we obtain that $Racd$ and therefore as a direct consequence of our premise, which holds for every x, y, we have that $d \models C$.

5) Let $A = B \circ C$. From $a \models A$ we get that there exist x, y, such that $Rxya$ and $x \models B$ and $y \models C$. Furthermore, from (iii), since $a \leq b$ and $Rxya$ we know that $Rxyb$. Thus $b \models B \circ C$, i.e., $b \models A$.

6) Finally, let $A = t : B$ and $a \models A$. That means that $B \in t_a^{\spadesuit}$ and since $a \leq b$, we have $t_a^{\spadesuit} \subseteq t_b^{\spadesuit}$, so $B \in t_b^{\spadesuit}$, i.e. $b \models A$ as well. □

In the following, since the majority of proofs are the same as in [19], we will give the proofs only for those cases that are new.

Lemma 3 (Entailment). *A entails B if and only if $A \to B$ is verified.*

5 Soundness and Completeness

5.1 Soundness

In order to prove soundness we need to prove that every instance of an axiom holds in arbitrary model and that inference rules preserve validity. We will consider only the axioms $(A13) - (A16)$.

Theorem 1 (Soundness). *Let* CS *be any constant specification. For each formula A we have*

$$\text{If} \quad \mathsf{RJ4}_{\mathsf{CS}} \vdash A \quad \text{then} \quad A \text{ is } \mathsf{CS}\text{-valid.}$$

Proof. Since our axioms are of the form $X \to Y$, using Lemma 3, it is enough to prove that for arbirtary $a \in K$, if $a \models X$ then $a \models Y$.

(A13) Suppose that $a \models t : (B \to C)$, i.e. $B \to C \in t_a^\spadesuit$. We need to show that $a \models s : B \to (t \cdot s) : C$. Suppose that $Rabc$ and $b \models s : B$, i.e. $B \in s_b^\spadesuit$. Since $B \to C \in t_a^\spadesuit$, we obtain that $C \in t_a^\spadesuit \cdot s_b^\spadesuit$ and, because $Rabc$, $t_a^\spadesuit \cdot s_b^\spadesuit \subseteq (t \cdot s)_c^\spadesuit$, $C \in (t \cdot s)_c^\spadesuit$, i.e. $c \models (t \cdot s) : C$.

(A14) Let $a \models t : B$, i.e. $B \in t_a^\spadesuit$. Then $t : B \in t : (t_a^\spadesuit) \subseteq (!t)_a^\spadesuit$. Therefore $a \models !t : t : B$.

(A15) Now, suppose that $a \models t : A \wedge s : B$. That means $A \in t_a^\spadesuit$ and $B \in s_a^\spadesuit$. Hence $A \wedge B \in t_a^\spadesuit \wedge s_a^\spadesuit \subseteq (t\tilde{\wedge}s)_a^\spadesuit$, so $a \models t\tilde{\wedge}s : (A \wedge B)$.

(A16) Finally, suppose $a \models t : A$, i.e., $A \in t_a^\spadesuit \subseteq t_a^\spadesuit \cup s_a^\spadesuit \subseteq (t + s)_a^\spadesuit$. Therefore $a \models (t + s) : A$. The other case is analoguous.

We need to prove that inference rules preserve the validity:

(MP) If a formula A is obtained from $B \to A$ and B, we have that $0 \models B \to A$ and $0 \models B$, hence $0 \models A$, since $R000$.

(ADJ) The case when a formula is obtained from adjunction is trivial.

(AN) If A is obtained by (AN), then $A = c : B$, where $(c, B) \in CS$. Therefore, $B \in c_0^\spadesuit$, so $0 \models c : B$. □

5.2 Completeness

In order to prove the completeness theorem, we will use a procedure based on [19]. That means that first we develop a calculus of intensional $\mathsf{RJ4_{CS}}$-theories, then a calculus of intensional T-theories, for a regular intensional theory T and at the end a calculus of prime intensional theories.

With all this machinery, we are able to define the canonical model, which gives us the completeness theorem. Below we state all definitions and lemmas that we need. The majority of the proofs are identical as in [19], so they will be omitted. We state only the original proofs.

Definition 6. *Let $T \subseteq$ For and CS any constant specification. We say that T is*

a) an intensional $\mathsf{RJ4_{CS}}$-theory *iff T is closed under adjunction and if $A \in T$ and $\mathsf{RJ4_{CS}} \vdash A \to B$, then $B \in T$;*

b) prime *iff it is an intensional $\mathsf{RJ4_{CS}}$-theory and if $A \vee B \in T$, then $A \in T$ or $B \in T$;*

c) regular iff it contains all theorems of $\mathsf{RJ4_{CS}}$*;*

d) consistent iff it does not contain the negation of some theorem of $\mathsf{RJ4_{CS}}$.

Lemma 4. *Let* $(\nu, \spadesuit)$ *be a valuation in a structure* $(K, 0, R, *)$ *and let* $a \in K$*. The set of all formulas* F *such that* $a \models F$*, denoted by* $T((\nu, \spadesuit), a)$*, is a prime theory. If* $0 \leq a$*, then* $T((\nu, \spadesuit), a)$ *is regular.*

Definition 7 (Calculus of intensional theories). *Let* CS *be an arbitrary constant specification. The calculus of intensional theories is the structure* $\mathcal{H} = (\mathcal{H}, \subseteq, \circ, 0)$*, where*

1) $\mathcal{H}$ *is the collection of all intensional* $\mathsf{RJ4_{CS}}$*-theories;*

2) $\subseteq$ *is set inclusion;*

3) $\circ$ *is a binary operation on* $\mathcal{H}$ *defined with*

$$S \circ T = \{C \mid \mathsf{RJ4_{CS}} \vdash A \circ B \to C, \text{ for some } A \in S \text{ and some } B \in T\};$$

4) 0 *is the set of all theorems of* $\mathsf{RJ4_{CS}}$.

Lemma 5. *The calculus* $\mathcal{H}$ *is a partially ordered commutative monoid, that means,* $\circ$ *is associative and commutative operation and* 0 *is an identity with respect to* $\circ$*. Also, the following holds for all* $a, b, c \in \mathcal{H}$:
if $a \subseteq b$ *then* $a \circ c \subseteq b \circ c$*;*
$a \circ a \subseteq a$ *(square decreasing).*

Definition 8 (Intensional T-theory). *An intensional* T*-theory is any set of formulas,* a*, which is an intensional* $\mathsf{RJ4_{CS}}$*-theory and whenever* $A \in a$ *and* $A \to B \in T$*, then* $B \in a$.

Now we define a calculus of intensional T-theories.

Definition 9 (Calculus of intensional T-theories). *The calculus of intensional* T*-theories is the structure* $\mathcal{H}_\mathbf{T} = (\mathcal{H}_T, \subseteq, \circ, 0_T)$*, where* T *is a regular theory,* $\mathcal{H}_T$ *is a set of all intensional* T*-theories,* $0_T = T$ *and* $\circ$ *and* $\subseteq$ *are defined as above.*

Lemma 6. *The calculus* $\mathcal{H}_\mathbf{T}$ *is a sub-semigroup of* $\mathcal{H}$.

Definition 10 (Positive relevant structure (PRS)). *The structure* $(K, 0, R)$*, where* K *is a set,* $0 \in K$ *and* R *is a ternary relation on* K*, which satisfies properties* (p1) – (p4) *will be called positive relevant structure (PRS).*

Let $\mathbf{M} = (M, \leq, \circ, 0)$ be commutative, partially ordered, square decreasing monoid satisfying that $a \leq b$ implies $a \circ c \leq b \circ c$. We say that PRS $(M, 0, R)$ is associated with $\mathbf{M}$ if M is equal to underlying set of $\mathbf{M}$, 0 is equal to identity of $\mathbf{M}$ and R is defined such that $Rabc$ iff $a \circ b \leq c$ in M.

Lemma 7. *If* $\mathbf{M}$ *is commutative, partially ordered, square decreasing monoid, then PRS* $(M, 0, R)$ *associated with* $\mathbf{M}$ *satisfies properties* $(p1) - (p4)$.
Furthermore, for all $a, b, c, d \in M$, $a \leq b$ *in* M *iff* $a \leq b$ *in* $\mathbf{M}$ *and* R^2abcd *iff* $a \circ b \circ c \leq d$ *in* $\mathbf{M}$.
The calculus $\mathcal{H}$ *is associated with PRS* $(\mathcal{H}, 0, R)$ *and the calculus* $\mathcal{H}_\mathbf{T}$ *is associated with PRS* $(\mathcal{H}_T, 0_T, R_T)$.

Let T be prime, regular, intensional $\mathsf{RJ4}_\mathsf{CS}$-theory. Let $(\mathcal{H}_T, 0_T, R_T)$ be the PRS associated with $\mathcal{H}_\mathbf{T}$ and let $\mathcal{H}'_T$ be the subset of $\mathcal{H}_T$ which consists of all prime intensional theories in $\mathcal{H}_T$. Let $0'_T = T$ and R'_T restriction of R_T to $\mathcal{H}'_T$.

Lemma 8. $(\mathcal{H}'_T, 0'_T, R'_T)$ *is a PRS, i.e., satisfies* $(p1) - (p4)$.

For a prime intensional theory a, we define $a^* = \{A \mid \neg A \notin a\}$.

Lemma 9. *Let* $(\mathcal{H}'_T, 0'_T, R'_T)$ *and* $*$ *be defined as above. Then* $(\mathcal{H}'_T, 0'_T, R'_T, *)$ *is a relevant structure (RS), i.e.,* $*$ *is an operation on* $\mathcal{H}'_T$ *and properties* $(p1) - (p6)$ *are satisfied.*

Definition 11 (Canonical model). *Let* CS *be any constant specification.* $\mathsf{RJ4}_\mathsf{CS}$-*model* $(\mathcal{H}'_T, 0'_T, R'_T, *, \nu, \spadesuit)$, *where* $(\mathcal{H}'_T, 0'_T, R'_T, *)$ *is RS from Lemma 9 and a valuation* $(\nu, \spadesuit)$ *defined with:*

a) $p \in \nu(a)$ *iff* $p \in a$;

b) $\spadesuit(t, a) = \{A \mid t : A \in a\}$,

will be called canonical T-model.

Lemma 10. *The canonical* T*-model is an* $\mathsf{RJ4}_\mathsf{CS}$*-model, i.e., it satisfies properties* $(p1) - (p13)$.

Proof. It follows from Lemma 9 that $(p1) - (p6)$ are satisfied. Let us show the others.

(p7) Suppose that $R'_T abc$, i.e., $a \circ b \subseteq c$ and suppose that $A \in t^\spadesuit_a \cdot s^\spadesuit_b$. Hence, there exist $B \in s^\spadesuit_b$ such that $B \to A \in t^\spadesuit_a$. By definition of $\spadesuit$, we have that $s : B \in b$ and $t : (B \to A) \in a$. Therefore,

$$(t : (B \to A)) \circ (s : B) \in a \circ b \subseteq c.$$

Also, note that, because of 3, the Axiom ($A13$) is equivalent to

$$(t : B \to A) \circ (s : B) \to (t \cdot s) : A, \tag{5}$$

so, since c is an intensional $\mathsf{RJ4}_{\mathsf{CS}}$-theory, and antecedent of 5 belongs to c, we obtain that $(t \cdot s) : A \in c$, i.e., $A \in (t \cdot s)^{\spadesuit}_{c}$.

(p8) Let $a \subseteq b$ and $A \in t^{\spadesuit}_{a}$. Then, $t : A \in a \subseteq b$, so, $A \in t^{\spadesuit}_{b}$.

(p9) Let $A \in s^{\spadesuit}_{a} \cup t^{\spadesuit}_{a}$. First, suppose that $A \in s^{\spadesuit}_{a}$, i.e., $s : A \in a$. Directly from the Axiom ($A16$) and intensionality of a we obtain the result. If $A \in t^{\spadesuit}_{a}$, the proof is analogous.

(p10) If $(t, A) \in \mathsf{CS}$, then, because of an axiom necessitation rule, we know that $\mathsf{RJ4}_{\mathsf{CS}} \vdash t : A$ and, since T is regular, $t : A \in T = 0'_T$ which implies that $A \in t^{\spadesuit}_{0}$.

(p11) Suppose that $A \in t : (t^{\spadesuit}_{a})$, i.e., exists $B \in t^{\spadesuit}_{a}$, such that $A = t : B$. Since $B \in t^{\spadesuit}_{a}$, we know that $t : B \in a$ and hence, by Axiom ($A14$), $!t : t : B \in a$ as well. That means $A = t : B \in (!t)^{\spadesuit}_{a}$.

(p12) Let $A \in s^{\spadesuit}_{a} \wedge t^{\spadesuit}_{a}$. There exist $B \in s^{\spadesuit}_{a}$ and $C \in t^{\spadesuit}_{a}$, such that $A = B \wedge C$. By definition of $\spadesuit$, we have that $s : B \in a$ and $t : C \in a$ and, because of adjunction, $s : B \wedge t : C \in a$. Again, intensionality gives us that $(s\tilde{\wedge}t) : (B \wedge C) \in a$, i.e., $A = B \wedge C \in (s\tilde{\wedge}t)^{\spadesuit}_{a}$.

(p13) If $a \subseteq b$ and $p \in \nu(a)$, we have that $p \in a$ and therefore $p \in b$ which concludes the proof. □

Lemma 11. *Let $A \in \mathsf{For}$. For all $a \in \mathcal{H}'_T$, $a \models A$ iff $A \in a$.*

Proof. We will prove only the case when $A = t : F$, for some $F \in \mathsf{For}$.
First suppose that $a \models t : F$. That means $F \in t^{\spadesuit}_{a}$, hence, by definition of $\spadesuit$, we obtain that $t : F \in a$.
For the other direction, suppose that $t : F \in a$. Again, directly by definition we obtain that $F \in t^{\spadesuit}_{a}$ and therefore $a \models t : F$. □

For the proof of the following lemma, Zorn's Lemma is necessary.

Lemma 12. *Let CS be an arbitrary constant specification. For every non-theorem, A, there exists a prime, regular $\mathsf{RJ4}_{\mathsf{CS}}$-theory which does not contain A.*

Theorem 2 (Completeness). *For an arbitrary constant specification CS, the system $\mathsf{RJ4}_{\mathsf{CS}}$ is semantically complete, i.e.,*

if A is CS-valid, then $\mathsf{RJ4}_{\mathsf{CS}} \vdash A$.

6 Conclusion

In this paper, the logic RJ4 is introduced and its models, which are combination of Kripke-style models for relevant logic R and basic modular models for justification logics, are developed. We propose an axiomatization and prove the soundness and completeness theorem.

As mentioned in the introduction, there is a close relationship between NR and our logic of relevant justifications. Let RLP be the system RJ4 plus the axiom $t : A \to A$ based on the total constant specification, i.e., every constant justifies every axiom (including $t : A \to A$). A *realization* is a mapping from modal formulas to formulas of justification logic that replaces each $\Box$ with some expression t: (different occurrences of $\Box$ may be replaced with different terms).

For further work, we plan to prove the realization theorem, i.e.:

Conjecture 1 (Realization). *There is a realization r such that for each modal formula A*

$$\mathsf{NR} \vdash A \quad \textit{implies} \quad \mathsf{RLP} \vdash r(A).$$

Acknowledgements

This work was supported by the Swiss National Science Foundation grant 200021_165549.

References

[1] A. R. Anderson and N. D. Belnap. *Entailment: The Logic of Relevance and Neccessity, Vol. I.* Princeton University Press, 1975.

[2] S. N. Artemov. Explicit provability and constructive semantics. *BSL*, 7(1):1–36, Mar. 2001.

[3] S. N. Artemov. The logic of justification. *RSL*, 1(4):477–513, Dec. 2008.

[4] S. N. Artemov. The ontology of justifications in the logical setting. *Studia Logica*, 100(1–2):17–30, Apr. 2012. Published online February 2012.

[5] S. N. Artemov and M. Fitting. Justification logic. In E. N. Zalta, editor, *The Stanford Encyclopedia of Philosophy.* Fall 2012 edition, 2012.

[6] S. N. Artemov and R. Kuznets. Logical omniscience as infeasibility. *APAL*, 165(1):6–25, 2014.

[7] K. Bimbo and J. M. Dunn. Larisa Maksimova's early contributions to relevance logic. *to appear in: Outstanding Contributions to Logic, edited by Sergei Artemov.*

[8] K. Bimbo and J. M. Dunn. The emergence of set-theoretical semantics for relevance logics. *IfCoLog Journal of Logics and their Applications*, 4(3):557–590, 2017.

[9] K. Bimbo, J. M. Dunn, and N. Ferenz. Two manuscripts, one by Routley, one by Meyer. *to appear in: Australasian Journal of Logic, special issue on the philosophy of Richard Sylvan*, 2018.

[10] S. Bucheli, R. Kuznets, and T. Studer. Justifications for common knowledge. *Applied Non-Classical Logics*, 21(1):35–60, Jan.–Mar. 2011.

[11] S. Bucheli, R. Kuznets, and T. Studer. Realizing public announcements by justifications. *Journal of Computer and System Sciences*, 80(6):1046–1066, 2014.

[12] M. Dunn and G. Restall. Relevance logic. In D. Gabbay and F. Guenthner, editors, *Handbook of Philosophical Logic*. Kluwer Academic Publishers, 2002.

[13] I. Kokkinis, P. Maksimović, Z. Ognjanović, and T. Studer. First steps towards probabilistic justification logic. *Logic Journal of IGPL*, 23(4):662–687, 2015.

[14] I. Kokkinis, Z. Ognjanović, and T. Studer. Probabilistic justification logic. In S. Artemov and A. Nerode, editors, *LFCS 2016*, volume 9537 of *LNCS*, pages 174–186. Springer, 2016.

[15] R. Kuznets and T. Studer. Justifications, ontology, and conservativity. In T. Bolander, T. Braüner, S. Ghilardi, and L. Moss, editors, *Advances in Modal Logic, Volume 9*, pages 437–458. College Publications, 2012.

[16] E. Mares. *Relevant Logic: A Philosophical Interpretation*. Cambridge University Press, 2004.

[17] M. Marti and T. Studer. Intuitionistic modal logic made explicit. *IfCoLog Journal of Logics and their Applications*, 3(5):877–901, 2016.

[18] R. K. Meyer. Entailment and relevant implication. *Logique et Analyse*, 11(44):472–479, 1968.

[19] R. Routley and R. Meyer. The semantics of entailment. *Studies in Logic and the Foundations of Mathematics*, 68:199–243, 1973. Truth, Syntax and Modality.

[20] R. Routley and R. K. Meyer. The semantics of entailment: II. *Journal of Philosophical Logic*, 1(1):53–73, 1972.

Received 28 February 2018

Dual-Belnap Logic and Anything but Falsehood

Yaroslav Shramko
Department of Philosophy, Kryvyi Rih State Pedagogical University, Kryvyi Rih, 50086, Ukraine
shramko@rocketmail.com

Abstract

This paper presents an inquiry into a proof system for a logic based on four Belnapian truth values, in which any truth value but the pure falsehood is designated. To this effect, I first implement a certain dualization of what Font terms 'Belnap's logic', and then show how it can be suitably extended. The resulting systems are of the Fmla-Set type dually to the standard formulation of Belnap's logic and the Exactly True Logic by Pietz and Rivieccio. I restate some philosophical motivation for the entailment relation of the Fmla-Set type by briefly comparing it with the usual Set-Fmla logical systems.

1 Preliminaries: Dunn and Belnap's four-valued semantics and designated truth values

J.Michael Dunn in his doctoral dissertation [9] initiated a strategy of semantic analysis, according to which sentences can systematically be considered not just true, or just false, but also neither true nor false, or both true and false simultaneously. This strategy has been technically implemented by constructing an 'intuitive semantics for first-degree entailment' in [10]; see also a comprehensive discussion (and generalization) of the subject in [11, 12]. Motivations for this approach may be various such as argumentative discourse, contradictory or incomplete theoretical systems, and philosophical paradoxes.

Following this strategy, Nuel Belnap [6, 7] introduced some weighty considerations from the computing field, in which sources and databases are often far from perfect, which forces the computers to deal with unreliable or corrupt information. Therefore, one arrives at four (generalized) truth values, according to the information that is 'told' to a computer with respect to a given sentence: 'just told True', 'just told False', 'told neither True nor False', 'told both True and False' (where 'True' and 'False' are ordinary classical truth values). It is most common to label these generalized truth values T, F, N, and B, respectively.

Let sentential language $\mathcal{L}$ be defined as follows:

$$\varphi ::= p \mid \varphi \wedge \varphi \mid \varphi \vee \varphi \mid {\sim}\varphi.$$

In line with the principles of semantic analysis sketched above, define valuation v as a map from the set of sentential variables to the *subsets* of the set of classical truth-values $\{t, f\}$. This valuation is extended to the whole language by the following conditions:

Definition 1.1.

(1) $t \in v(\varphi \wedge \psi) \Leftrightarrow t \in v(\varphi)$ and $t \in v(\psi)$,
$f \in v(\varphi \wedge \psi) \Leftrightarrow f \in v(\varphi)$ or $f \in v(\psi)$;

(2) $t \in v(\varphi \vee \psi) \Leftrightarrow t \in v(\varphi)$ or $t \in v(\psi)$,
$f \in v(\varphi \vee \psi) \Leftrightarrow f \in v(\varphi)$ and $f \in v(\psi)$;

(3) $t \in v({\sim}\varphi) \Leftrightarrow f \in v(\varphi)$,
$f \in v({\sim}\varphi) \Leftrightarrow t \in v(\varphi)$.

Belnapian four truth values (being ascribed to a sentence φ) are then explicated as follows:

$v(\varphi) = B$ (told both True and False) $\Leftrightarrow t \in v(\varphi)$ and $f \in v(\varphi)$,

$v(\varphi) = T$ (just told True) $\Leftrightarrow t \in v(\varphi)$ and $f \notin v(\varphi)$,

$v(\varphi) = F$ (just told False) $\Leftrightarrow t \notin v(\varphi)$ and $f \in v(\varphi)$,

$v(\varphi) = N$ (told neither True nor False) $\Leftrightarrow t \notin v(\varphi)$ and $f \notin v(\varphi)$.

One therefore obtains an elegant semantic construction built on the "four values and three connectives" system, which can further be employed to determine entailment relation as a tool for "evaluating inferences", and finally, to obtain "logic, that is, a canon of inference" [6, p. 15]. This is normally done by marking certain truth values as *designated*, and by defining entailment as a relation that *preserves* this designated status in the course of reasoning.

Which values should be taken as designated among the Belnapian B, T, F, and N? This question can be answered differently depending on the underlying philosophical intuitions and the goals of logical analysis. Clearly, the pure falsehood F should be disqualified from the very outset. Four options are possible for the remaining three.

One option is to pick out $\{T, B\}$ as the set of designated truth values, which is the mainstream choice for Dunn and Belnap's four-valued semantics. A truth value is considered then to be designated if and only if it contains t (the classical True), being thus *at least true*. This choice is founded on the idea that entailment relation "never leads us from told True to the absence of told True (preserves Truth)" [3, p. 519] and brings about the system of 'tautological entailments' of relevant logic, see [2, Chapter III].

Alternatively, one can follow a dual intuition that "implication, entailment, validity, etc. should have as much to do with falsity preservation as with truth preservation—it is just that the direction is reversed" [10, p. 165]. Here it is important that valid entailment "never leads us from the absence of told False to told False (preserves non-Falsity)" [3, p. 519]. From this perspective a truth value is considered non-designated if and only if it is *at least not false*, containing thus f (the classical False), and the corresponding set of designated truth values will be $\{T, N\}$.

In the framework of the four-valued semantics, the above two choices are equivalent in the sense that they determine one and the same entailment relation (when defined on the same sets of premises and conclusions), see [10, p. 165], [12, p. 10], [13, Proposition 2.3]. Still, formally we have here two different choices of two different sets of designated truth values.

Andreas Pietz[1] and Umberto Rivieccio in [21] investigated a logic based on Belnapian four truth values, but with only T as designated. They give an informal motivation for such an 'exactly true logic', ensuring thus "a consequence relation that preserves *truth-and-non-falsity*" [21, p. 128]. This logic validates certain principles that are not valid in the original Dunn-Belnap's semantics, but is still not collapsed into classical logic.

One remaining option deserves attention: to allow as designated *any* truth value, *except the worst one*. According to Belnap, "the worst thing to be told is that something you cling to is false, *simpliciter*" [3, p. 516]. So, T is the "best of all" [ibid], N and B still hold out a hope of a better outcome, and only F is irrecoverable. Hence, it is reasonable to pose a question using the logic with $\{N, T, B\}$ as the subset of distinguished elements among the four Belnapian truth values. Such a logic should ensure preservation of everything but the (outright) falsehood.

João Marcos in [20] differentiates between entailment relations based on the sets of designated truth values $\{T\}$, $\{T, B\}$ and $\{N, T, B\}$. He shows how these relations

[1] After the name change—Andreas Kapsner.

can be explicated by "uniform classic-like semantical and proof-theoretical frameworks" in terms of bivaluations and the corresponding two-signed tableau systems. It is observed that "the inner structure of the four-valued formalism could be seen as a result from a natural combination of classical logic with itself" [20, p. 290].

The present paper is a companion article to [24] extending it by a detail examination of some characteristic features of the entailment relation based on the set of designated truth values $\{N, T, B\}$, and addresing the problem of its deductive formalization. Section 2 recalls a specific proof-theoretic characterization of the Dunn-Belnap semantics, considered by Josep Maria Font in [13] under the name 'Belnap's logic' in the form of a 'Hilbert-style calculus'. Section 3 briefly reviews a way of extending Belnap's logic, proposed by Pietz and Rivieccio to get their system ETL for a four-valued logic with T as the sole designated truth-value. Following this, Section 4 describes a dualization of Belnap's logic obtained by inverting its inference rules, and the corresponding definition of the entailment relation. Section 5 proceeds to certain extension of dual Belnap's logic resulting in logical system NFL ('non-falsity logic') for grasping the entailment relation for a backward preservation of the pure falsity (F). This system is proved to be sound and complete with respect to the intended semantics, and thus presents a solution of the stated problem. The paper is concluded with some philosophical explanations of the logics under consideration.

2 A Hilbert-style presentation of Belnap's logic

Font in [13, p. 5] associates 'Belnap's four-valued logic' with an entailment relation of the SET-FMLA[2] type, that is, "a relation $\models_{\mathcal{B}}$ between arbitrary sets of sentences and a sentence". Consider the following definition of $\models_{\mathcal{B}}$:

Definition 2.1. *Let* Γ *be any set of formulas, and* ψ *be any formula. Then* $\Gamma \models_{\mathcal{B}} \psi =_{df} \forall v : (\forall \varphi \in \Gamma : t \in v(\varphi)) \Rightarrow t \in v(\psi)$.[3]

This definition implies an acceptance of $\{T, B\}$ as the set of designated truth values, stating explicitly the preservation of classical truth (t) from premises to the conclusion. Moreover, the following lemma ensures the preservation of classical falsity (f) in a backward direction:

Lemma 2.2. $\Gamma \models_{\mathcal{B}} \psi \Leftrightarrow \forall v : f \in v(\psi) \Rightarrow (\exists \varphi \in \Gamma : f \in v(\varphi))$.

Proof. See proof of Lemma 2.2 in [24]. □

[2]Cf. the classification of logical frameworks in [18, p. 198].

[3]Generally, Γ may be infinite, but in view of the well-known compactness property it is enough to consider some finite subset of Γ, cf. Definition 2.1 in [13].

This well-known result reinforces the point that Belnap's logic could equivalently be defined through the set of designated truth values $\{T, N\}$. Because both ways of characterizing $\models_{\mathcal{B}}$ are equivalent, we take the set $\{T, B\}$ determined by Definition 2.1 to be the canonical set of designated truth values for Belnap's logic.

For a proof-theoretic characterization of $\models_{\mathcal{B}}$, Font considers a specific system, which he describes as a "Hilbert-style axiomatization" of Belnap's logic [13, p. 10], denoting it $\vdash_H$. This system comprises only the so-called direct rules of inferences of the form $\Gamma \vdash \psi$ (organized vertically in a two-level shape), and has no axioms. The set of rules for $\vdash_H$ is as follows:

(R1) $\dfrac{\varphi \wedge \psi}{\varphi}$ (R2) $\dfrac{\varphi \wedge \psi}{\psi}$ (R3) $\dfrac{\varphi, \psi}{\varphi \wedge \psi}$

(R4) $\dfrac{\varphi}{\varphi \vee \psi}$ (R5) $\dfrac{\varphi \vee \psi}{\psi \vee \varphi}$ (R6) $\dfrac{\varphi \vee \varphi}{\varphi}$

(R7) $\dfrac{\varphi \vee (\psi \vee \chi)}{(\varphi \vee \psi) \vee \chi}$ (R8) $\dfrac{\varphi \vee (\psi \wedge \chi)}{(\varphi \vee \psi) \wedge (\varphi \vee \chi)}$ (R9) $\dfrac{(\varphi \vee \psi) \wedge (\varphi \vee \chi)}{\varphi \vee (\psi \wedge \chi)}$

(R10) $\dfrac{\varphi \vee \psi}{\sim\sim\varphi \vee \psi}$ (R11) $\dfrac{\sim\sim\varphi \vee \psi}{\varphi \vee \psi}$ (R12) $\dfrac{\sim(\varphi \vee \psi) \vee \chi}{(\sim\varphi \wedge \sim\psi) \vee \chi}$

(R13) $\dfrac{(\sim\varphi \wedge \sim\psi) \vee \chi}{\sim(\varphi \vee \psi) \vee \chi}$ (R14) $\dfrac{\sim(\varphi \wedge \psi) \vee \chi}{(\sim\varphi \vee \sim\psi) \vee \chi}$ (R15) $\dfrac{(\sim\varphi \vee \sim\psi) \vee \chi}{\sim(\varphi \wedge \psi) \vee \chi}$

Let us take a closer look at some deductive features of $\vdash_H$. Remarkably, it has no theorems (which is no surprise—no axioms, no theorems). Thus, this system is designed to establish (non-degenerate) valid consequences of the form $\Gamma \vdash \psi$, where Γ is non-empty. Elements of Γ can be called *assumption formulas*, and ψ is a *conclusion* derivable from Γ. Accounting for Font's characterization of $\vdash_H$ as a "Hilbert-style presentation", an inference (or derivation) of ψ from Γ in $\vdash_H$ should be defined as a finite *consecutive* list of (occurrences of) formulas, each of which either belongs to Γ or comes by an inference rule from some formulas preceding it in the list, and the last formula of which is ψ (cf. [19, p. 35]). If there is an inference of ψ from Γ in $\vdash_H$, then ψ is derivable from Γ in $\vdash_H$, and consequence $\Gamma \vdash \psi$ is said to be *valid* in $\vdash_H$.

Let $\Gamma \vdash_H \psi$ means that consequence $\Gamma \vdash \psi$ is valid in $\vdash_H$. By way of illustration, consider inferences for the following consequences: (a) $\varphi \vdash_H \sim\sim\varphi$ and (b) $\varphi \wedge \psi \vdash_H \sim\sim\varphi \wedge \psi$.

(a):

1. φ	(assumption)
2. $\varphi \vee {\sim}{\sim}\varphi$	1: (R4)
3. ${\sim}{\sim}\varphi \vee {\sim}{\sim}\varphi$	2: (R10)
4. ${\sim}{\sim}\varphi$	3: (R6)

(b):

1. $\varphi \wedge \psi$	(assumption)
2. φ	1: (R1)
3. ${\sim}{\sim}\varphi$	2: (*a*)
4. $\varphi \wedge \psi$	(assumption)
5. ψ	4: (R2)
6. ${\sim}{\sim}\varphi \wedge \psi$	3, 5: (R3)

For more examples of this inferential technique in systems like $\vdash_H$ one may wish to consult [15, pp. 125-126]. Observe, that taken literary (b) presents an inference $\varphi \wedge \psi, \varphi \wedge \psi \vdash_H {\sim}{\sim}\varphi \wedge \psi$. However, since Γ is considered to be a genuine set, consequence with a *contracted* assumptions set holds with no additional structural adjustments.

Interestingly Font, despite of his "Hilbert-style" characterization of $\vdash_H$, suggests also a construction of its inferences in a tree-like form resembling natural deduction, see [13, p. 11]. This suggestion can be exemplified by the following inferences of the consequences (a) and (b) above:

$$\text{(a)}\quad \dfrac{\dfrac{\dfrac{\varphi}{\varphi \vee {\sim}{\sim}\varphi}\,\text{(R4)}}{{\sim}{\sim}\varphi \vee {\sim}{\sim}\varphi}\,\text{(R10)}}{{\sim}{\sim}\varphi}\,\text{(R6)} \qquad\qquad \text{(b)}\quad \dfrac{\dfrac{\dfrac{\varphi \wedge \psi}{\varphi}\,\text{(R1)}}{{\sim}{\sim}\varphi}\,\text{(a)} \qquad \dfrac{\varphi \wedge \psi}{\psi}\,\text{(R2)}}{{\sim}{\sim}\varphi \wedge \psi}\,\text{(R3)}$$

As one can see, through such a construction inferences are evolving as direct derivations in the form of trees, possibly branching upwards. The derived formula constitutes the root of a tree, whereas its leaves stand for the formulas from which the root is derived (assumptions). Such form of inferences could be rather convenient and illustrative for explaining the main point of a proof-theoretic dualization of Belnap's logic, considered in Section 4 below.

The following fact helps to simplify inferences in $\vdash_H$ by eliminating extraneous disjunctions and turning disjunctions into conjunctions if required:

Lemma 2.3. *For every rule (R10)–(R15) of the form* $\dfrac{\varphi \vee \chi}{\psi \vee \chi}$ *the following rules are derivable in* $\vdash_H$*: (a)* $\dfrac{\varphi}{\psi}$*; (b)* $\dfrac{\varphi \wedge \chi}{\psi \wedge \chi}$.

Proof. See Proposition 3.2 in [13]. □

In particular, this lemma allows to establish all the properties of De Morgan negation for $\sim$. System $\vdash_H$ is sound and complete with respect to Definition 2.1:

Theorem 2.4. $\Gamma \vdash_H \psi \Leftrightarrow \Gamma \models_{\mathcal{B}} \psi$.

Proof. See Theorem 3.11 in [13]. □

3 Disjunction elimination and exactly true logic

As already observed, inference rules in $\vdash_H$ are all *direct* regulations ensuring a straightforward transition from premise(s) to conclusion. The first three rules deliver a complete inferential characterization of conjunction: (R1), (R2) for conjunction elimination and (R3) for conjunction introduction. The situation with disjunction is more intricate because the property of disjunction elimination is inexpressible within the SET-FMLA framework by a direct inference rule. Considering such an inexpressibility, this property is compensated in $\vdash_H$ by certain additional rules, most crucially, rules (R10)–(R15) with an additional disjunctive context attached to the usual double negation and De Morgan laws.

However, the property of disjunction elimination holds in $\vdash_H$ in a form of a *meta-principle* (or an admissible *meta-rule*), as is stated in the following lemma:

Lemma 3.1. *If $\varphi \vdash_H \chi$ and $\psi \vdash_H \chi$, then $\varphi \vee \psi \vdash_H \chi$.*

Proof. As observed in the proof of Proposition 3.3 in [13], if $\varphi \vdash_H \psi$, then $\varphi \vee \chi \vdash_H \psi \vee \chi$. Now, assume $\varphi \vdash_H \chi$ and $\psi \vdash_H \chi$. By the above observation, and using (R5), we obtain: (*) $\varphi \vee \psi \vdash_H \chi \vee \psi$ and (**) $\chi \vee \psi \vdash_H \chi \vee \chi$. The following inference completes the proof:

1. $\varphi \vee \psi$	(assumption)
2. $\chi \vee \psi$	1: (*)
3. $\chi \vee \chi$	2: (**)
4. χ	3: (R6)

□

Analogously, the property of contraposition is an admissible meta-principle in $\vdash_H$:

Lemma 3.2. *If $\varphi \vdash_H \psi$, then $\sim\psi \vdash_H \sim\varphi$.*

Proof. It is enough to show that the contrapositive versions of all the rules (R1)–(R15) are also the rules of $\vdash_H$. □

The absence of disjunction elimination among the derivable principles of $\vdash_H$ allows for some interesting extensions that would otherwise be impossible. For example, Pietz and Rivieccio [21] employ it to obtain a deductive characterization for their 'exactly true logic', which accepts T as the only designated truth value. Namely, consider the following definition:

Definition 3.3. $\Gamma \models_{\mathcal{T}} \psi =_{df} \forall v : (\forall \varphi \in \Gamma : v(\varphi) = T) \Rightarrow v(\psi) = T$.

The corresponding proof-system ETL can be obtained by extending $\vdash_H$ with the following rule of inference:[4]

$$\text{(R16)} \quad \frac{\varphi \wedge (\sim\varphi \vee \psi)}{\psi}$$

Some properties of ETL are worthy of note. First, it validates *ex contradictione quodlibet*, that is, $\varphi \wedge \sim\varphi \vdash_{\text{ETL}} \psi$ holds. However, contraposition and disjunction elimination are *not* admissible meta-principles of ETL; therefore, it does not collapse to classical or Kleene's logic. In particular, the classically valid consequence $\sim\psi \vdash \sim(\varphi \wedge \sim\varphi)$ (and more generally $\psi \vdash \varphi \vee \sim\varphi$) fails in ETL, which is evidence for the non-admissibility of contraposition. To see that disjunction elimination is also not admissible, it is sufficient to observe that $(\varphi \wedge \sim\varphi) \vee (\psi \wedge \sim\psi) \vdash \chi$ (valid in strong Kleene) fails in ETL, even though both $\varphi \wedge \sim\varphi \vdash_{\text{ETL}} \chi$ and $\psi \wedge \sim\psi \vdash_{\text{ETL}} \chi$ hold. Pietz and Rivieccio dub the latter property "anti-primeness" [21, p. 129].

ETL is sound and complete with respect to $\models_{\mathcal{T}}$:

Theorem 3.4. $\Gamma \vdash_{\text{ETL}} \psi \Leftrightarrow \Gamma \models_{\mathcal{T}} \psi$.

Proof. See Theorem 3.4 in [21]. □

4 A dualization of Belnap's logic

Now, it is time to look at the four-valued consequence relations from a somewhat different (in fact, dual) perspective. Heinrich Wansing rightly remarks that "[t]he term 'duality' has several meanings even in mathematics" [27, p. 486]. However, as Michael Atiyah once noted, "[f]undamentally, duality gives *two different points of view of looking at the same object*" [4, p. 69].

Proceeding from the basic logical duality between Fregean *the True* and *the False* (to wit, classical t and f) one can first arrive at a duality between sentences of the object language $\mathcal{L}$ (with $\wedge$ and $\vee$), cf. [19, pp. 21-25], and then at a duality between expressions about consequence (most generally conceived as a relation between *arbitrary* sets of sentences of $\mathcal{L}$), based on the use of the concept of duality as "related to order reversal" [27, p. 486].

[4]In [21, p. 133] this rule is called "disjunctive syllogism", which is not quite accurate because the latter name is generally reserved for a slightly different principle, saying $\sim\varphi \wedge (\varphi \vee \psi) \vdash \psi$. In fact, (R16) presents an ordinary rule of *modus ponens* for a material conditional standardly defined through a disjunction in which the antecedent is negated, and in such a form this rule is often referred to in the literature as 'Ackermann's rule γ', see [1, p. 119].

Definition 4.1. *Let φ, ψ be any sentences of $\mathcal{L}$, and let φ^d be obtained from φ by interchanging between $\wedge$ and $\vee$, and replacing every atomic sentence with its negation (and likewise for ψ^d). Let $\Gamma^d = \{\varphi^d : \varphi \in \Gamma\}$, and $\Delta^d = \{\psi^d : \psi \in \Delta\}$ where Γ, Δ are non-empty sets of sentences of $\mathcal{L}$. Then $\Delta^d \vdash \Gamma^d$ is said to be* dual *to $\Gamma \vdash \Delta$.*

An easy induction on the length of a formula gives for any formula φ, and for any valuation v:

Lemma 4.2. *$t \in v(\varphi) \Leftrightarrow f \in v(\varphi^d)$, and $f \in v(\varphi) \Leftrightarrow t \in v(\varphi^d)$.*

Next step is to extend the notion of duality to logical systems (formulated in language $\mathcal{L}$) in general:

Definition 4.3. *Logical system L is said to be* self-dual *if $\Gamma \vdash_L \Delta \Leftrightarrow \Delta^d \vdash_L \Gamma^d$; logical systems L_1 and L_2 are said to be* mutually dual *if $\Gamma \vdash_{L_1} \Delta \Leftrightarrow \Delta^d \vdash_{L_2} \Gamma^d$.*

Notice again that Definitions 4.1 and 4.3 generally involve consequence expressions of the SET-SET framework. Expressions of Belnap's logic can be viewed as a special case of SET-SET consequence expressions with the singleton restriction in the succedent. Clearly, by definition, neither $\vdash_H$ nor ETL are self-dual, and cannot be such, precisely because they deal with the asymmetric consequence expressions of the SET-FMLA type.

This suggests a way of a *structural dualization* of $\vdash_H$ (and ETL) by constructing the corresponding logical system of a FMLA-SET framework. Namely, a 'Hilbert-style axiomatization' of the *dual Belnap logic* $\vdash_{dH}$ can be formulated as follows:

$$(\mathrm{R1}_d)\ \frac{\varphi}{\varphi \vee \psi} \qquad (\mathrm{R2}_d)\ \frac{\psi}{\varphi \vee \psi} \qquad (\mathrm{R3}_d)\ \frac{\varphi \vee \psi}{\varphi, \psi}$$

$$(\mathrm{R4}_d)\ \frac{\varphi \wedge \psi}{\varphi} \qquad (\mathrm{R5}_d)\ \frac{\varphi \wedge \psi}{\psi \wedge \varphi} \qquad (\mathrm{R6}_d)\ \frac{\varphi}{\varphi \wedge \varphi}$$

$$(\mathrm{R7}_d)\ \frac{(\varphi \wedge \psi) \wedge \chi}{\varphi \wedge (\psi \wedge \chi)} \qquad (\mathrm{R8}_d)\ \frac{(\varphi \wedge \psi) \vee (\varphi \wedge \chi)}{\varphi \wedge (\psi \vee \chi)} \qquad (\mathrm{R9}_d)\ \frac{\varphi \wedge (\psi \vee \chi)}{(\varphi \wedge \psi) \vee (\varphi \wedge \chi)}$$

$$(\mathrm{R10}_d)\ \frac{\sim\sim\varphi \wedge \psi}{\varphi \wedge \psi} \qquad (\mathrm{R11}_d)\ \frac{\varphi \wedge \psi}{\sim\sim\varphi \wedge \psi} \qquad (\mathrm{R12}_d)\ \frac{(\sim\varphi \vee \sim\psi) \wedge \chi}{\sim(\varphi \wedge \psi) \wedge \chi}$$

$$(\mathrm{R13}_d)\ \frac{\sim(\varphi \wedge \psi) \wedge \chi}{(\sim\varphi \vee \sim\psi) \wedge \chi} \qquad (\mathrm{R14}_d)\ \frac{(\sim\varphi \wedge \sim\psi) \wedge \chi}{\sim(\varphi \vee \psi) \wedge \chi} \qquad (\mathrm{R15}_d)\ \frac{\sim(\varphi \vee \psi) \wedge \chi}{(\sim\varphi \wedge \sim\psi) \wedge \chi}$$

This system manipulates consequence expressions of the FMLA-SET type, i.e., is designed to establish valid consequences of the form $\varphi \vdash \Delta$. Every inference in

$\vdash_{dH}$ has only one assumption, and a non-empty set of conclusions. Intuitively, an expression $\varphi \vdash \Delta$ means that at least one sentence among the elements of Δ is derivable from φ.[5]

To put it formally, an inference (or derivation) of Δ from φ in $\vdash_{dH}$ is a finite *consecutive* list of (occurrences of) formulas, the first of which is φ. All other formulas of the list are formed by applying the inference rules to formulas that precede these in the list, with Δ being the set of *terminating formulas* of the inference. A formula is terminating if and only if it has such an occurrence in the list, that is never used later as a premise of an inference rule applied in this inference. If there is an inference of Δ from φ in $\vdash_{dH}$, then Δ is derivable from φ in $\vdash_{dH}$, and consequence $\varphi \vdash \Delta$ is said to be *valid* in $\vdash_{dH}$.

By way of example consider the inferences in $\vdash_{dH}$ of the dual versions of formulas (a) and (b) above: (a_d) $\sim\sim\varphi \vdash_{dH} \varphi$ and (b_d) $\sim\sim\varphi \vee \psi \vdash_{dH} \varphi \vee \psi$.

(a_d):

1. $\sim\sim\varphi$	(assumption)
2. $\sim\sim\varphi \wedge \sim\sim\varphi$	1: ($R6_d$)
3. $\varphi \wedge \sim\sim\varphi$	2: ($R10_d$)
4. φ	3: ($R4_d$), termination

(b_d):

1. $\sim\sim\varphi \vee \psi$	(assumption)
2. $\sim\sim\varphi$	1: ($R3_d$)
3. ψ	1: ($R3_d$)
4. $\varphi \vee \psi$	3: ($R2_d$), term.
5. φ	2: (a_d)
6. $\varphi \vee \psi$	5: ($R1_d$), term.

Note, that we had to infer the formula $\varphi \vee \psi$ twice, since without steps 5–6 the formula $\sim\sim\varphi$ would be terminating, and we would had the inference of $\sim\sim\varphi \vee \psi \vdash_{dH} \sim\sim\varphi, \varphi \vee \psi$ instead of (b_d).

One can also construct inferences in $\vdash_{dH}$ in a form of derivation trees, which—dually to the trees in $\vdash_H$—may branch downwards. Any derivation tree in $\vdash_{dH}$ has only one leaf, but can have many roots. The leaf of a tree stands for the formula from which its conclusions (roots) are derived. As an illustration consider the following derivation tree for (b_d):

$$\dfrac{\sim\sim\varphi \vee \psi}{\dfrac{\dfrac{\sim\sim\varphi}{\varphi}\,(a_d)}{\varphi \vee \psi}\,(R1_d) \qquad \dfrac{\psi}{\varphi \vee \psi}\,(R2_d)}\,(R3_d)$$

It is not difficult to obtain the dual version of Lemma 2.3:

[5]It is notable, that the reading of $\varphi \vdash \Delta$ as "φ entails ψ_1, or ψ_2, ..., or ψ_n, where $\psi_1, \psi_2, \ldots, \psi_n = \Delta$" is possible, but not inevitable. Rather the reading "φ entails ψ_i, for some $\varphi_i \in \Delta$" seems preferable.

Lemma 4.4. *For each rule* (Ri_d) *(*$10 \leqslant i \leqslant 15$*) of the form* $\dfrac{\varphi \wedge \chi}{\psi \wedge \chi}$*, the following rules hold: (a)* $\dfrac{\varphi}{\psi}$*, and (b)* $\dfrac{\varphi \vee \chi}{\psi \vee \chi}$.

The duality between $\vdash_H$ and $\vdash_{dH}$ are established by the following theorem:

Theorem 4.5. $\Gamma \vdash_H \psi \Leftrightarrow \psi^d \vdash_{dH} \Gamma^d$.

Proof. To prove this theorem it is enough to observe that the dual version of every rule of $\vdash_H$ is derivable in $\vdash_{dH}$ and vice versa. □

It is most natural to define entailment relation of the FMLA-SET type as a relation that backwardly preserves classical falsity (f) from conclusions to the assumption:

Definition 4.6. $\varphi \models_{\mathcal{DB}} \Delta =_{df} \forall v : (\forall \psi \in \Delta : f \in v(\psi)) \Rightarrow f \in v(\varphi)$.

One can observe a semantical duality between $\models_{\mathcal{B}}$ and $\models_{\mathcal{DB}}$:

Theorem 4.7. *For any* Γ*, for any* $\psi : \Gamma \models_{\mathcal{B}} \psi \Leftrightarrow \psi^d \models_{\mathcal{DB}} \Gamma^d$.

Proof. Consider arbitrary Γ and ψ. Let $\forall v : (\forall \varphi \in \Gamma : t \in v(\varphi)) \Rightarrow t \in v(\psi)$. Assume, $\exists v : (\forall \varphi^d \in \Gamma^d : f \in v(\varphi^d))$ and $f \notin v(\psi^d)$. By using Lemma 4.2 one very quickly gets a contradiction. The proof of the converse is similar. □

Definition 4.6 explicitly suggests $\{T, N\}$ as the set of designated truth values (and hence, dually to Definition 2.1, $\{F, B\}$ as the set of *non-designated* values). Still, just like $\models_{\mathcal{B}}$, relation $\models_{\mathcal{DB}}$ preserves classical truth in the forward direction, as the following lemma states, being obtained by a simple dualization of Lemma 2.2:

Lemma 4.8. $\varphi \models_{\mathcal{DB}} \Delta \Leftrightarrow \forall v : t \in v(\varphi) \Rightarrow (\exists \psi \in \Delta : t \in v(\psi))$.

Proof. For every valuation v define its dual v^*, such that $t \in v^*(p) \Leftrightarrow f \notin v(p)$, and $f \in v^*(p) \Leftrightarrow t \notin v(p)$. A direct induction extends this valuation to any formula of the language. Now, assume $\varphi \models_{\mathcal{DB}} \Delta$. Consider an arbitrary valuation v, such that $\forall \psi \in \Delta : t \notin v(\psi)$. We have then $\forall \psi \in \Delta : f \in v^*(\psi)$, and hence, $f \in v^*(\varphi)$. Thus, $t \notin v(\varphi)$. The proof of the converse is similar. □

Dually to $\vdash_H$, conjunction introduction and contraposition are inexpressible in $\vdash_{dH}$ as direct inference rules. Nevertheless, admissibility of the corresponding meta-principles can be obtained in $\vdash_{dH}$ by dualization of Lemma 3.1 and Lemma 3.2:

Lemma 4.9. *If* $\varphi \vdash_{dH} \psi$ *and* $\varphi \vdash_{dH} \chi$*, then* $\varphi \vdash_{dH} \psi \wedge \chi$.

Lemma 4.10. *If $\varphi \vdash_{dH} \psi$, then $\sim\psi \vdash_{dH} \sim\varphi$.*

System $\vdash_{dH}$ is sound with respect to Definition 4.6:

Theorem 4.11. *If $\varphi \vdash_{dH} \Delta$, then $\varphi \models_{\mathcal{DB}} \Delta$.*

Proof. For every rule (R1$_d$), (R2$_d$), and (R4$_d$)–(R15$_d$) of the form $\frac{\alpha}{\beta}$ assume $f \in v(\beta)$. Then an assumption that $f \notin v(\alpha)$ will lead to a contradiction. For (R3$_d$) assume $f \in v(\varphi)$ and $f \in v(\psi)$. Then we obtain a contradiction from the assumption that $f \notin v(\varphi \vee \psi)$. □

To obtain completeness of $\vdash_{dH}$ with respect to Definition 4.6 one can employ a technique of bringing any formula of $\mathcal{L}$ to a *normal form*. In what follows I dualize appropriately the definitions and proofs from [13, pp. 12-14]. Let *Var* be the set of propositional variables of $\mathcal{L}$, and $Lit = Var \cup \{\sim p : p \in Var\}$ be the set of *literals*. Let *Cl* be the set of *clauses*—the least set of formulas containing *Lit* and closed under $\wedge$. Let $var(\varphi)$ be the set of variables of φ, and $var(\Gamma)$ be the set of variables of formulas from Γ. For any clause φ the set of its literals $lit(\varphi)$ is defined inductively by: $lit(\varphi) = \{\varphi\}$ if $\varphi \in Lit$, and $lit(\varphi \wedge \psi) = lit(\varphi) \cup lit(\psi)$. For $\Gamma \subseteq Cl$, $lit(\Gamma)$ is the set of literals of formulas from Γ. As usual, $\varphi \dashv\vdash_{dH} \psi$ means $\varphi \vdash_{dH} \psi$ and $\psi \vdash_{dH} \varphi$, and likewise for $\dashv\models_{\mathcal{DB}}$.

Lemma 4.12. *For all $\varphi \in (\mathcal{L})$ there is a finite $\Gamma \subseteq Cl$, such that $var(\varphi) = var(\Gamma)$, and for any $\psi \in \mathcal{L}$, for all $\gamma \in \Gamma : \varphi \wedge \psi \dashv\vdash_{dH} \gamma \wedge \psi$.*

Proof. By induction on the length of φ. If $\varphi \in Var$, then we can put $\Gamma = \{\varphi\}$. Let $\varphi = \varphi_1 \wedge \varphi_2$, and Γ_1, Γ_2 correspond to φ_1, φ_2 by inductive hypothesis. Then $\Gamma = \{\gamma_1 \wedge \gamma_2 : \gamma_1 \in \Gamma_1, \gamma_2 \in \Gamma_2\}$ satisfies $var(\Gamma) = var(\varphi)$, and we have: $(\varphi_1 \wedge \varphi_2) \wedge \psi \dashv\vdash_{dH} \varphi_1 \wedge (\varphi_2 \wedge \psi)$ (by R7$_d$) $\dashv\vdash_{dH} \gamma_1 \wedge (\varphi_2 \wedge \psi)$ (for all $\gamma_1 \in \Gamma_1$, by inductive hypothesis) $\dashv\vdash_{dH} \varphi_2 \wedge (\gamma_1 \wedge \psi)$ (for all $\gamma_1 \in \Gamma_1$, by a principle derivable in $\vdash_{dH}$) $\dashv\vdash_{dH} \gamma_2 \wedge (\gamma_1 \wedge \psi)$ (for all $\gamma_1 \in \Gamma_1, \gamma_2 \in \Gamma$, by inductive hypothesis) $\vdash_{dH}$) $\dashv\vdash_{dH} (\gamma_2 \wedge \gamma_1) \wedge \psi)$ (for all $\gamma_1 \in \Gamma_1, \gamma_2 \in \Gamma$, by converse of R7$_d$ derivable in $\vdash_{dH}$). The cases with $\varphi = \varphi_1 \vee \varphi_2$ and $\varphi = \sim\varphi_1$ are analogous. □

Lemma 4.13. *For any $\varphi \in \mathcal{L}$ there is a finite $\Gamma \subseteq Cl$, such that $var(\varphi) = var(\Gamma)$ and $\varphi \dashv\vdash_{dH} \bigvee \Gamma$.*

Proof. Similarly as above, by induction on the length of φ. Let $\varphi \in Var$. Then $\Gamma = \{\varphi\}$. Let $\varphi = \varphi_1 \wedge \varphi_2$, and Γ_1, Γ_2 correspond to φ_1, φ_2 by inductive hypothesis. Then $\Gamma = \{\gamma_1 \wedge \gamma_2 : \gamma_1 \in \Gamma_1, \gamma_2 \in \Gamma_2\}$ satisfies $var(\Gamma) = var(\varphi)$, and we have: $\varphi \dashv\vdash_{dH} \gamma_1 \wedge \varphi_2$ (for all $\gamma_1 \in \Gamma_1$, by Lemma 4.12) $\dashv\vdash_{dH} \varphi_2 \wedge \gamma_1$ (for all $\gamma_1 \in \Gamma_1$, by R5$_d$)$\dashv\vdash_{dH}$ $\gamma_2 \wedge \gamma_1$ (for all $\gamma_1 \in \Gamma_1, \gamma_2 \in \Gamma_2$, by Lemma 4.12) $\dashv\vdash_{dH}$ $\gamma_1 \wedge \gamma_2$ (for all $\gamma_1 \in \Gamma_1, \gamma_2 \in \Gamma_2$, by R5$_d$). Hence, $\varphi \vdash_{dH} \bigvee\{\gamma_1 \wedge \gamma_2 : \gamma_1 \in \Gamma_1, \gamma_2 \in \Gamma_2\}$ (by R1$_d$). To get the converse, assume $\bigvee\{\gamma_1 \wedge \gamma_2 : \gamma_1 \in \Gamma_1, \gamma_2 \in \Gamma_2\}$. By R3$_d$, $\{\gamma_1 \wedge \gamma_2 : \gamma_1 \in \Gamma_1, \gamma_2 \in \Gamma_2\}$, and since φ is equivalent through $\dashv\vdash_{dH}$ to each $\gamma_1 \wedge \gamma_2$, we thus obtain φ. The cases with $\varphi = \varphi_1 \vee \varphi_2$ and $\varphi = {\sim}\varphi_1$ are analogous. □

Lemma 4.14. *Every $\varphi \in \mathcal{L}$ is equivalent both through $\dashv\vdash_{dH}$ and $\mathrel{=\!\mid\!\models}_{\mathcal{DB}}$ to a disjunction of clauses with the same variables.*

Proof. For $\dashv\vdash_{dH}$ the lemma holds by Lemma 4.13. Due to Theorem 4.11 it holds for $\mathrel{=\!\mid\!\models}_{\mathcal{DB}}$ as well. □

Lemma 4.15. *Let $\varphi \in Cl$, $\Delta \subseteq Cl$. Then $\varphi \models_{\mathcal{DB}} \Delta \Rightarrow \varphi \vdash_{dH} \Delta$.*

Proof. Assume $\varphi \models_{\mathcal{DB}} \Delta$. For a fixed $\varphi \in Cl$ define valuation v^l on literals by putting for every $p \in Var$: $t \in v^l(\varphi) \Leftrightarrow p \in lit(\varphi)$; $f \in v^l(\varphi) \Leftrightarrow {\sim}p \in lit(\varphi)$. By Definition 4.6, $(\forall\psi \in \Delta : {\sim}p \in lit(\psi)) \Rightarrow {\sim}p \in lit(\varphi)$, and by Theorem 4.8 $p \in lit(\varphi) \Rightarrow \exists\psi \in \Delta : p \in lit(\psi)$, for every p. Thus, $lit(\varphi) \subseteq lit(\psi)$. Since both φ and ψ are clauses, ψ is a conjunction of the same literals appearing in φ, and maybe other ones, modulo some associations, permutations, repetitions, etc. By using (R4$_d$), (R5$_d$), (R6$_d$) and (R7$_d$), we get $\varphi \vdash_{dH} \psi$, and thus, $\varphi \vdash_{dH} \Delta$. □

Theorem 4.16. *For any φ and Δ : if $\varphi \models_{\mathcal{DB}} \Delta$, then $\varphi \vdash_{dH} \Delta$.*

Proof. By Lemma 4.15 and Lemma 4.14. □

I finish this section with a brief review of some well-known notions and results from *abstract algebraic logic*, as displayed, e.g., in [13, 16, 17], adjusted to a FMLA-SET framework. Assume a standard notion of *logical matrix* for language $\mathcal{L}$ as a pair $\langle \mathbf{A}, D\rangle$, where $\mathbf{A}$ is an algebra of type $\mathcal{L}$ with universe A, and $D \subseteq A$. If A forms a *lattice*, then D is a *lattice filter* on A which can also be prime. The *Leibniz congruence* $\Omega_{\mathbf{A}}(D)$ of the matrix $\langle \mathbf{A}, D\rangle$ is defined as the largest congruence of $\mathbf{A}$, such that if any two elements $a, b \in \mathbf{A}$ are connected by the congruence relation and $a \in D$, then $b \in D$ as well. A matrix is said to be *reduced* if its Leibniz congruence is the identity relation.

Consider a structural consequence relation of a FMLA-SET type, i.e. a relation $\vdash \subseteq \mathcal{L} \times P(\mathcal{L})$ satisfying the following properties for all $\varphi, \psi \in \mathcal{L}$ and all $\Gamma, \Delta \subseteq \mathcal{L}$:

Reflexivity: $\varphi \vdash \{\varphi\} \cup \Delta$.
Monotonicity: if $\varphi \vdash \Gamma$, then $\varphi \vdash \Gamma \cup \Delta$.
Transitivity: if $\varphi \vdash \Gamma$ and $\psi \vdash \{\varphi\} \cup \Delta$, then $\psi \vdash \Gamma \cup \Delta$.
Structurality: if $\varphi \vdash \Delta$, then $\sigma\varphi \vdash \sigma\Delta$, for every uniform substitution σ on $\mathcal{L}$.

A logic L in a FMLA-SET framework can be then defined as a pair $\langle \mathcal{L}, \vdash \rangle$. A logical matrix is considered to be a *model* of a logic L when $\varphi \vdash_L \Delta$ implies for any valuation v on A (a homomorphism from $\mathcal{L}$ to $\mathbf{A}$) $v(\psi) \in D$ (for some $\psi \in \Delta$), whenever $v(\varphi) \in D$. In such a case the set D is called a filter for L or an L-filter. A logic L is said to be *complete* relative to a class of its matrix models iff for every $\Delta \cup \{\varphi\} \subseteq \mathcal{L}$, such that $\varphi \nvdash_L \Delta$, there is a logical matrix $\langle \mathbf{A}, D \rangle$ (which is a model of L) and a valuation $v \in Hom(\mathcal{L}, A)$, such that $v(\varphi) \in D$ but $v(\psi) \notin D$, for every $\psi \in \Delta$. It is well-known that every logic is complete with respect to the class of all its reduced models, see, e.g., [14, p. 207].

Observe, that the set of Belnapian truth values $\{T, B, N, F\}$ constitute a lattice with operations of meet, join and involution that correspond to the connectives determined by Definition 1.1. This lattice labeled in [13, p. 3] as $\mathfrak{M}_4$ generates the variety of *De Morgan lattices* **DM**. Famously, $\mathfrak{M}_4$ has exactly two prime filters $D_b = \{T, B\}$ and $D_n = \{T, N\}$.

Theorem 4.11 in fact demonstrates that matrix $\langle \mathfrak{M}_4, D_n \rangle$ is a model of $\vdash_{dH}$. By using Theorem 4.16, it can also be shown that $\vdash_{dH}$ is complete with respect to the class of logical matrices $\langle \mathbf{A}, D \rangle$, where $\mathbf{A}$ is **DM** and D is the set of filters generated by D_n (or equivalently by D_b). Note, that D is closed under intersections, being thus itself a complete lattice.

In the next section I will need the following lemma, which can be obtained from Theorem 3.14 in [13]:

Lemma 4.17. *If* $\mathbf{A}$ *is a non-trivial (i.e. not one-element) algebra, then* $\langle \mathbf{A}, D \rangle$ *is a reduced matrix for* $\vdash_{dH}$ *iff* $\mathbf{A} \in \mathbf{DM}$ *and* D *is a lattice filter of* $\mathbf{A}$.

5 The non-falsity logic and dual γ

Definition 4.6 explicates the entailment relation of the dual Belnap logic as essentially preserving classical falsity (f) in a *backward direction.* We can strengthen this property, and consider a relation that is backwardly hereditary with respect to Belnapian *exact falsity*:

Definition 5.1. $\varphi \models_{\mathcal{F}} \Delta =_{df} \forall v : (\forall \psi \in \Delta : v(\psi) = F) \Rightarrow v(\varphi) = F.$

Dually to Definition 3.3, this relation takes F as *the only non-designated* truth value, and thus, is based on the set of designated values $\{T, B, N\}$. As explained in [24, p. 1308], such a choice may be suitable if we wish "to allow as designated all the truth values *except the worst one*", and thus, to consider "anything but the (outright) falsehood".

The following theorem establishes semantical duality between $\models_{\mathcal{T}}$ and $\models_{\mathcal{F}}$:

Theorem 5.2. *For any* Γ, *for any* $\psi : \Gamma \models_{\mathcal{T}} \psi \Leftrightarrow \psi^d \models_{\mathcal{F}} \Gamma^d$.

Proof. Similarly as the proof of Theorem 4.7. □

This duality suggests a deductive formalization of the non-falsity logic (NFL) on the basis of the dual Belnap logic obtained by extending system $\vdash_{dH}$ by the *dual Ackermann's rule* γ:[6]

$$(\mathrm{R16}_d) \quad \frac{\varphi}{\psi \vee (\sim\psi \wedge \varphi)}$$

Theorem 5.3. $\Gamma \vdash_{\mathrm{ETL}} \psi \Leftrightarrow \psi^d \vdash_{\mathrm{NFL}} \Gamma^d$.

Proof. In addition to the proof of Theorem 4.5 one has to consider (R16) and to state that its dual version is derivable in NFL, and analogously with derivability of the dual version of ($\mathrm{R16}_d$) in ETL. □

NFL is sound with respect to Definition 5.1:

Theorem 5.4. *If* $\varphi \vdash_{\mathrm{NFL}} \Delta$, *then* $\varphi \models_{\mathcal{F}} \Delta$.

Proof. A simple check confirms the fact that every rule ($\mathrm{R1}_d$)–($\mathrm{R16}_d$) preserves the truth value F from conclusions to the premise. □

NFL is a paraconsistent system, since $\varphi \wedge \sim\varphi \vdash \psi$ is not NFL-derivable. Indeed, assume $v(\psi) = F$, and take $v(\varphi) = B$. Then $v(\varphi \wedge \sim\varphi) = B$, and hence, $\varphi \wedge \sim\varphi \not\models_{\mathcal{F}} \psi$. By Theorem 5.4 $\varphi \wedge \sim\varphi \nvdash_{\mathrm{NFL}} \psi$.

But NFL is not paracomplete as the following derivation shows:

1. φ	(assumption)
2. $\psi \vee (\sim\psi \wedge \varphi)$	1: ($\mathrm{R16}_d$)
3. ψ	2: ($\mathrm{R3}_d$)
4. $\sim\psi \wedge \varphi$	2: ($\mathrm{R3}_d$)
5. $\sim\psi$	4: ($\mathrm{R4}_d$)
6. $\psi \vee \sim\psi$	3: ($\mathrm{R1}_d$), termination
7. $\psi \vee \sim\psi$	5: ($\mathrm{R2}_d$), termination

[6]In [24] we used the rule of *dual disjunctive syllogism* in the form $\frac{\varphi}{\sim\psi \vee (\psi \wedge \varphi)}$.

Moreover, the principle of conjunction introduction is not admissible in NFL. Indeed, we have both $\varphi \vdash_{\text{NFL}} \psi \vee {\sim}\psi$ and $\varphi \vdash_{\text{NFL}} \chi \vee {\sim}\chi$. But if we take $v(\varphi) = T$, $v(\psi) = B$ and $v(\psi) = N$, we obtain $v((\psi \vee {\sim}\psi) \wedge (\chi \vee {\sim}\chi)) = F$, thus $\varphi \not\models_{\mathcal{F}} (\psi \vee {\sim}\psi) \wedge (\chi \vee {\sim}\chi)$, and hence, $\varphi \nvdash_{\text{NFL}} (\psi \vee {\sim}\psi) \wedge (\chi \vee {\sim}\chi)$.

To prove the completeness of NFL with respect to Definition 5.1 one can dualize an algebraic technique employed in [21]. Consider again lattice $\mathfrak{M}_4$ with the lattice order $\leqslant$. Let $x \preceq y$ generally stands for the lattice equation $x \sqcap y \approx x$. Since $\mathfrak{M}_4$ generates the variety of De Morgan lattices **DM**, it satisfies an equation $x \approx y$ iff this equation is satisfied in all the lattices from **DM**. Once $\langle \mathcal{L}, \wedge, \vee, {\sim} \rangle$ is known to form a De Morgan lattice, we have:

Lemma 5.5. *For every $\varphi, \psi_1, \ldots, \psi_n \in \mathcal{L}$ the following are equivalent:*

(i) $\varphi \models_{\mathcal{F}} \psi_1, \ldots, \psi_n$;

(ii) $\mathfrak{M}_4$ *satisfies* ${\sim}(\psi_1 \vee \ldots \vee \psi_n) \wedge \varphi \preceq \psi_1 \vee \ldots \vee \psi_n$.

Proof. $(i) \Rightarrow (ii)$: Assume (i), and consider an arbitrary valuation v. If $v(\psi_1 \vee \ldots \vee \psi_n) = T$, the lemma holds. If $v(\psi_1 \vee \ldots \vee \psi_n) = N$, then $v({\sim}(\psi_1 \vee \ldots \vee \psi_n)) = N$, and $N \sqcap x \leqslant N$ holds for any $x \in \{T, F, B, N\}$. The same argument holds for $v(\psi_1 \vee \ldots \vee \psi_n) = B$. If $v(\psi_1 \vee \ldots \vee \psi_n) = F$, then by (i) $v(\varphi) = F$, and (ii) holds as well.

$(ii) \Rightarrow (i)$: Assume (ii), and consider a valuation v, such that $v(\psi_1 \vee \ldots \vee \psi_n) = F$. Then $T \wedge v(\varphi) \leqslant F$, and thus, $v(\varphi) = F$. □

Combining this lemma with the algebraic implications of the completeness result for $\vdash_{dH}$ from the previous section, we get the desired theorem:

Theorem 5.6. *If $\varphi \models_{\mathcal{F}} \Delta$, then $\varphi \vdash_{\text{NFL}} \Delta$.*

Proof. Assume $\varphi \nvdash_{\text{NFL}} \Delta$. This implies that there is some reduced matrix $\langle \mathbf{A}, D \rangle$, and a valuation $v \in Hom(\mathcal{L}, A)$, such that $v(\varphi) \in D$ but $v(\psi) \notin D$, for every $\psi \in \Delta$. Since NFL is an extension of $\vdash_{dH}$, this matrix will also be a model of $\vdash_{dH}$, and by Lemma 4.17, $\mathbf{A} \in \mathbf{DM}$ and D is a lattice filter of $\mathbf{A}$. Note, that D is closed under (R16$_d$). Now, suppose $\varphi \models_{\mathcal{F}} \Delta$. By Lemma 5.5, $\langle \mathbf{A}, D \rangle$ satisfies ${\sim}(\psi_1 \vee \ldots \vee \psi_n) \wedge \varphi \preceq \psi_1 \vee \ldots \vee \psi_n$, were $\psi_1, \ldots, \psi_n = \Delta$. Hence, $v({\sim}(\psi_1 \vee \ldots \vee \psi_n) \wedge \varphi) \leqslant v(\psi_1 \vee \ldots \vee \psi_n)$, and thus, $v({\sim}(\psi_1 \vee \ldots \vee \psi_n)) \wedge \varphi)) \notin D$. But since D is closed under (R16$_d$), we should have $v({\sim}(\psi_1 \vee \ldots \vee \psi_n)) \wedge \varphi)) \in D$, a contradiction. □

6 Concluding remarks: feasibility of Fmla-Set entailment

This paper elaborates a general method of dualizing the proof systems of certain kind, which can be referred to as 'degenerated' Hilbert-style axiomatic systems, which have in fact no axioms, but only direct inference rules of the SET-FMLA type. Syntactically the dualization in question consists just in reversing all the inference rules of the system to be dualized (together with the proper dualization of all the involved sentences), and in switching thus to a system that deals now with FMLA-SET consequences. The semantic definition of the entailment relation is subject to the analogous dualization, which reflects a general duality between truth and falsity.

The described procedure was performed on two systems—Font's formulation of Belnap's logic and Pietz and Rivieccio's formulation of exactly true logic, both based on a four-valued Belnapian semantics, but with different choices of designated truth values. As a result we obtain two new systems, formalizing the dual Belnap logic and the non-falsity logic, which belong to the FMLA-SET framework. Soundness and completeness of these systems with respect to the corresponding four-valued semantics were established.

In view of the technical considerations of the present paper a reader may not feel comfortable with the very idea of a logic formulated in the FMLA-SET framework. If we agree that "logic is the science of argument" [23, p. ix], what kind of argument could comprise logical systems of such type?

It may be noted that Belnap's logic, and specifically system $\vdash_H$ defined in a SET-FMLA framework, is an exemplar of what can be called *a logic of proof*, where an argument is conceived as a procedure of *proving some sentence* by inferring it from a collection of premises. In such a setting an argument is just a logical device that ultimately "leads to a conclusion, *one* conclusion, or so one would think" [25, p. 333]. In this sense, as observed in [23, p. ix], "ordinary arguments are lopsided: they can have any number of premisses but only one conclusion". But is any kind of logical deduction necessarily such?

Shoesmith and Smiley in their now classic book [23], drawing on pioneering insights by Gerhard Gentzen, Rudolf Carnap and William Kneale, advance a *multiple-conclusion logic* that allows "any number of conclusions as well, regarding them ... as setting out the field within which the truth must lie if the premisses are to be accepted" [23, p. ix]. Since then, the subject of a multiple-conclusion logic has been taken up by various authors, see [22] for a prominent example of this.

However, Florian Steinberger recently challenged the very idea of multiple conclusions by appealing to standards of logical inferentialism, "the position that the meanings of the logical constants are determined by the rules of inference they obey" [25, p. 333]. This position, he argues, is incompatible with multiple-conclusion proof systems because such systems are supposedly not "connected to our ordinary deductive inferential practices", and thus, he says, "constitute a departure from our ordinary forms of inference and argument" [25, pp. 335, 340].

Even leaving a dubious issue of finding logical structures "in nature" (see [25, pp. 339]) aside, one can point out serious limitations of an inferentialist conception based on the notion of proof (to wit, *assertion*) only. Wansing, for instance, pays particular attention to the speech act of *denial* "in the context of a use-based, inferentialist account of linguistic meaning" [27, p. 483], distinguishing then between provability, disprovability, and their duals, where "the dual of provability is reducibility to non-truth", and "the dual of disprovability is reducibility to non-falsity" [27, p. 486]. He also considers the corresponding 'inferential relations'[7], and moreover, supplements the well-known *Brouwer-Heyting-Kolmogorov (BHK) interpretation* of the logical (intuitionistic) connectives formulated in terms of canonical proofs by "interpretations in terms of canonical disproofs, canonical reductions to absurdity (alias non-truth), and canonical reductions to non-falsity" [27, p. 493].

Luca Tranchini by constructing a natural deduction system for dual-intuitionistic logic observes a close correspondence between introduction rules of the natural deduction system for intuitionistic logic and Brouwer-Heyting-Kolmogorov 'proof-interpretation' of the logical constants. He suggests a dual 'refutation-interpretation' for the logical constants through the 'dual-BHK' clauses, which in turn correspond to elimination rules of the natural deduction system for dual-intuitionistic logic, see [26, p. 645-646]. Note, incidentally, that the natural deduction system for dual-intuitionistic logic constructed by Tranchini is "a single-premise multiple-conclusions system in which derivation trees branch downward" [26, p. 632].

The latter observation not only supports the justifiability of multiple-conclusion systems, but also highlights the relevance of an entailment relation *with only one premise*, and particularly the systems considered in Sections 4 and 5 above. Namely, dual Belnap logic, as well as non-falsity logic, belonging to the FMLA-SET logical framework, can be most naturally considered a kind of what Kosta Došen once called *logics of refutation*:

> A refutation would be a deduction where we have at most one premise; from this premise we try to deduce a number of conclusions, with the intent to show that all these conclusions are refutable, so that the premise

[7] In particular, the consequence of the form $\varphi \vdash \psi_1, \ldots, \psi_n$ stands for an inferential relation "φ is reducible to absurdity from *counterassumptions* $\psi_1, \ldots, \psi_n$".

must be refutable too. Sequents are read backwards: if all sentences on the right are refutable, a sentence on the left is refutable. [8, p. 111]

In this way, systems $\vdash_{dH}$ and NFL restore an essential lopsidedness of the 'ordinary proof-arguments', but in a dual fashion, as the pure 'refutation-arguments', which can have any number of conclusions but only one premise. The set of conclusions forms then a *refutation set* for a given premise, whereas the premise stands for a hypothesis to be tested for refutability.

Systems $\vdash_{dH}$ and NFL differ in their selection criteria to possible refutations. A sentence can generally be considered refutable if it can take a non-designated truth value. Then, in dual Belnap logic, it is sufficient for a sentence to be not (classically) true, *or* to be at least (classically) false, depending on the chosen set of designated truth values (either $\{T, B\}$, or $\{T, N\}$—both options are possible on an equal footing), to qualify as a refutation in a given inference. By contrast, in the non-falsity logic, the criterion is much stronger—here a genuine refutation must be both false *and* not true.

Both these systems, being interesting in their own right, indicate a general usefulness of the FMLA-SET logical framework for certain logico-methodological purposes, and thus, its worthiness for further elaboration.

References

[1] Ackermann, W. Begründung einer strengen Implikation, *Journal of Symbolic Logic* 21 (1956), 113-128.

[2] Anderson, A.R. and Belnap N.D. *Entailment: The Logic of Relevance and Necessity*, Vol. I, Princeton, NJ: Princeton University Press, 1975.

[3] Anderson, A.R., Belnap, N.D., and Dunn, J.M. *Entailment: The Logic of Relevance and Necessity*, Vol. II, Princeton, NJ: Princeton University Press, 1992.

[4] Atiyah, M. Duality in mathematics and physics, in: *Conferències FME (Facultat de Mathemàtiques i Estadística, Universitat Politècnica de Catalunya), Centre de Recerca Matemàtica (CRM), Institut de Matemàtica de la Universitat de Barcelona (IMUB)*, Vol. 15, 2008, 69-91.

[5] Belnap, N.D. Tautological entailments (abstract), *Joumal of Symbolic Logic* 24 (1959), 316.

[6] Belnap, N.D. A useful four-valued logic, in: J.M. Dunn and G. Epstein (eds.), *Modern Uses of Multiple-Valued Logic*, Dordrecht: D. Reidel Publishing Company, 1977, 8-37.

[7] Belnap, N.D. How a computer should think, in: G. Ryle (ed.), *Contemporary Aspects of Philosophy*, Oriel Press, 1977, 30-55.

[8] Došen, K. *Logical Constants: an Essay in Proof Theory*, D. Phil. thesis, University of Oxford, 1980.

[9] Dunn, J.M. *The Algebra of Intensional Logics*, Doctoral Dissertation, University of Pittsburgh, Ann Arbor, 1966 (University Microfilms).

[10] Dunn, J.M. Intuitive semantics for first-degree entailment and coupled trees, *Philosophical Studies* 29 (1976), 149-168.

[11] Dunn, J.M. A comparative study of various model-theoretic treatmets of negation: a history of formal negation. In: D.M. Gabbay and H. Wansing (eds.), *What is Negation?*, Dordrecht/Boston/London: Kluwer Academic Publishers, 1999, 23-51.

[12] Dunn, J.M. Partiality and its dual, *Studia Logica* 66 (2000), 5-40.

[13] Font, J.M. Belnap's four-valued logic and De Morgan lattices, *Logic Journal of the IGPL* 5 (1997) 413-440.

[14] Font, J.M. *Abstract Algebraic Logic. An Introductory Textbook*, London: College Publications, 2016.

[15] Font, J.M., Guzmán, F., and Verdú, V. Characterization of the reduced matrices for the $\{\wedge, \vee\}$-fragment of classical logic, *Bulletin of the Section of Logic* 20 (1991), 124-128.

[16] Font, J.M., and R. Jansana, *A General Algebraic Semantics for Sentential Logics*, vol. 7 of Lecture Notes in Logic: Springer-Verlag, 1996.

[17] Font, J.M., Jansana, R. and Pigozzi, D. A survey of abstract algebraic logic, *Studia Logica* 74 (2003), 13-97.

[18] Humberstone, L. *The Connectives*, Cambridge, MA: MIT Press, 2011.

[19] Kleene, S.C. *Mathematical Logic*. New York: Wiley, 1967.

[20] Marcos, J. The value of the two values, in: J.-Y. Beziau, M.E. Coniglio (eds.) *Logic without Frontiers: Festschrift for Walter Alexandre Carnielli on the Occasion of his 60th Birthday* (Tributes), College Publications, 2011, 277-294.

[21] Pietz, A. and Rivieccio, U. Nothing but the Truth, *Journal of Philosophical Logic* 42 (2013), 125-135.

[22] Restall, G. Multiple conclusions, in: P. Hájek, L. Valdes-Villanueva, and D. Westerstahl (eds.), *Logic Methodology and Philosophy of Science. Proceedings of the Twelfth International Congress*, London: King's College Publications, 2005, 189-205.

[23] Shoesmith, D.J. and T.J. Smiley, *Multiple Conclusion Logic*, Cambridge: Cambridge University Press, 1978.

[24] Shramko, Y., Zaitsev, D., Belikov, A. First-degree entailment and its relatives, *Studia Logica* 105 (2017), 1291-1317.

[25] Steinberger, F. Why conclusions should remain single, *Journal of Philosophical Logic* 40 (2011), 333-355.

[26] Tranchini, L. Natural deduction for dual-intuitionistic logic, *Studia Logica* 100 (2012), 631-648.

[27] Wansing, H. Proofs, disproofs, and their duals, in: V. Goranko, L. Beklemishev and V. Shehtman (eds.), *Advances in Modal Logic*. Volume 8, London, College Publications, 2010, 483-505.

Received 18 February 2018

www.ingramcontent.com/pod-product-compliance
Lightning Source LLC
Chambersburg PA
CBHW080514090426
42734CB00015B/3047